Selke / Culter / Hernandez
Plastics Packaging

Susan E.M. Selke
John D. Culter
Ruben J. Hernandez

Plastics Packaging

Properties, Processing, Applications,
and Regulations

2nd Edition

HANSER

Hanser Publishers, Munich • Hanser Gardner Publications, Cincinnati

The Authors:
Professor Dr. Susan E.M. Selke, Michigan State Universitiy, School of Packaging, East Lansing, MI 48824-1223, USA
John D. Culter, President Advanced Materials Engineering Inc., Napels, Fl 34114-9232, Silver Lake Blvd. 1231, USA
Professor Ruben J. Hernandez (†)

Distributed in the USA and in Canada by
Hanser Gardner Publications, Inc.
6915 Valley Avenue, Cincinnati, Ohio 45244-3029, USA
Fax: (513) 527-8801
Phone: (513) 527-8977 or 1-800-950-8977
Internet: http://www.hansergardner.com

Distributed in all other countries by
Carl Hanser Verlag
Postfach 86 04 20, 81631 München, Germany
Fax: +49 (89) 98 48 09
Internet: http://www.hanser.de

The use of general descriptive names, trademarks, etc., in this publication, even if the former are not especially identified, is not to be taken as a sign that such names, as understood by the Trade Marks and Merchandise Marks Act, may accordingly be used freely by anyone.

While the advice and information in this book are believed to be true and accurate at the date of going to press, neither the authors nor the editors nor the publisher can accept any legal responsibility for any errors or omissions that may be made. The publisher makes no warranty, express or implied, with respect to the material contained herein.

Library of Congress Cataloging-in-Publication Data
Selke, Susan E. M.
Plastics packaging: properties, processing, application and regulations /
Susan E.M. Selke, John D. Culter, Ruben J. Hernandez.-- 2nd ed.
p. cm.
Ruben J. Hernandez appears first on the previous edition.
Includes bibliographical references and index.
ISBN 1-56990-372-7 (hardcover)
1.   Plastics in packaging. 2.   Polymers. I. Culter, John D., 1937- II.
Hernandez, Ruben J., 1945- III. Hernandez, Ruben J., 1945- Plastics
packaging. IV. Title.
TS198.3.P5H47 2004
688.8--dc22   2004020052

Bibliografische Information Der Deutschen Bibliothek
Die Deutsche Bibliothek verzeichnet diese Publikation in der Deutschen Nationalbibliografie;
detaillierte bibliografische Daten sind im Internet über <http://dnb.ddb.de> abrufbar.
ISBN 3-446-22908-6

© Carl Hanser Verlag, Munich 2004
Production Management: Oswald Immel
Typeset by Susan E.M. Selke, USA
Coverconcept: Marc Müller-Bremer, Rebranding, München, Germany
Coverdesign: MCP • Susanne Kraus GbR, Holzkirchen, Germany
Printed and bound by Druckhaus "Thomas Müntzer" GmbH, Bad Langensalza, Germany

In memory of

Ruben J. Hernandez
1945-2002

Professor
School of Packaging
Michigan State University

Preface

This book is intended to provide a basic understanding of plastic packaging materials. It covers the properties of common packaging plastics, and relates these properties to the chemical structure of the polymers. Common processing methods for transforming plastic resins into packages are covered.

In this book we discuss the uses of plastics in packaging. Although this is not a course in chemistry nor in material science, we attempt to stress the relationship between chemical structure and packaging material properties. We expect the reader to have some knowledge of chemistry and physics. The major purpose of this book is to provide the students in the School of Packaging with reading material on plastics for packaging; however, we hope that it can also be useful to packaging professionals responsible for writing specifications, designing, fabricating, testing, and controlling the quality of plastic materials. We also hope to trigger the reader's curiosity to pursue further studies in the exciting world of packaging materials.

This second edition fixes some of the errors that, despite our best efforts, found their way into the first edition. Unfortunately, we're sure that we have not yet found them all! We have also expanded and updated the discussion of PET bottle production, retort pouches, polylactides, plastics recycling, and a variety of other topics.

We have deliberately included some information that goes well beyond what would normally be included in an introductory level packaging course, in order that it will be available for the more advanced student and for the practitioner. In the first edition, we attempted to identify this material by putting it inside grey boxes. We have abandoned that convention in this edition for a variety of reasons, the most compelling being that instructors vary in what they choose to incorporate in an introductory course. We decided that it is better to leave it up to individual instructors to tell students what sections, or portions of sections, to omit in their study of this text.

Contents

1   Introduction

1.1   Historic Note

The first man-made plastic, a form of cellulose nitrate, was prepared in 1838 by A. Parker and shown at the Great International Exhibition in London in 1862. It was intended to be a replacement for natural materials such as ivory and was called parkesine. In 1840, Goodyear and Hancock developed the "vulcanization" procedure that eliminated tackiness and added elasticity to natural rubber. The change in the properties of the natural rubber was obtained by the addition of sulfur powder that produced additional chemical bonds in the bulk of the rubber.

In 1851, hard rubber, or ebonite, was commercialized. In 1870 a patent was issued to J. Hyatt, of New York, for celluloid, a type of cellulose nitrate with low nitrate content produced at high temperature and pressure. This was the first commercially available plastic, and the only one until the development of Bakelite by Baekeland in 1907. Bakelite is the oldest of the purely synthetic plastics and consisted of a resin obtained by the reaction of phenol and formaldehyde.

The exact nature of plastics, rubber, and similar natural materials was not known until 1920, when H. Staudinger proposed a revolutionary idea: all plastics, rubber, and materials such as cellulose were polymers, or macromolecules. Before Staudinger's theory, the scientific community was very confused about the exact nature of plastics, rubbers and other materials of very high molecular weight. To most research workers in the 19th century, the finding that some materials had a molecular weight in excess of 10,000 g/mol appeared to be untrustworthy. They confused such substances with colloidal systems consisting of stable suspensions of small molecules.

Staudinger rejected the idea that these substances were organic colloids. He hypothesized that the high molecular weight substances known as polymers were true macromolecules formed by covalent bonds. Staudinger's macromolecule theory stated that polymers consist of long chains in which the individual monomers (or building blocks) are connected with each other by normal covalent bonds. The unique polymer properties are a consequence of the high molecular weight and long chain nature of the macromolecule. While at first his hypothesis was not readily accepted by most scientists, it eventually became clear that this explanation permitted the rational interpretation of experiments and so gave to industrial chemists a firm guide for their work. An explosion in the number of polymers followed. Staudinger was awarded the Nobel Prize in 1953. It is well established now that plastics, as well as many other substances such as rubber, cellulose, and DNA, are macromolecules.

Since 1930, the growth in the number of polymers and their applications has been immense. During the 1930s, industrial chemical companies initiated fundamental research programs that had a tremendous impact on our society. For example, Wallace Carothers, working at DuPont de Nemours and Co., developed diverse polymeric materials of defined structures and investigated how the properties of these materials depend on their structure. In 1939 this program resulted in the commercialization of nylon.

A commercial process for the synthesis of polyethylene was successfully developed in the 1930s by ICI (Imperial Chemical Industries), in England. In 1955, K. Ziegler in Germany and J. Natta in Italy developed processes for making polyethylene at low pressure and temperature using special catalysts. They were awarded the Nobel Prize, Ziegler in 1964 and Natta in 1965, for their contributions in the development of new polymerization catalysts with unique stereo-regulating powers. Linear polyethylene produced using solution and gas technologies was introduced in the 1970s. The continuous development of new polymers resulted in additional breakthroughs in the mid-1980s and early 1990s. Single-site catalysts, which were originally discovered by Natta in the mid-1950s, were commercialized for syndiotactic polystyrene in 1954, polypropylene in 1984, and polyethylenes in the early 1990s.These catalysts permit much greater control over the molecular weight and architecture of polyolefins such as polyethylene and polypropylene. Table 1.1 shows the approximate introduction dates for some common plastics.

Today, dozens of different synthetic plastics are produced throughout the world by hundreds of companies. In North America alone, plastics production in 2003 totaled about 49 million tonnes (107 billion lbs). Packaging is the largest single market for plastics, amounting to about 13 million tonnes (28 billion lbs) per year, about a quarter of total U.S. plastics production, in 2003.

1.2   Role of Plastics in Packaging

The term plastics is used instead of polymer to indicate a specific category of high molecular weight materials that can be shaped using a combination of heat, pressure, and time. All plastics are polymers, but not all polymers are plastics. In this text, we will discuss the major plastics that are useful as packaging materials. To a limited extent, we will discuss cellophane, which is a wood-based material that is a polymer, but not a plastic. We will also discuss adhesives, which are polymers and may or may not be plastics, but which are very useful in the fabrication of plastic and other types of packaging.

Table 1.1 Approximate Dates of Introduction for Some Common Plastics

Date	Polymer	Date	Polymer
1907	Phenol-formaldehyde resins	1947	Epoxies
1927	Polyvinyl chloride	1948	ABS resins
1927	Cellulose Acetate	1952	Polyethylene, linear
1930	Styrene-butadiene rubber	1955	Polypropylene
1936	Polymethyl methacrylate	1957	Polycarbonate
1936	Polyvinyl acetate	1957	LLDPE
1938	Polystyrene	1964	Ionomer resins
1938	Nylon 66	1965	Polyimides
1939	Polyvinylidene chloride	1970	Moldable elastomers
1941	Polytetrafluoroethylene	1972	Acrylonitrile copolymers
1942	Polyesters, unsaturated	1972	Ethylene vinyl alcohol
1942	Polyethylene, branched	1974	Aromatic polyamides
1943	Butyl rubber	1978	PET
1943	Nylon 6	1985	Liquid crystal polymers
1943	Fluoropolymers	1992	Metallocene polymers
1943	Silicones	1994	PEN

Packaging started with natural materials such as leaves. From there, it progressed to fabricated materials such as woven containers and pottery. Glass and wood have been used in packaging for about 5000 years. In 1823, Durand in England patented the "cannister," the first tin-plate metal container. The double seamed three-piece can was in use by 1900. Paper and paperboard became important packaging materials around 1900. As soon as plastic materials were discovered, they were tried as packaging materials, mainly to replace paper packaging. Use of cellophane, which is a polymer but not truly a plastic, predated much of the use of plastics.

The use of plastics in packaging applications began, for the most part, after World War II. Polyethylene had been produced in large quantities during the war years, and it became commercially available immediately after the war. Its first application had been as insulation for wiring in radar and high frequency radio equipment. It was soon found that it could be formed easily into various shapes useful for packaging. An early application was in bread bags, replacing waxed paper. Polyethylene coatings replaced wax in heat-sealable paperboard. As a coating, it was also combined with paper to replace waxed paper and cellophane. The driving force behind the expansion of polyethylene use was to obtain a resealable package as well as a transparent material that allowed the product to be visible. Polyethylene remains the leading packaging plastic because of its low raw material price, versatile properties, and its ease of manufacture and fabrication.

The growth of plastics packaging has accelerated rapidly since the 1970s, in large part because of one of the main features of plastics - low density. This low density made the use of plastics attractive because of the weight savings, which translates into energy savings for transportation of packaged goods. In addition, plastic packages are usually thinner than their counterparts in glass, metal, paper, or paperboard. Therefore, conversion to plastic packaging often permits economies of space as well as of weight. Savings in the amount of distribution packaging needed may also result. Another important property is the relatively low melting temperatures of plastics compared to glass and metals. Lower melting temperatures mean less energy is required to produce and fabricate the materials and packages. While use of plastics in all applications has grown rapidly during this period, the growth in packaging has outpaced the growth in other sectors. Packaging is the largest single market for plastics, as illustrated in Figure 1.1. In 2003, packaging accounted for about 28% of all thermoplastics and engineering resins used in North America. According to The Association of Plastics Manufacturers in Europe, APME, packaging accounted for 37.2 percent of all plastics used in Europe in 2002 and 2003 (www.apme.org).

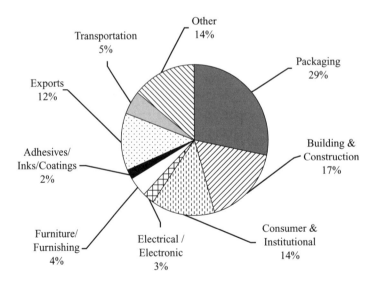

Figure 1.1 Major North American markets for plastics, 2003 (American Plastics Council, www.americanplasticscouncil.org).

Many of the early applications of plastics were in food packaging. The substitution of plastic films for paper in flexible packaging led to the development of many new combinations of materials, and to the use of several polymers together to gain the benefit of their various attributes. The development of flexible packaging for foods picked up speed in the late 1940s and 1950s as the prepared foods business began to emerge. Milk cartons using polyethylene coated paperboard were introduced in the 1950s. Here the driving force was economics: glass is more expensive in a systems sense, breakage of glass on line requires extensive cleaning, and returnable bottles brought all sorts of foreign objects into an otherwise clean environment.

In industrial packaging, plastics were used early as a part of multiwall shipping sacks that replaced bulk shipments, drums, and burlap sacks. Again, polyethylene film was the predominant material used. Cement in 110 kg (50 lb) bags became a major application of polyethylene film in the industrial sector. The polyethylene liner protects the cement from moisture that would cause it to solidify. Another large use of plastics in industrial packaging is as cushioning to protect goods from vibration and impact during shipping. Polystyrene, polyurethane and polyethylene foams, along with other polymers, are used as cushioning, competing against paper-based cushioning materials.

Medical packaging has been another big user of plastics. As converting techniques improved, so that accurate molding of small vials could be accomplished at low cost, and as new polymers became available with the necessary characteristics, plastics have been substituted for glass in many applications. As medical procedures became more complex, more disposable kits were introduced, designed to have complete sets of equipment for specific procedures. These kits require special packaging to keep the parts organized and easily usable. Here thermoformed trays became standard, so that kits of pre-sterilized, disposable instruments and supplies, in the proper varieties and amounts, can be readily assembled. The plastic packaging allows the sterilization to occur after the package is sealed, thus eliminating the possibility of recontamination after sterilization, as long as the package remains intact. Sterilization with ethylene oxide is facilitated by the use of spun-bonded polymeric fabrics. Radiation sterilization depends on the use of polymers that retain their integrity after exposure to ionizing radiation.

The energy crisis in the 1970s, while at first leading to attacks on plastics as users of precious petroleum, actually accelerated the movement to plastic packaging because of the weight reduction possible. Many metal cans and glass bottles were replaced by plastic cans and bottles, and in many cases changes in package design moved the product out of rigid packaging altogether, into flexible packaging, which more often than not was made of plastic. Similarly, some metal drums were replaced by plastic drums. A major driving force was to reduce the fuel used for transportation of both packages and packaged goods by reducing the weight of the package. One important example is the introduction of the plastic beverage bottle.

Environmental concerns of the 1980s and early 1990s, caused by littering issues and a perceived lack of landfill space, caused a major rethinking of the plastic packaging in use. Companies that used plastics had to defend the uses that were in place, and justify new applications. The result was a more responsible approach to packaging in general, by most companies. As politicians and the public became more informed about the truth concerning plastics and the environment, the issues receded from the forefront, although they have not disappeared altogether. Today, plastic packaging has earned its position as one of the choices of the package designer. Decisions about which material(s) should be used require consideration of (1) product protection requirements, (2) market image, (3) cost, and (4) environmental issues.

1.3   Book Structure

This book is intended to provide (1) an introduction to the plastics used in packaging, (2) discussion of how their use relates to their properties, and (3) explanation of how these properties relate to their chemical structure, along with (4) an introduction to converting these plastic resins into useful packages. We have used much of the material in this book in our undergraduate course on plastics packaging at the School of Packaging, Michigan State University.

Chapter 2 provides some introductory concepts and definitions. Chapter 3 looks at the relationship between the chemical and physical structure and the properties of plastics. Chapter 4 provides a description of the plastics commonly used in packaging, except for biodegradable plastics, which are covered in Chapter 16. Chapter 5 looks at the other ingredients that go into a plastic resin. Chapter 6 examines adhesion, adhesives, and heat sealing. Chapter 7 covers conversion of plastic resins into film and sheet forms. Chapter 8 examines how film and sheet can be modified by lamination and by coating. Chapter 9 discusses flexible packaging. Chapter 10 covers thermoforming. Chapter 11 discusses injection molding of plastics, with a special look at closures; rotational and compression molding; and tubes. Chapter 12 looks at formation of plastics into bottles and other containers by blow molding. Chapter 13 looks at distribution packaging, with an emphasis on foams and cushioning. Chapter 14 looks at the barrier characteristics, and other mass transfer characteristics, of packaging and how they relate to the shelf life of products. In Chapter 15, we examine at various laws and regulations impacting packaging choices. Finally, Chapter 16 looks at environmental issues associated with plastic packaging, including biodegradable and biobased plastics.

Throughout the text, long examples are placed in boxes. Most chapters end with a set of questions. In many cases, the answers can be found (or calculated) from the material in the chapter. In other cases, answering the questions requires the student to put together information from several previous chapters. Sometimes, the questions are intended to stimulate thinking in preparation for what will be discussed in subsequent chapters, and cannot be answered completely with only the information that has already been presented. The correct solutions to quantitative questions are included.

2   Basic Concepts and Definitions

2.1   Terminology

The plastic materials we use in packaging are polymers and macromolecules. Before beginning the discussion and description of polymer structures and properties, we should define these and some related terms. Mastering these terms will provide the reader with a terminology useful for understanding many of the polymer properties described in the rest of the book. Some of these concepts were learned in basic chemistry courses, but now will be applied to polymers. Additional terms will be defined in subsequent chapters.

2.1.1   Macromolecule

Polymers belong to the group of materials called macromolecules, which simply means very large molecules, and can be defined as compounds made of a very large number of atoms chemically connected by covalent bonds. The relative molecular mass (also referred to as molecular weight) of macromolecules ranges from several hundred to millions of daltons (g/mol). Therefore, the chemical and mechanical properties of macromolecules are in large part determined simply by the very large size of the molecules. Polymers form a large group of materials that include plastics, adhesives, rubbers, fibers, and surface coatings, as well as cellulose, deoxyribonucleic acid (DNA), and ribonucleic acid (RNA).

In a macromolecule, the atoms that are linked together by covalent bonds and run through the whole molecule form the backbone, or main chain. The backbones of a large number of polymers are formed entirely by carbon atoms. However, there are polymers whose molecular backbones include carbon and oxygen (polyesters); silicon, carbon and oxygen (silicone rubbers and resins); or nitrogen, carbon and oxygen (nylons, also called polyamides) (Fig. 2.1).

2.1.2 Polymer

A polymer is defined by the International Union of Pure and Applied Chemistry, IUPAC, as a substance made of large molecules that is characterized by the multiple repetition of one or more species of atoms or group of atoms (called monomers or constitutional units) linked to each other covalently in amounts sufficient to provide a set of properties that do not vary markedly with the addition or removal of one or a few of the constitutional units. IUPAC and many other authorities consider macromolecule and polymer to be synonyms; others define polymer as a macromolecule which consists of multiple repetitions of monomer(s). With this definition, enzymes are macromolecules but not polymers, since they are very large molecules made by the non-repetitive combination of 20 amino acids.

Although the possible number of polymers is theoretically limitless, the economics of their production and processing, as well as the physical and chemical properties that they have, restrict the number of commercial importance to a few dozen (see Fig. 2.2), and in packaging applications the number of polymers used is even smaller. The polymers most commonly used in packaging are polyolefins, specifically polyethylene and polypropylene. Polystyrene, polyvinyl chloride, and polyethylene terephthalate (PET) are also among the most commonly used packaging polymers.

2.1.3 Plastic

Plastics are a special group of polymers with characteristics that differentiate them from fibers, rubbers, adhesives and other materials that are also polymers. The main characteristic of plastics is their ability, while solid in the finished state, to be made to flow and be molded using controlled heat and pressure at relatively low temperatures, compared to glass and metals. Plastics, then, are materials capable of being deformed continually without rupture by a stress which exceeds a yield value during processing. The plastic is easily shaped into the desired form by increasing the temperature and pressure to soften it and mold it into a new shape. On cooling, the plastic becomes hard and is able to retain the new shape. The distinctions between plastics, fibers, adhesives and rubbers are not always clear. The same type of polymer may belong to more than one group, depending on its molecular weight and on how it is processed. Currently, for example, new catalyst technology is causing a blurring of the boundaries between rubbers and plastics.

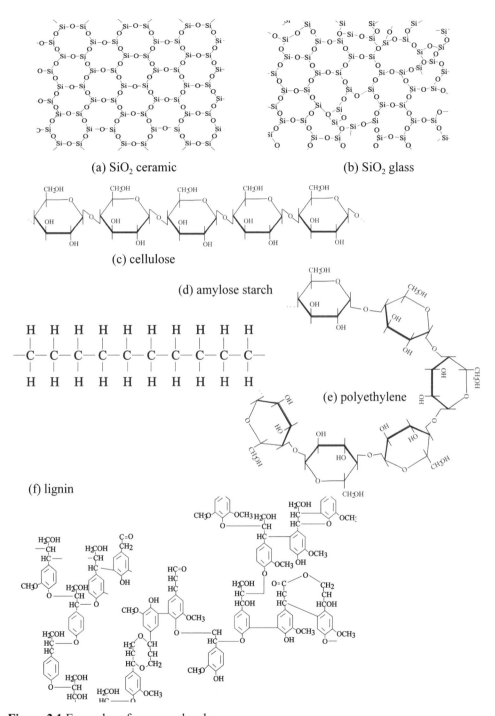

(a) SiO$_2$ ceramic

(b) SiO$_2$ glass

(c) cellulose

(d) amylose starch

(e) polyethylene

(f) lignin

Figure 2.1 Examples of macromolecules.

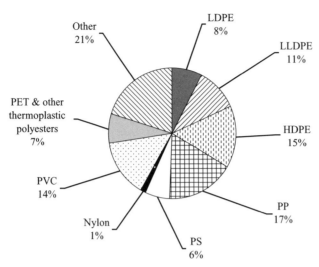

Figure 2.2 Plastics use in North America, 2003, million metric tons (Society of the Plastics Industry, www.societyplasticsindustry.org).

Some plastics, called *thermoplastics*, are able to repeatedly undergo this shape-changing treatment. Others, known as *thermoset* materials, can be shaped only once, because they form irreversible covalent bonds between chains during the "setting" process and consequently will no longer melt and flow. Vulcanized rubber is an example of a thermoset polymer. Thermoset polymers tend to have relatively high chemical resistance and excellent mechanical properties. However, the vast majority of plastics used in packaging are thermoplastics.

2.1.4   Monomer

A monomer is the unit, made of a simple molecule, or a compound constituted of such molecules, from which polymers are produced by a chemical polymerization reaction. For example, ethylene ($CH_2=CH_2$) is the monomer used to produce polyethylene. Propylene ($CH_3-CH=CH_2$), is polymerized to form polypropylene, and vinyl chloride ($CH_2=CHCl$) is polymerized to form polyvinyl chloride. As illustrated, polymer names are often formed by prefixing the term "poly" to the monomer name.

2.1.5 Constitutional Unit

The constitutional unit is the smallest chemical unit whose repetition completely describes the main chain structure. For example, in polyethylene (PE)

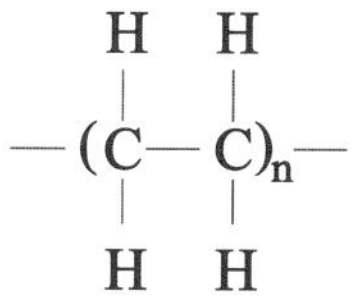

$$-\left(\underset{\underset{\text{H}}{|}}{\overset{\overset{\text{H}}{|}}{\text{C}}} - \underset{\underset{\text{H}}{|}}{\overset{\overset{\text{H}}{|}}{\text{C}}}\right)_n-$$

$CH_2=CH_2$ is the monomer, since polyethylene is formed by the polymerization of ethylene molecules. However, the group CH_2 is the smallest unit that completely describes PE, and therefore is the constitutional unit.

It will be noted that in this and other depictions of the structure of polymer molecules, we have not indicated the species at the ends of the molecules. The polymers, of course, do not have dangling bonds at the ends. However, because the molecules are so large, the groups at the end generally are insignificant in determining behavior. Further, there are a variety of possible ending groups, including hydrogens, groups with double bonds, and initiator residues. Therefore we are following the common convention in polymer science of leaving the end structures unspecified.

2.1.6 Homopolymer

A homopolymer is a polymer formed from only one type of monomer. For example, homopolymer polypropylene is formed only from propylene monomers, so it has the following structure:

$$-\left(CH_2 - \underset{\overset{|}{CH}}{\overset{\overset{CH_3}{|}}{}}\right)_n-$$

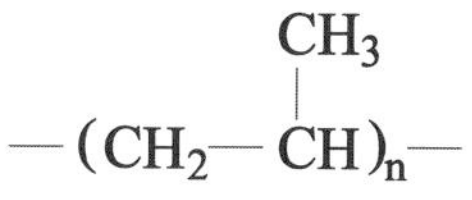

2.1.7 Copolymer

A copolymer is a polymer composed of two or more different types of monomers. For example, if E represents ethylene and P propylene, a copolymer of E and P might have the following structure:

-PPEEEPPPPEPEEEE-

Copolymers can have a variety of structures, depending on the number and arrangement of the monomeric units. A random structure, illustrated above, typically results when the comonomers are introduced into the polymerization reactor at the same time and allowed to react randomly. Other important structures are alternating, block and graft copolymers. The structure has a significant effect on the final copolymer properties. We will discuss this topic in Section 3.3. Many plastics used in packaging are copolymers.

In condensation polymerization, the reaction starts with two monomers but these polymers are not copolymers. The reason is that the first reaction generally makes an intermediate - an ester or an amide - which is then polymerized into a poly*ester* or a poly*amide*. Copolyesters and copolyamides incorporate a third monomer.

2.2   Polymer Nomenclature

Precise chemical names for polymers are frequently long and unwieldy, so a system of generally accepted abbreviations has arisen. Table 2.1 lists the names of a number of common homopolymers and copolymers, along with the abbreviations commonly used for them. Polymer chemists usually use parentheses in the name to contain the monomer, or repeating unit; for example poly(acrylonitrile) or poly(ethylene). However, in many fields, including packaging, it is common to omit the parentheses, and refer simply to polyacrylonitrile or polyethylene. Similarly, chemists would typically denote a copolymer as poly(monomer 1- co - monomer 2), such as poly(ethylene-co-vinyl acetate), but in other fields it is common to drop both the co- and the poly- and refer to the polymer simply as ethylene vinyl acetate. In some cases, a slash is used to separate the monomers, so the name becomes ethylene/vinyl acetate, as shown in Table 2.1.

The names of some plastics, particularly condensation polymers (see Section 3.8), do not follow this customary pattern. For example, as we shall see in Chapter 4, nylons (or polyamides) have a unique numbering system related to their structure. Polycarbonate is another polymer whose name does not accurately reflect the monomer used to produce it. In this and some other cases, a generic family name is so closely associated with a particular member of a polymer family, that this family name is often used as if it represented only that particular plastic. For example, the term polyester is often used interchangeably with PET, and vinyl interchangeably with PVC. It is also common, for some copolymers, to totally ignore the comonomer in naming the plastic, or to not specify what comonomer is used. For example, we refer to high impact polystyrene, or HIPS, when it is actually a copolymer. Similarly, we talk about polyvinylidene chloride, PVDC, as a packaging material, although we never use the homopolymer, and the name, in fact, represents a whole family of copolymers.

Table 2.1 Common Plastics

Homopolymer	Symbol	Copolymer	Symbol
Cellulose Acetate	CA	Acrylonitrile/butadiene/styrene	ABS
Epoxy	EP	Ethylene/ethyl acrylate	E/EA
Polyamide	PA	Ethylene/propylene	E/P
Polyacrylonitrile	PAN	Ethylene/vinyl acetate	EVA
Polybutylene acrylate	PBA	Ethylene/vinyl alcohol	EVOH
Polybutylene terephthalate	PBT	Linear low density	LLDPE
Polycarbonate	PC	polyethylene	
Polyethylene	PE	Vinyl chloride/ethylene	VC/E
High density polyethylene	HDPE	Vinyl chloride/ethylene/methyl	VC/E/MA
Low density polyethylene	LDPE	acrylate	
Polyethylene terephthalate	PET	Vinyl chloride/vinyl acetate	VC/VA
Polyethylene naphthalate	PEN	Vinyl chloride/vinylidene	VC/VDC
Polymethyl methacrylate	PMMA	chloride	
Polypropylene	PP		
Polystyrene	PS		
Polytetrafluoroethylene	PTFE		
Polyurethane	PU		
Polyvinyl acetate	PVA		
Polyvinyl alcohol	PVOH		
Polyvinyl chloride	PVC		
Polyvinylidene chloride	PVDC		

2.3   Interatomic and Intermolecular Forces in Polymers

Because of the very large size of polymer molecules, both interatomic and intermolecular forces play a crucial role in their formation and properties. A brief review of the most important atomic forces follows.

2.3.1   Interatomic Forces

Interatomic forces are the forces that bond atoms together, forming molecules. These forces connect atoms like carbon C, Hydrogen H, Oxygen O and Nitrogen N, making possible the construction of the very large polymer molecules. In polymeric materials, we find two types of interatomic bonds: covalent bonds, which are very common, and ionic bonds, which are very rare.

2.3.1.1 Covalent Bonds

A covalent bond consists of two electrons shared by two adjacent atoms, with each atom contributing one electron. The bond is formed when the orbitals of the two atoms overlap and form a hybrid. We represent this bond by a dash between the symbols for the two elements, for instance, C-C. The covalent bond is the predominant bond in polymers, linking the atoms of carbon (and also oxygen and nitrogen when present) to form the backbone of the polymer chain. Covalent bonds also link hydrogen, oxygen, carbon, and other atoms in side groups onto the backbone chain. The angle formed by three carbon atoms C-C-C in a linear chain measures 109.5°, as is predicted by orbital hybridization theory. The C-C-C angle is partially responsible for the elastic behavior of polymers, as the bond angle, and hence the dimensions of the molecular segments, can be affected by stress.

Atoms can be joined by one or more covalent bonds. If two pairs of electrons are shared, the bond is referred to as a double bond, for instance C=C. Three shared pairs of electrons results in a triple bond. Many polymers contain six member carbon rings that are often described as containing alternating single and double bonds that can switch positions with one another. While this description is not totally accurate, it is useful in understanding some of the unique properties of these materials. This type of bond is referred to as aromatic, and polymers containing these structures are called aromatic polymers. A common example is polystyrene:

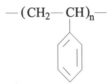

When atoms are joined by a single covalent bond, they are not "locked in place," but can rotate around the bond. Ease of rotation is one of the factors that influences the properties of the polymer, as will be discussed later. Rotation is not possible about double or triple bonds, or within ring structures, because of the characteristics of the electron orbitals.

Electron sharing between dissimilar atoms is not always equal. When one atom in the pair has a higher affinity for the shared electrons than the other atom, the bond is known as a polar covalent bond. For example, in a carbon-chlorine bond, such as in polyvinyl chloride, the chlorine has a higher electron affinity than carbon, so it gets a disproportionate share of the electrons, resulting in the C-Cl bond being a polar covalent bond. The importance of this polarity will be discussed in Section 2.3.2. Electron sharing between identical atoms, such as C-C or O-O, is nonpolar.

Table 2.2 shows the energy necessary to break covalent bonds between two atoms. As the energy increases, the molecule becomes stronger and more stable. Conversely, the least energetic bonds indicate the points where the molecule is most susceptible to degradation or other chemical change.

Table 2.2 Energies of Covalent Bonds

Bond	Energy (kJ/mol)	Bond	Energy (kJ/mol)
C–C	347	C≡N	891
C=C	614	C–H	414
C–O	360	C–F	439
C=O	715	C–Cl	326
C–N	293	O–O	146
C=N	614	O–H	460

2.3.1.2 *Ionic Bonds*

An ionic bond results when the affinity of two atoms for their shared electrons is so unequal that it results in actual transfer of one or more electrons from the atom with the lowest attraction, the electropositive atom, to the atom with the highest attraction, the electronegative atom. This transfer results in the formation of a positive ion and a negative ion. Oppositely charged ions in the substance are then attracted to each other by electrostatic attractions. Ionic bonds are common in many substances (salts, for example) but are rarely found in plastics. They are present in the side chains in certain polymers, however, and convey some unique and useful properties. For example, Na^+ and Zn^{++} when used to neutralize carboxylic groups, $-COO^-$, bonded to hydrocarbon chains such as polyethylene, produce polymers called ionomers.

2.3.2   Intermolecular and Intramolecular Forces

When atoms and molecules approach one another closely, they begin to exert forces on one another that do not result from the sharing or transfer of electrons. These forces are known as intermolecular forces, if the forces are between two neighboring molecules, and intramolecular forces if they are between different parts of the same molecule. These forces are much weaker than covalent or ionic bonds but are very important to the properties of polymers because the huge size of polymer molecules results in a proportionally huge number of these attractions.

Intermolecular forces are often referred to as *secondary bonds*, or secondary forces, while covalent and ionic bonds are referred to as *primary bonds*. The primary bonds determine the molecular structure of the material. The secondary forces are responsible for its physical nature. Gases have relatively weak secondary forces, liquids have stronger ones, and solids have the strongest secondary forces. When a solid is heated and melts, the heat is providing the energy necessary to disrupt the secondary forces sufficiently to allow flow. An "ideal gas" can be described as a (hypothetical) substance with no ability to form secondary bonds.

There are three primary categories of secondary forces, or van der Waals forces: dispersion forces, induction forces, and dipole forces - from weakest to strongest. These forces are highly sensitive to the distance between the molecules, with an approximate range of action between 3 and 5 Å (1 Å = 1 x 10^{-10} m). This distance range, as we will see, is crucial in the adhesion of polymers and in polymer surface preparation for adhesion, which are very important in packaging. The strength of secondary forces decreases proportionally to the 6th power of the distance separating the molecules. The cohesive energy density, which indicates the amount of energy required, per unit volume, to move a molecule far enough away from its neighbors that no significant intermolecular forces remain, is a useful parameter for measuring secondary forces. Several important polymer properties, such as solubility, miscibility, viscosity, surface tension, and friction, as well as adhesion between two materials, are largely determined by the strength and type of intermolecular forces. Table 2.3 summarizes the relationship between intermolecular forces and the mechanical characteristics of polymers.

Table 2.3 Importance of Intermolecular Forces in the Mechanical Characteristics of Polymers

Intermolecular Forces	Cohesive Energy Density	Typical Polymer Characteristics
Small	Low	Relatively flexible rubbery behavior, high permeability. Ex: PE, PP
Medium	High	Stiffer, plastic behavior. Ex: PET
Large	Higher	High resistance to stress, high strength, good mechanical properties, low permeability. Ex: Nylon, EVOH

2.3.2.1 Dispersion Forces

Dispersion forces, also called London forces, or London dispersion forces, are the most prevalent type of intermolecular forces. They are found in all substances, resulting from the natural fluctuation of the electron cloud in an atom that causes time-varying partial positive and negative charges. These charges sum to zero over time, but at any instant lead to electrostatic attractions between neighboring atoms with opposite charges. Dispersion forces are typically about 0.4-0.8 kJ/mol.

2.3.2.2   Induction Forces

Induction forces are found only when polar atoms are present, such as in ethylene/vinyl acetate copolymer. Polar atoms are those with a partial, or full, positive or negative charge resulting from polar or ionic bonds (as discussed in Section 2.2.1). The positive and negative charges in polar atoms can induce a corresponding fluctuation of the electrons in neighboring nonpolar atoms, thereby creating temporary polarity in these neighbors, and resulting in electrostatic attraction between the atoms. Induction forces are intermediate in strength between dispersion forces and dipole forces.

2.3.2.3   Dipole Forces

A permanent electrical dipole is formed wherever there is a polar bond, resulting in an asymmetric distribution of positive and negative charges in a molecule. The attractions between the permanent partial positive charges and partial negative charges in neighboring molecules are the strongest of the secondary forces, reaching up to 8 kJ/mol. The characteristics of PVC, for example, are largely determined by its strong secondary forces, which result from the C-Cl dipoles.

2.3.2.4   Hydrogen Bonds

Hydrogen bonds are the strongest type of secondary forces. While they can be regarded as a subcategory of dipole forces, their strength and unique characteristics generally results in classifying these forces in a special group - distinct from the ordinary van der Waals forces. Hydrogen bonds can reach strengths as high as 40 kJ/mol, and result from the attraction between a strongly electronegative atom and a hydrogen that is covalently bonded to a strongly electronegative atom. Fluorine, oxygen, nitrogen, and to a lesser extent chlorine, have very high electron affinities. A hydrogen atom attached to one of these atoms has an unusually high excess positive charge, and the electronegative atom itself has an unusually high excess negative charge. The attraction between such a hydrogen and a neighboring electronegative atom is thus unusually strong, and can take on some of the characteristics of a very weak covalent bond. In fact, in 1999, evidence of actual electron-sharing in some hydrogen bonds was reported. Hydrogen bonds play an important role in the intermolecular forces of polymers such as polyamides (nylons) and ethylene/vinyl alcohol.

2.4   Properties Determined by Chemical Composition

The properties of polymers are strongly influenced both by their chemical composition and their physical state. The chemical composition includes the atoms that make up the polymer molecules, as well as the number and arrangement of those atoms. It is determined by both the chemical composition of the monomers making up the polymer, and the polymerization conditions. Once determined, the chemical composition is fixed unless chemical reactions occur.

The physical state of the polymer includes such variables as the crystallinity of the material and the degree of orientation of the molecules. It is strongly dependent on the processing history of the material, and can be changed by changes in the environmental conditions to which the polymer is exposed, like heat and pressure, as will be discussed later.

The following properties are determined primarily by the chemical structure of the polymer:

- Density
- Thermal properties (enthalpy, thermal conductivity)
- Thermal expansion
- Chemical reactivity
- Melting and softening temperature
- Solubility, diffusion and permeability
- Friction

The long chain nature of the polymers is particularly important in determining:

- Strength
- Viscosity
- Elasticity
- Stress relaxation, creep and other time-dependent properties
- Melting temperature

Note that for some properties, such as melting temperature, both the chemical structure and the long chain nature of polymers play important roles. As will be seen later, the stiffness and flexibility characteristics of polymers are largely determined by the degree of hindrance to free rotation about single bonds in the backbone chain.

2.5 Categorization of Plastics

The many types of plastics are often grouped into various categories. Commodity plastics are those used in high volumes, and are characterized by modest prices and reasonable performance. This group includes polyethylene, polypropylene, polystyrene, and polyvinyl chloride. Engineering polymers are sometimes defined as those that maintain their mechanical properties at temperatures above 100°C and that command a higher price than the commodity plastics. These include nylon, polyesters, and polycarbonate (although the most common polyester, PET, is moving into the commodity plastics category). Other categories sometimes used include intermediate, for those polymers that lie between the commodity and engineering resins (such as PET), and advanced, for those that have very high prices and unique properties. The term specialty resins is sometimes used to refer to packaging polymers that are relatively high in cost and therefore used for their unique properties, such as ethylene vinyl alcohol and polyvinylidene chloride.

Study Questions

1. Cellulose, the major ingredient of paper, is a polymer. Is it a plastic? Why or why not?

2. What, if any, are the differences between the intermolecular forces in polymers and those in small molecules? How does this impact polymer properties?

3. What is the major difference between thermoplastics and thermosets? Which category is used more often in packaging? Why?

4. If we polymerized pentene, what would be the name of the resulting polymer? What if we copolymerized pentene and ethylene?

5. What monomer is used to produce polystyrene? Draw its chemical structure.

6. What monomer, or monomers, are used to produce ABS?

7. Why is chemical structure more important than physical structure in determining the density of a polymer?

8. Name a polymer that has a backbone chain made of: (1) only carbon atoms, (2) carbon and oxygen, (3) silicon, carbon and oxygen, and (4) nitrogen, carbon and oxygen.

9.   Define the following terms: macromolecule, polymer, and plastic. When are the constitutional unit and monomeric unit of a polymer the same?

10. What are the main similarities/differences between plastics, fibers, rubber, and adhesives?

11. Give an example of a polymer whose monomeric unit is different from its constitutional unit. Give an example where the monomeric and constitutional units are the same. Draw all the structures.

12. What are "intermolecular forces" in polymers? What are the main types of intermolecular forces? Which is the most common one? Why are intermolecular forces important? What is the typical range of intermolecular forces, in Å? How might we measure the intermolecular forces in a polymer?

13. Write the full names of the polymers represented by the following abbreviations: PVC, PAN, PVDC, PE, PP, EVA, PS, HDPE, PC, EVOH, LDPE, and PET.

14. Suggest an explanation for the behavior of polymers in Table 2.3. In other words, why do polymers with small, medium, or large intermolecular forces have the cohesive energy density and polymer characteristics listed?

15. Why is the cohesive energy density of a polymer related to its solubility?

3 Polymer Structure and Properties

3.1 Introduction

Polymers are produced by polymerization reactions that chemically bond large numbers of monomers together. Monomer(s), catalysts, solvents or fluid carriers and other secondary chemicals are continually fed into polymerization reactors. The reaction process is maintained under controlled temperature and pressure. Common categories of polymerization reactions are bulk, in which monomers are fed as a pure gas or liquid; solution, in which monomers are dissolved in a solvent; suspension, in which monomers are suspended in an immiscible medium; and emulsion, in which monomers are dispersed as very tiny particles in the carrier.

3.2 Molecular Architecture

Polymers are produced by chemically linking together a very large number of monomeric units in a chain-like or network structure. Because the monomers can react at different locations of the molecule, the monomers can be linked in the chain in different ways. Homopolymers can be linear, branched, or even cross-linked, while copolymers may have even more complex architectures such as block and graft. The molecular architecture of each polymer, in addition to the polymer's chemical composition and molecular weight, is an important determinant of the polymer properties. For example, a linear structure in general yields a more compact arrangement of atoms than a branched structure, which translates into higher density, and often into higher crystallinity, as well. If the polymer is more crystalline, it will have improved chemical resistance, increased tensile strength, and will be a better barrier to the permeation of gases.

3.2.1   Linear Polymers

In a linear polymer molecule, the monomers form a single long backbone chain. A linear polymer can be schematically represented as shown in Figure 3.1. As the figure illustrates, the polymer may not be totally linear, but branches, if any exist, are quite rare. Examples of linear polymers are HDPE, LLDPE, PET, nylons, PVC, and PP.

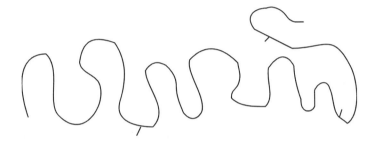

Figure 3.1 Representation of a linear polymer.

One determinant of the polymer architecture is the functionality of the monomers from which it is formed. Functionality refers to the number of bonds that a monomer can form with other monomeric molecules during the polymerization process. If the monomer is bifunctional (has a functionality of two), it will generally form a linear polymer. A molecule with a functionality of one cannot form a polymer at all. It can react with one other molecule to form a dimer, but since each of the two molecules has "used up" its ability to react, it cannot grow further into a polymer. Trifunctional (or higher) monomeric units can produce either branched or cross-linked polymers, depending on the functionality of the monomer and the stoichiometry and conversion of the reaction.

Examples of bifunctional monomers are $H_2C = CH_2$, $HOOC - C_6H_4 - COOH$, and $H_2C = CHCl$.

3.2.2   Branched Polymers

In a branched polymer, some monomers become part of side chains, branching off the main chain or off other branches. Monomers with a functionality of 3 or greater may form branched polymers, since three parts of the polymer molecule can extend from a single monomer wherever a tri-functional monomer is located. The degree of branching of the polymer will depend on the number of multifunctional monomers present during the reaction.

Branches can also arise during the polymerization reaction when reaction conditions convert what would normally be a stable structure into a reactive entity. This is very common with polyethylene. During the polymerization, a free radical can abstract (remove) a hydrogen atom from the backbone of the polymer and start a new growth point. From this growth point a whole new chain can grow. For LDPE, the branches (side chains) can be nearly as long as the backbone. (In fact, the backbone is defined as the longest pathway through the molecule, and parts of it may have originated as branches that grew longer than a portion of the molecule that used to be the backbone.) In polyethylene we talk about short chain branching (SCB) and long chain branching (LCB). LCD usually refers to side chains of 50 carbon atoms or more. Both long and short chain branches are illustrated in Figure 3.2.

Figure 3.2 Representation of a branched polymer.

In polymers with side groups that are a part of the regular structure of one of the monomers, as in the case of polystyrene, polypropylene, polybutyl methacrylate, etc., we generally do not consider these as branches or side chains. Instead, they are called side groups. An exception to this rule is for polyethylene/α-olefin copolymers, the LLDPE family, where the side group is referred to as a side chain. The reason for this exception is that the polymers look exactly like those that could be made by a side chain growth reaction. In fact, it is common for books to list LLDPE as a branched polymer, even though its name, linear low density polyethylene, clearly and correctly describes it as a linear polymer.

The effects of SCB and side groups are similar. They disrupt the ability of the polymer to crystallize. If the disruption is not complete, the added bulkiness will make the rate of crystallization slow down. SCB has little effect on the flow properties of a polymer, but LCB has a profound effect. We will discuss this in a later chapter. For now, we can illustrate the effects of branching by comparing LDPE with HDPE. The densities differ, the tensile properties differ, and the elastic character of the polymers differs greatly, even through both are made from the same monomer.

3.2.3   Cross-Linked Polymers

As indicated above, multifunctional monomers, instead of merely forming branches, can link main chains together to such an extent that the molecules are transformed into network type structures, known as cross-linked polymers, as shown in Figure 3.3. This is especially likely if one of the monomers has a functionality of 4, such as divinylbenzene. These cross-linked networks can be two dimensional, but most often are three-dimensional.

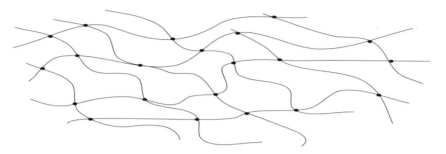

Figure 3.3 Representation of a cross-linked polymer.

Cross-linked polymers, although not widely used in packaging, have important applications. In addition to cross-linking generated during the polymerization process, cross-linking can result from degradation reactions occurring during extrusion or other processing, or as a consequence of exposure to UV or gamma radiation. In film extrusion, if polymer with excessive exposure to high temperatures is used, it can result in cross-linked regions called *gels*, which cause film weakness and can create runnability problems in packaging equipment. Excessive use of regrind is one source of such problems.

The vast majority of polymers used in packaging are thermoplastics, and most have a linear or nearly linear structure. Important packaging polymers with a branched structure include low density polyethylene and ethylene vinyl acetate.

Cross-linking is characteristic of thermoset materials, and is important in adhesives, many polymer coatings for cans, and very strong polymers for machine parts. In thermosets, linear or branched polymers react under the heat and pressure associated with the forming operation to produce a large number of cross-links which, once formed, no longer permit the polymer to flow. Thus these materials are set into a permanent shape during processing and do not soften upon reapplication of heat and pressure. If only a very small number of cross-links are formed, a polymer may still exhibit typical plastic behavior, including the ability to be softened and reshaped.

3.3   Copolymer Structure

As indicated before, a copolymer is a polymer which is composed of two or more different types of monomers. For example, ethylene and propylene can be polymerized together in the same chemical reactor to produce polymeric chains containing units contributed by each the two monomers. A common way of designating such a copolymer is E/P. Depending on the molar ratio of ethylene and propylene in the reacting gas mixture, the copolymer produced may have properties very similar to PE, or to PP, or properties intermediate between PE and PP. The goal of copolymerization is to control the properties of the resulting polymer through the ratio and type of co-reactants and the reactor conditions (temperature, pressure and catalyst). This allows the processing temperature, permeability, glass transition temperature, thermoforming temperature, toughness, elastic modulus, and other properties to be varied in ways not possible with homopolymers. Copolymers can have linear, branched, or cross-linked structures just like homopolymers, and have additional variations related to the relative positioning of the monomer units within the structure. The monomers can be arranged to produce random, alternating, block, or graft architectures.

3.3.1   Random Copolymers

In a random copolymer, the monomers are randomly distributed along the chain, so there is no pattern to their arrangement (Fig. 3.4). For example, random polypropylene copolymers are a type of polypropylene in which the basic structure of the polymer chain is modified by the incorporation of ethylene comonomer during the polymerization process. This results in changes in the physical properties compared to homopolymer PP such as increased clarity, improved impact resistance, increased flexibility, and a decrease in the melting point and heat sealing temperature.

E----EEEEPEEEPPEEPPPPEPPPEEEEEEPPEEPP------PEEE

Figure 3.4 Diagram of a random copolymer. Capital letters represent different monomers.

In general, the properties of random copolymers which are affected mostly by chain flexibility and intermolecular forces tend to be a weighted average of those of the respective homopolymers. Properties affected primarily by crystallinity, however, such as melt temperature, tend to be decreased by copolymerization, due to the irregularity that copolymerization produces in the chemical structure of the polymer molecules. This will be discussed further in Section 3.10.

Most of the copolymers used in packaging are random copolymers. In many cases, if only small amounts of the comonomer are used, the modifying monomer is not reflected in the name used for the polymer. For example, as we saw above, polypropylene copolymers are commonly referred to just in that way, rather than as poly(propylene-co-ethylene), which would be the chemically accurate designation.

3.3.2   Alternating Copolymers

In an alternating copolymer, the backbone chain is produced by the systematic alternation of two monomers in the main chain (Fig. 3.5). For example, vinyl acetate and maleic anhydride form a copolymer with such an alternating arrangement.

A----ABABABABABABABABABABABABABABABABABABA----B

Figure 3.5 Diagram of an alternating copolymer. Capital letters represent different monomers.

Properties of an alternating AB copolymer tend to be what would be predicted from a homopolymer with repeating unit C, where C = AB. These copolymers are quite unusual, since they can be produced only if the reaction rate for a growing polymer chain ending in unit A is much faster with monomer B than with monomer A, and the reaction rate for a growing polymer chain ending in unit B is much faster with monomer A than with monomer B. Only in this way can random collisions result in a predominantly alternating structure.

3.3.3   Block Copolymers

In a block copolymer, long groups of one type of monomer are followed by long groups of the other, within the backbone chain (Fig. 3.6). For example, some styrene-butadiene copolymers have a block copolymer structure.

A-----AAAAAAAAAAAABBBB----BBBBBBAAAAAAAAA---A

Figure 3.6 Diagram of block copolymer. Capital letters represent different monomers.

One way of forming block copolymers is to first polymerize each monomer separately to a low degree of polymerization, and then combine these small polymer molecules with each other (Fig. 3.7).

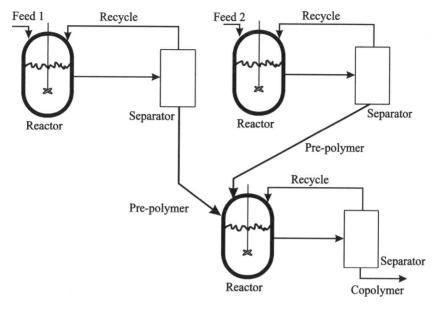

Figure 3.7 Block copolymerization reaction system using pre-polymerization of monomers.

Another option is to introduce the monomers into the reactor in an alternating sequence (Fig. 3.8).

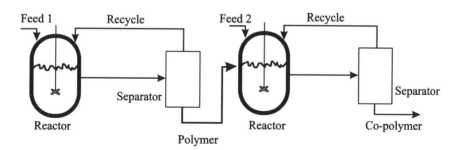

Figure 3.8 Block copolymerization reaction system utilizing sequential polymerization of monomers.

The properties of block copolymers are quite different from those of random or alternating copolymers. The segments of each type of monomer will be attracted more strongly to other segments of the same structure than to segments of the comonomer, so the polymer tends to have small regions which are predominantly one or the other type of monomer. Within these regions, the behavior tends to be very similar to the respective homopolymers. For example, in an ethylene/propylene block copolymer, there would be polyethylene crystallites and polypropylene crystallites, each exhibiting a melt temperature similar to that of homopolymer PE and PP. Similarly, there would be polyethylene and polypropylene amorphous areas, with glass transition temperatures in each area also corresponding to the homopolymers.

Block copolymers tend to behave more like blends than like random or alternating copolymers. However, they have a distinct advantage over blends, since the regions will be connected to each other by covalent bonds, and therefore cannot easily be separated. These polymers are often able to preserve some of the advantages of each of the homopolymers to a greater extent than can random copolymers. For example, styrene-butadiene block copolymers have much greater impact strength than random styrene-butadiene copolymers, since the butadiene blocks are much more effective in absorbing impacts than are randomly placed butadiene monomers.

3.3.4   Graft Copolymers

A graft copolymer has a backbone consisting of one type of monomer, and branches of another (Fig. 3.9). For example, high impact polystyrene is formed from a polystyrene backbone with grafted polybutadiene branches.

```
A---AAAAAAA--AAAAAAAA--AAAAAAAAA--AAAAAA---A
       B                    B            B
       B                    B            B
       B                    B            B
       B                    ⋮            B
       B                    B            B
       ⋮                                 ⋮
       B                                 B
```

Figure 3.9 Diagram of a graft copolymer. Capital letters represent different monomers.

The properties of graft copolymers are similar to those of block copolymers. They have the same tendency to arrange themselves in microregions consisting predominantly of each of the types of monomer units. Both block and graft copolymers have special usefulness in adhesives designed for joining together two dissimilar polymers. Since polymers are most strongly attracted to other materials like themselves, a block copolymer produced with units similar to each of the plastics to be joined will tend to arrange itself with the substrate A types of units next to substrate A, and the substrate B types of units next to substrate B. Since the copolymer units are joined by covalent bonds, the adhesive effectively ties the whole structure together.

Graft copolymers are commonly produced by building reactive sites into a linear polymer. Then in a subsequent reaction, polymerization by the comonomer is carried out at these reactive sites. For example, the incorporation of random vinyl bromide units in polystyrene provides sites for subsequent production of a graft polystyrene-polymethyl methacrylate copolymer.

3.3.5 Combinations of Copolymer Types

Copolymers can also be a combination of types. For example, acrylonitrile/butadiene/styrene (ABS) is a two-phase polymer system that combines a random copolymer of styrene and acrylonitrile (SAN) and a dispersed graft copolymer made of butadiene rubber grafted onto the SAN backbone (Fig. 3.10).

```
A---AAABBABBAAAABBBBAABABABBBBBBAAABABBBABA---BA
         C                          C
          C                          C
           C                          C--CC
            C--CC
```

Figure 3.10 Diagram of a graft copolymer with a random copolymer backbone. Capital letters represent different monomers.

3.4 Chain Polymerization, Addition Polymers

In 1929, Wallace H. Carothers classified polymers as *addition polymers* or *condensation polymers* based on their structure. Addition polymers are those that are formed by "adding" the whole monomer into the chain, resulting in a polymer in which the constitutional unit is the same as the monomeric unit (or one in which the monomer is a multiple of the constitutional unit, such as in the case of polyethylene). Condensation polymers, on the other hand, are produced by a condensation reaction in which, usually, a small byproduct molecule is formed as each unit is added into the growing chain. Therefore the polymer's constitutional unit is different from the monomer.

Later, in 1953, P. J. Flory divided the polymers by their reaction mechanism into *chain-reaction* and *step-reaction* polymers, rather than by comparing the polymer's constitutional unit and the monomer. The addition polymers are generally produced by a chain reaction mechanism, and the condensation polymers produced by a step-reaction mechanism. Currently it is customary, though not scientifically correct, to refer to *addition or chain-reaction polymerization*, and to *condensation or step-reaction* polymerization. Some have suggested that the classification of polymers should also include another categorization: whether the polymers are produced by incorporating monomers only, or whether large oligomers take part in the reaction, with or without the formation of byproducts, so that reactions such as $P_z + P_x \rightarrow P_n$ occur.

We will first discuss addition or chain-reaction polymerization, and will discuss condensation or step-reaction polymers in Section 3.8. Addition polymers used in packaging include, among others, polyethylene, polypropylene, polyvinyl chloride, and polystyrene. Polyesters, nylons, and polycarbonate are condensation polymers.

3.4.1   Addition or Chain Polymerization

Chain polymers are synthesized by bonding *unsaturated* monomers, most commonly containing one double bond, together to form the polymer chain. Monomers containing more than one double bond or even triple bonds can also be polymerized by this mechanism. Because the double bond "opens up," the polymerization process occurs by simply adding the monomers without producing any molecular byproducts. The process can be represented as

$$P_n + M \rightarrow P_{n+1}$$

where P_n is a growing chain which already incorporates n monomer units, reacting with a monomer M to yield a larger chain, P_{n+1}. For example, the synthesis of polyethylene from ethylene is shown in Figure 3.11.

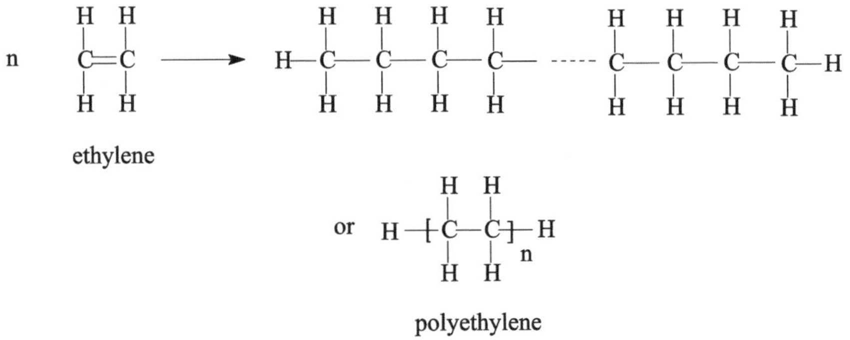

Figure 3.11 Synthesis of polyethylene.

Similarly, the formation of many other addition polymers can be represented by Figure 3.12.

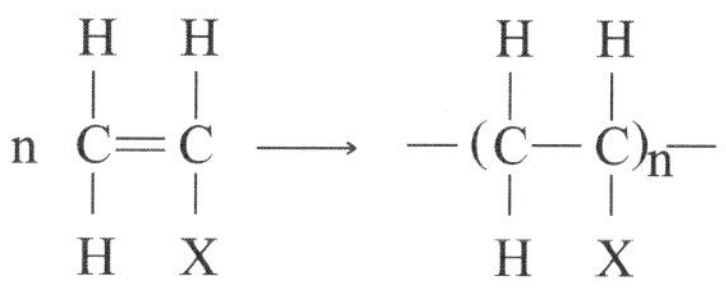

$$n \quad \underset{\underset{H}{|}}{\overset{\overset{H}{|}}{C}} = \underset{\underset{X}{|}}{\overset{\overset{H}{|}}{C}} \longrightarrow -(\underset{\underset{H}{|}}{\overset{\overset{H}{|}}{C}} - \underset{\underset{X}{|}}{\overset{\overset{H}{|}}{C}})_n-$$

Figure 3.12 Addition polymerization

3.4.2 Vinyl Polymers

Addition polymers formed from monomers with the structure $CH_2=CHX$, as shown in Figure 3.12, are called *vinyl* polymers. Table 3.1 shows a number of common packaging addition polymers that differ in the chemical group appearing in the X position in the vinyl molecule.

A second class of vinyl monomers has the structure $CH_2=CXY$. An example is the vinylidene chloride monomer, where both X and Y are chlorine molecules:

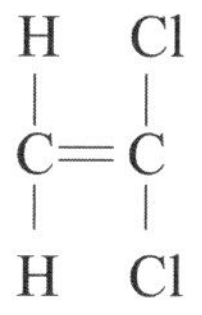

$$\underset{\underset{H}{|}}{\overset{\overset{H}{|}}{C}} = \underset{\underset{Cl}{|}}{\overset{\overset{Cl}{|}}{C}}$$

which produces poly(vinylidene chloride) (PVDC):

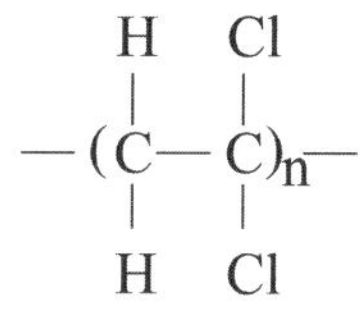

$$-(\underset{\underset{H}{|}}{\overset{\overset{H}{|}}{C}} - \underset{\underset{Cl}{|}}{\overset{\overset{Cl}{|}}{C}})_n-$$

Monomers in which F is substituted for H are also regarded as vinyl polymers. In tetrafluoroethylene $CF_2=CF_2$, all four hydrogens have been replaced by fluorine. The resultant polymer is poly(tetrafluoroethylene) (PTFE), or Teflon. PTFE is a polymer characterized by very low friction, giving it non-stick properties, and very high chemical resistance.

Table 3.1 Common Vinyl Addition Polymers of the Form $CH_2 = CHX$

X	Monomer	Polymer Name
$-CH_3$	Propylene	Poly(propylene) (PP)
	Styrene	Poly(styrene) (PS)
$-Cl$	Vinyl chloride	Poly(vinylchloride) (PVC)
	Vinyl acetate	Poly(vinyl acetate) (PVA)
$-C \equiv N$	Acrylonitrile	Poly(acrylonitrile) (PAN)

Propylene:

$$\begin{array}{cc} H & H \\ | & | \\ C & = C \\ | & | \\ H & CH_3 \end{array}$$

Poly(propylene) (PP):

$$-(\overset{\displaystyle H}{\underset{\displaystyle H}{C}} - \overset{\displaystyle H}{\underset{\displaystyle CH_3}{C}})_n-$$

Styrene:

$$\begin{array}{cc} H & H \\ | & | \\ C & = C \\ | & | \\ H & C_6H_5 \end{array}$$

Poly(styrene) (PS):

$$-(\overset{\displaystyle H}{\underset{\displaystyle H}{C}} - \overset{\displaystyle H}{\underset{\displaystyle C_6H_5}{C}})_n-$$

Vinyl chloride:

$$\begin{array}{cc} H & Cl \\ | & | \\ C & = C \\ | & | \\ H & H \end{array}$$

Poly(vinylchloride) (PVC):

$$-(\overset{\displaystyle H}{\underset{\displaystyle H}{C}} - \overset{\displaystyle Cl}{\underset{\displaystyle H}{C}})_n-$$

Vinyl acetate:

$$\begin{array}{cc} H & H \\ | & | \\ C & = C \\ | & | \\ H & O \\ & | \\ & C = O \\ & | \\ & CH_3 \end{array}$$

Poly(vinyl acetate) (PVA):

$$-(\overset{\displaystyle H}{\underset{\displaystyle H}{C}} - \overset{\displaystyle H}{\underset{\displaystyle O}{C}})_n-$$
$$C = O$$
$$CH_3$$

Acrylonitrile:

$$\begin{array}{cc} H & H \\ | & | \\ C & = C \\ | & | \\ H & C \equiv N \end{array}$$

Poly(acrylonitrile) (PAN):

$$-(\overset{\displaystyle H}{\underset{\displaystyle H}{C}} - \overset{\displaystyle H}{\underset{\displaystyle C \equiv N}{C}})_n-$$

3.4.3 Free-Radical Polymerization

Addition polymers are most often produced by a free radical polymerization mechanism. A free radical is a molecular fragment that contains an unpaired electron. It can result from the decomposition of a molecule, or the reaction of a molecule with another free radical. Free radicals are energetic species which are able to attack the relatively weak C=C bond, resulting in incorporation of the monomer in the growing chain, and generation of a larger free radical. These reactions produce very long chains very quickly.

Three basic steps are involved in free-radical polymerization: initiation, which begins the chain growth; propagation, which increases the size of the polymer molecule; and termination, which ends the growth of the molecule.

3.4.3.1 Initiation

Initiation has two steps, the formation of the free radical and the addition of the first monomer. The initial generation of a free radical is usually accomplished by the decomposition of an *initiator*, which is a relatively unstable compound such as a peroxide (R-O-O-R'), hydroperoxide (R-O-O-H) or azo compound (R-N=N-R'). Sometimes UV radiation and high temperatures are used to generate free radicals. After the initiator free radical is formed, the addition of the first monomer unit to the chain is accomplished, as shown in Figure 3.13.

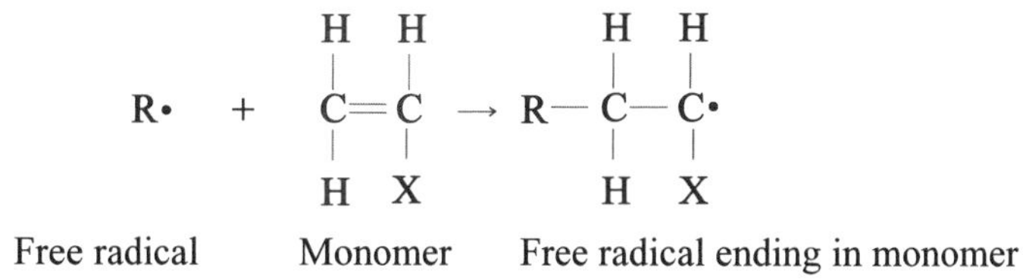

Figure 3.13 The initiation step in free-radical polymerization.

3.4.3.2 Propagation

During propagation, monomer is added to the growing chain, producing an increase in the size of the polymer molecule, as shown in Figure 3.14.

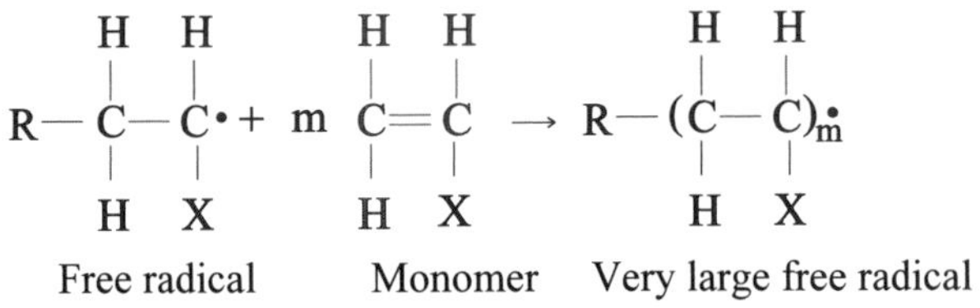

$$R-\overset{\overset{\displaystyle H}{|}}{\underset{\underset{\displaystyle H}{|}}{C}}-\overset{\overset{\displaystyle H}{|}}{\underset{\underset{\displaystyle X}{|}}{C}} \bullet + \; m \; \overset{\overset{\displaystyle H}{|}}{\underset{\underset{\displaystyle H}{|}}{C}}=\overset{\overset{\displaystyle H}{|}}{\underset{\underset{\displaystyle X}{|}}{C}} \;\rightarrow\; R-(\overset{\overset{\displaystyle H}{|}}{\underset{\underset{\displaystyle H}{|}}{C}}-\overset{\overset{\displaystyle H}{|}}{\underset{\underset{\displaystyle X}{|}}{C}})_m^{\bullet}$$

Free radical   Monomer   Very large free radical

Figure 3.14 The propagation step in free-radical polymerization.

3.4.3.3 Termination

Termination occurs when something happens to end the growth of the molecule. There are a number of termination mechanisms. The one illustrated in Figure 3.15 shows two large free radicals recombining.

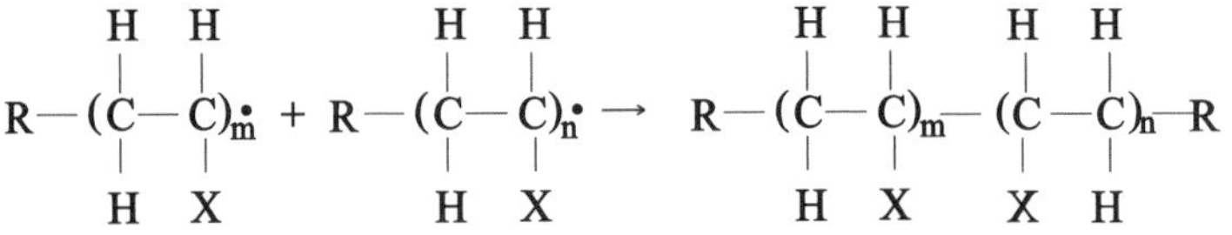

$$R-(\overset{\overset{\displaystyle H}{|}}{\underset{\underset{\displaystyle H}{|}}{C}}-\overset{\overset{\displaystyle H}{|}}{\underset{\underset{\displaystyle X}{|}}{C}})_m^{\bullet} + R-(\overset{\overset{\displaystyle H}{|}}{\underset{\underset{\displaystyle H}{|}}{C}}-\overset{\overset{\displaystyle H}{|}}{\underset{\underset{\displaystyle X}{|}}{C}})_n^{\bullet} \;\rightarrow\; R-(\overset{\overset{\displaystyle H}{|}}{\underset{\underset{\displaystyle H}{|}}{C}}-\overset{\overset{\displaystyle H}{|}}{\underset{\underset{\displaystyle X}{|}}{C}})_m-(\overset{\overset{\displaystyle H}{|}}{\underset{\underset{\displaystyle X}{|}}{C}}-\overset{\overset{\displaystyle H}{|}}{\underset{\underset{\displaystyle H}{|}}{C}})_n-R$$

Figure 3.15 An example of termination by recombination in free-radical polymerization.

3.4.4   Polyethylene Polymerization Processes

Polyethylene, the simplest addition polymer, is also the most important packaging polymer, and we will briefly describe its polymerization process. Polyethylene, as discussed above, is made by opening the double bond in the ethylene molecule, and chemically bonding the monomers together in a reactor. That reactor can involve an autoclave (stirred tank) process or a tubular process. It can be done at low pressure (about 300 psi) or at pressures as high as 50,000 psi. Temperatures are controlled at some elevated level such as 125-250°C, but the temperature needed is very specific to the type of polymer structure desired.

Autoclaves are large stirred vessels, as shown in Figure 3.16(a). In high pressure polymerization of ethylene in an autoclave, ethylene itself acts as a solvent for the forming polymer. At the extreme pressures used, 15,000-50,000 psi, ethylene is above its critical point, and therefore acts as both a gas and a liquid at the same time. Highly branched low density polyethylene, such as that used for blow molding, is made this way. Autoclave polymers are highly branched materials because the free radical initiators and the monomer are being randomly mixed with the growing polymer chain at high energy conditions. Under these circumstances, a free radical can react with a monomer to increase the size of the growing molecule, as shown in Figure 3.14, or it can abstract a proton from a polymer chain to create a reaction site internal to that chain, as shown in Figure 3.17. If the site of the attack is on the same chain near the free radical, as happens most often, a short chain branch is produced. When the site is far from the end of the molecule, a long chain branch is produced.

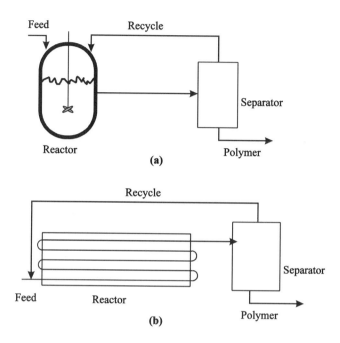

Figure 3.16 Polyethylene reactors: (a) autoclave reactor, (b) tubular reactor.

Tubular reactors, Figure 3.16(b), are used to make film grade LDPE. The flow through the reactor pipe is turbulent, so there is good mixing in the radial direction; however, the blunt velocity profile limits the back mixing that occurs. Free radical initiators are injected into the flow at various spots along the path to cause the polymerization to occur, and to control the branching. Because the reaction zone is limited in extent, the amount of branching is more limited than in an autoclave.

For high density polyethylene, polymerization is usually carried out using Ziegler-Natta (ZN) catalysts that control the combining of the monomers. Figure 3.18 shows the general concept, and Figure 3.19 gives details of the addition process. Because the combining only occurs on the catalyst surface, at relatively low energy conditions, there is almost no branching.

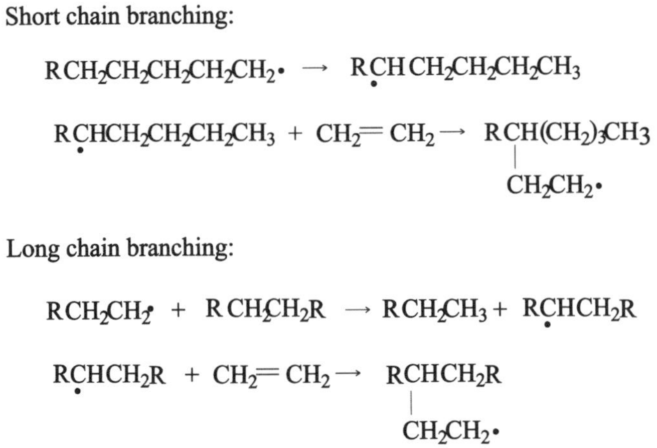

Short chain branching:

$$RCH_2CH_2CH_2CH_2CH_2\cdot \;\rightarrow\; R\overset{\cdot}{C}HCH_2CH_2CH_2CH_3$$

$$R\overset{\cdot}{C}HCH_2CH_2CH_2CH_3 + CH_2\!\!=\!\!CH_2 \rightarrow RCH(CH_2)_3CH_3$$
$$\qquad\qquad\qquad\qquad\qquad\qquad\qquad | $$
$$\qquad\qquad\qquad\qquad\qquad\qquad\quad CH_2CH_2\cdot$$

Long chain branching:

$$RCH_2CH_2^\cdot + RCH_2CH_2R \;\rightarrow\; RCH_2CH_3 + R\overset{\cdot}{C}HCH_2R$$

$$R\overset{\cdot}{C}HCH_2R + CH_2\!\!=\!\!CH_2 \rightarrow RCHCH_2R$$
$$\qquad\qquad\qquad\qquad\qquad\qquad\quad | $$
$$\qquad\qquad\qquad\qquad\qquad\qquad CH_2CH_2\cdot$$

Figure 3.17 Chain branching in polyethylene, illustrating how chain transfer reactions can yield short or long chain branches.

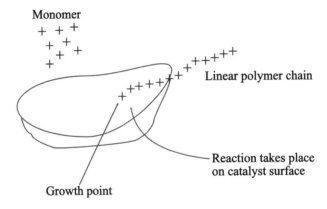

Figure 3.18 The effect of the catalyst surface in shaping or "tailoring" the molecule is of considerable importance.

To provide the effect of limited branching, which helps the processing characteristics of HDPE, a small quantity of hexene is typically introduced. Hexene is also used to control the crystallinity and reduce the density of HDPE to meet various end-use demands. For film grades, density is usually about 0.956 g/cm^3, which means that there are less than 10 branches/1000 carbon atoms in the polymer.

Because the polymerization is aided by a catalyst, it can be carried out at low pressures, usually around 300 psi. The polymerization can be done in slurry, solution, or gas phase. Solution phase polymerization is like that described previously, in that the polymer is dissolved in the monomer. In the slurry phase process, a carrying solvent is used to dissolve the monomer and suspend the catalyst. Once the polymer reaches a high molecular weight, it becomes insoluble and creates the slurry as it drops out of solution. Gas phase polymerization is carried out in a fluidized bed of catalyst in a vertical tower. The ethylene percolates up through the catalyst bed, fluidizing it, polymerizes, and precipitates out as a powder that is drawn off the bottom of the tower. Molecular weight distribution is controlled by the choice of catalyst and reactor conditions. It can also be dramatically changed by using two, or more, reactors in series with different reaction conditions. This approach is used to create bimodal HMW-HDPE, used for grocery sacks and other applications requiring high stiffness.

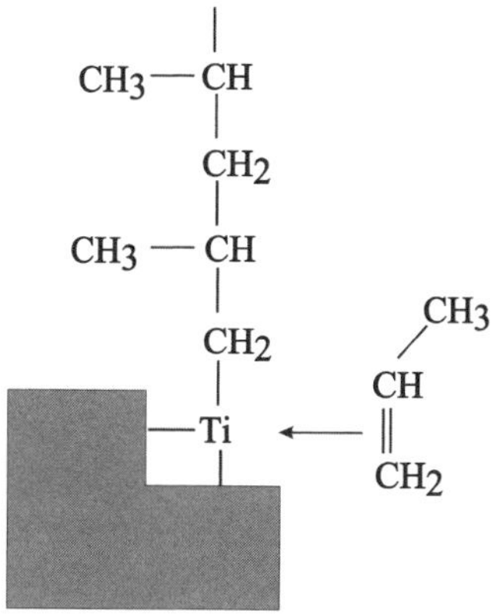

$$
\begin{array}{c}
| \\
CH_3 \!-\! CH \\
| \\
CH_2 \\
|
\end{array}
$$

CH$_3$ — CH
|
CH$_2$
|
— Ti ← CH
||
CH$_2$

CH$_3$

Figure 3.19 The titanium atom is a key in facilitating the polymerization reaction.

Chromium-based Ziegler-Natta catalysts leave a terminal unsaturation in some of the molecules; that is, there is a double bond left between the last two carbon atoms in the chain. This unsaturation is a reactive site, and it can react during further processing, such as in compounding of additives or pelletizing, with free radicals or other active species which may be generated. The result is the formation of a limited amount of long chain branching, LCB. This LCB causes the polymers made from chromium catalysts to have some unique properties. We will discuss the effects of LCB further in later parts of the book.

As has been indicated, the catalyst particle sites control the structural arrangement of the polymer. Unfortunately, not all of the sites on a ZN catalyst behave the same. There are always some sites that produce a low molecular weight, highly branched material. In HDPE, this waxy material must be removed to produce a polymer with the desired characteristics.

If the reaction is to produce linear low density PE, using a comonomer to produce the side groups that interfere with crystallization, the site characteristics also affect the resulting polymer. On ZN catalysts, there are sites that tend to produce the highly branched, low molecular weight material, while others produce mainly a high molecular weight polymer with very little branching - an HDPE material. The fact that the sites do not all incorporate the comonomer with equal efficiency affects the end-use properties of the polymer. These facts do not make the process useless, but they do complicate the interchangeability of grades of LLDPE from different suppliers or even within the same supplier, when using difference catalyst systems causes the grade differences. The branching characteristics can be measured by using Temperature Rising Elution Fractionation (TREF) to characterize the materials. The best characterization is the measurement of end-use properties necessary for the packaging material. The point is that LLDPE polymers may not be interchangeable, even if the melt index (MI) and the density are the same.

Additional discussion of these catalyst systems can be found in Section 4.2.

3.4.5   Other Addition Polymerization Mechanisms

Addition polymers can also be produced by mechanisms that involve ionic intermediates, reacting in much the same way as illustrated for free radical intermediates. These types of polymerization are used much less frequently than the free radical types, but are very important in some specialty reactions, such as preparation of block copolymers.

3.5   Molecular Configuration and Conformation

The configuration of a molecule refers to the fixed arrangement of the atoms in the molecule, which is determined by the chemical bonds that have been formed. The configuration of a polymer chain *cannot* be altered unless chemical bonds are broken and reformed. The linear, branched, or cross-linked architecture of polymer molecules, and the different types of copolymers discussed in Section 3.3 are examples of different molecular configurations. Even within a linear homopolymer, there can be different configurations of molecules, as will be explained in Sections 3.6 and 3.7.

In contrast, the conformation of a molecule is its arrangement in space. Different conformations can arise from rotation about single bonds. Thus, one conformational state of a molecule can be changed to another conformational state without requiring the breaking of primary chemical bonds. When a polymer molecule has sufficient energy, such as in the melt or in solution, it rapidly changes from one conformation to another. In the solid phase, the ability of a molecule to change its conformation is a function of temperature and of degree of crystallinity, as will be discussed later.

3.6 Head-to-Head and Head-to-Tail Configurations of Vinyl Polymers

The addition of a vinyl monomer to a free radical during the polymerization process (propagation phase) can take place in either of two ways:

(1)

$$
\begin{array}{cccccccc}
& H & X & & H & X & & & H & X & H & X \\
& | & | & & | & | & & & | & | & | & | \\
R- & C- & C\bullet & + & C= & C & \rightarrow & R- & C- & C- & C- & C\bullet \\
& | & | & & | & | & & & | & | & | & | \\
& H & H & & H & H & & & H & H & H & H
\end{array}
$$

or

(2)

$$
\begin{array}{cccccccc}
& H & X & & X & H & & & H & X & X & H \\
& | & | & & | & | & & & | & | & | & | \\
R- & C- & C\bullet & + & C= & C & \rightarrow & R- & C- & C- & C- & C\bullet \\
& | & | & & | & | & & & | & | & | & | \\
& H & H & & H & H & & & H & H & H & H
\end{array}
$$

The CHX group is considered the "head" of the monomer, and the CH_2 group the tail. If only reaction (1) occurs, the resulting polymer is said to have a head-to-tail configuration, where the X group is attached on every other carbon in the main chain:

$$
\begin{array}{cccc}
X & X & X & X \\
| & | & | & | \\
-CH_2-CH-CH_2-CH-CH_2-CH-CH_2-CH-
\end{array}
$$

If both reactions (1) and (2) take place, both head-to-head and tail-to-tail configurations occur randomly in the molecule, along with head-to-tail configurations:

$$
\begin{array}{cccc}
X & X & X & X \\
| & | & | & | \\
-CH_2-CH-CH-CH_2-CH_2-CH-CH_2-CH-
\end{array}
$$

If only reaction (2) occurs, the molecule would have a head-to-head/tail-to-tail configuration. However, this generally does not occur.

In most cases, free radical polymerization of vinyl polymers results in predominantly head-to-tail addition. The major reason is the increased stability of the free radical that results from reaction (1) compared to the free radical resulting from reaction (2). Carbon free radicals are classified according to the number of carbon atoms attached to the carbon with the unpaired electron (Fig. 3.20). The carbon in a primary free radicals is attached to only one other carbon atom, secondary free radicals are attached to two, and tertiary free radicals are attached to three. Tertiary free radicals are the most stable, secondary radicals the next, and primary free radicals the least stable. The more unstable a free radical is, the more energy it takes to form it from a stable molecule. Therefore, the energy conditions are most favorable for producing tertiary free radicals, followed by secondary, and lastly by primary. Since reaction (1) results in a secondary free radical, it is favored energetically over reaction (2), which results in a primary free radical.

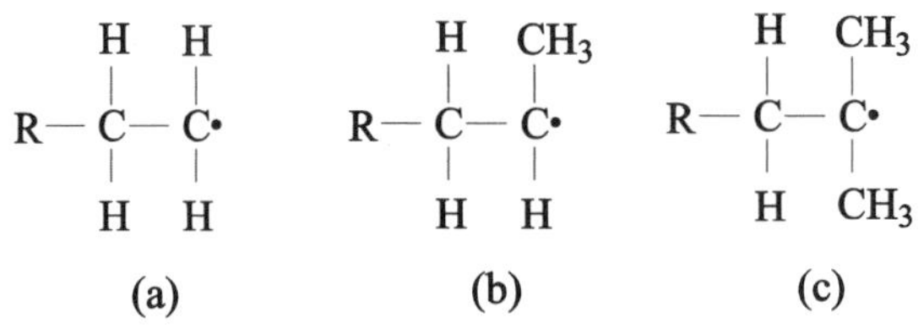

$$
\begin{array}{ccc}
\quad\quad \text{H}\quad\text{H} & \quad\quad \text{H}\quad\text{CH}_3 & \quad\quad \text{H}\quad\text{CH}_3 \\
\quad\quad |\quad\quad| & \quad\quad |\quad\quad\;\; | & \quad\quad |\quad\quad\;\; | \\
\text{R}-\text{C}-\text{C}\bullet & \text{R}-\text{C}-\text{C}\bullet & \text{R}-\text{C}-\text{C}\bullet \\
\quad\quad |\quad\quad| & \quad\quad |\quad\quad| & \quad\quad |\quad\quad\;\; | \\
\quad\quad \text{H}\quad\text{H} & \quad\quad \text{H}\quad\text{H} & \quad\quad \text{H}\quad\text{CH}_3 \\
(a) & (b) & (c)
\end{array}
$$

Figure 3.20 Examples of (a) a primary free radical, (b) a secondary free radical, and (c) a tertiary free radical.

The configuration affects the properties of the polymer. For example, occasional head-to-head addition sites in predominantly head-to-tail polyvinyl chloride are known to be more susceptible to the initiation of thermal degradation of the polymer than the regular head-to-tail locations.

3.7   Stereochemistry

The configuration of a polymer molecule is dependent not only on which atoms are attached to which others, but also on the three-dimensional arrangement of the atoms. These different arrangements in three-dimensional space are referred to as the stereochemistry of the molecule. For tactic polymers like polypropylene, three different patterns, called symmetry configurations or stereoregularity, exist, as illustrated in Figure 3.21.

In order to distinguish one configuration from another, we must standardize the conformation, so that we can separate differences in appearance caused by bond rotation from differences in the chemical structure of the molecule. Therefore, we look at the molecule in its most extended form. For a polymer with a main chain composed only of carbon atoms, this is a zig-zag, with the carbon atoms all in one plane, having their normal bond angle of 109.5°. Then we can look at the positioning of the groups attached to the chain. In particular, for polypropylene, we look at the positioning of the methyl groups. If all the methyl groups are on the same side of the chain, we have isotactic PP.

If they alternate sides in a regular pattern, we have syndiotactic PP. If there is no order to the side of the chain they appear on, we have atactic PP.

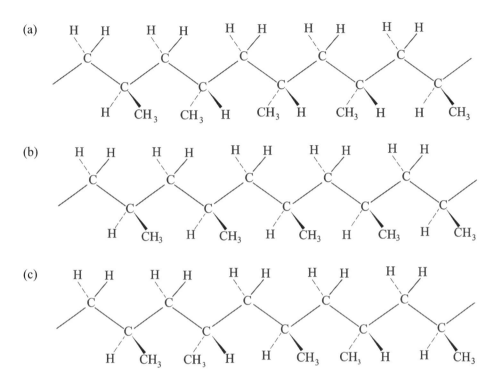

Figure 3.21 Types of polypropylene: (a) atactic, (b) isotactic, (c) syndiotactic.

The stereochemistry, or tacticity, of the molecule is important in determining whether or not the polymer is able to crystallize, as will be discussed in Section 3.10.1.

The existence of different stereochemical configurations, which by definition cannot be converted from one to another by rotation of bonds, is dependent on the inclusion of a *chiral* carbon in the molecule. A chiral carbon is one which has four different substituents attached, and which therefore can exist in two mirror-image forms, as illustrated in Figure 3.22. In a polymer molecule, any main-chain carbon is attached to at least three different substituents - each end of the molecule and a hydrogen or other side group. Therefore, determining whether a carbon is chiral requires looking only at the two side groups. If they are identical, such as in -CH_2-, the carbon is not chiral. If they are different, as in:

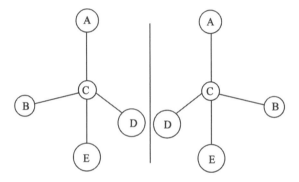

then the carbon is chiral. A polymer that contains chiral carbons can be either atactic, isotactic, or syndiotactic. An isotactic polymer is one in which the two different substituent groups always occur on the same side of the fully extended polymer chain. In a syndiotactic polymer, the groups alternate sides of the chain. In an atactic polymer, the side of the chain on which each group appears is random - there is no pattern.

Figure 3.22 The two mirror-image forms of a chiral carbon.

3.8   Step Polymerization, Condensation Polymers

Condensation polymers are formed from monomers with functional (reactive) groups such as:

$$\underset{\text{acids}}{-\overset{\overset{\displaystyle O}{\|}}{C}OH} \qquad \underset{\text{alcohols}}{-COH} \qquad \underset{\text{amines}}{-NH_2}$$

The first step in a step reaction mechanism is the formation of esters or amides from the diols and diacids, or diacids and diamines, respectively. From these intermediates, the polymerization reaction (second step) proceeds. Because the first step is a faster reaction than the second, the monomer is used up quickly. During the two steps of the reaction, small molecules such as H_2O or CH_3OH are eliminated. Water is the most frequent byproduct molecule, for example from the reaction between a diacid and a dialcohol. Unlike addition polymers, condensation polymers, because they incorporate functional groups, generally have non-carbon atoms (hetero-atoms) as part of the main backbone chain. Examples are shown in Figure 3.23.

(a)

$$-O-\overset{\overset{\textstyle O}{\|}}{C}-R_1-\overset{\overset{\textstyle O}{\|}}{C}-O-R_2-O-\overset{\overset{\textstyle O}{\|}}{C}-R_1-\overset{\overset{\textstyle O}{\|}}{C}-O-$$

(b)

$$-R_1-\overset{\overset{\textstyle O}{\|}}{C}-\overset{\overset{\textstyle H}{|}}{N}-R_2-\overset{\overset{\textstyle H}{|}}{N}-\overset{\overset{\textstyle O}{\|}}{C}-R_1-\overset{\overset{\textstyle O}{\|}}{C}-\overset{\overset{\textstyle H}{|}}{N}-R_2-\overset{\overset{\textstyle H}{|}}{N}-$$

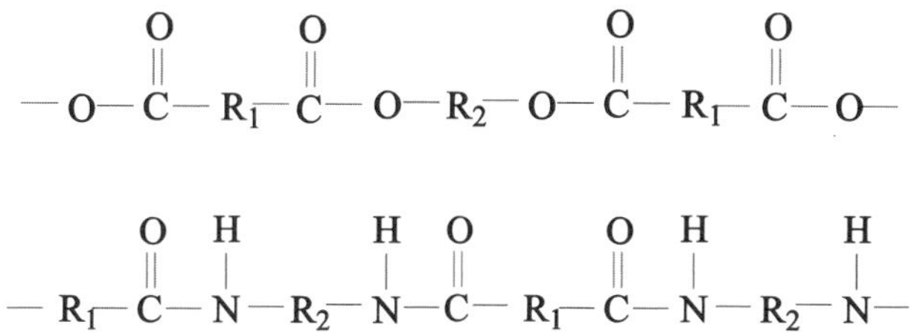

Figure 3.23 Examples of condensation polymers: (a) polyester, (b) polyamide.

The formation of condensation polymers requires that the monomers contain groups with appropriate functionality, such as:

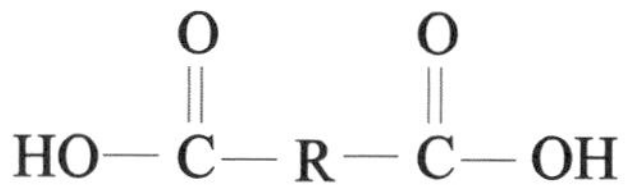

$$HO-\overset{\overset{\textstyle O}{\|}}{C}-R-\overset{\overset{\textstyle O}{\|}}{C}-OH$$
di-acid

$$HO-CH_2-R-CH_2-OH$$
di-alcohol

$$H_2N-CH_2-R-CH_2-NH_2$$
di-amine

The functionality of the monomers must be at least 2 to form a polymer. If the functionality is 2, a linear polymer will result. For a cross-linked polymer, the functionality must be 3 or more.

The reaction between a di-alcohol and a di-acid, for example, proceeds as follows:

Step 1

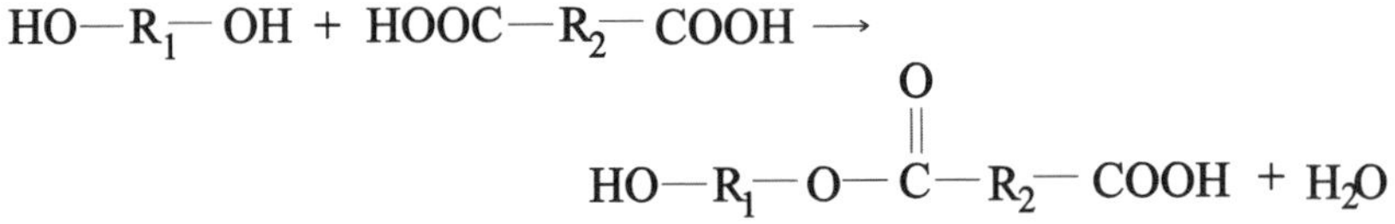

$$HO{-}R_1{-}OH + HOOC{-}R_2{-}COOH \longrightarrow$$

$$HO{-}R_1{-}O{-}\overset{\overset{\displaystyle O}{\|}}{C}{-}R_2{-}COOH + H_2O$$

Step 2

$$2\ HO{-}R_1{-}O{-}\overset{\overset{\displaystyle O}{\|}}{C}{-}R_2{-}COOH \longrightarrow$$

$$HO{-}R_1{-}O{-}\overset{\overset{\displaystyle O}{\|}}{C}{-}R_2{-}\overset{\overset{\displaystyle O}{\|}}{C}{-}O{-}R_1{-}O{-}\overset{\overset{\displaystyle O}{\|}}{C}{-}R_2{-}COOH + H_2O$$

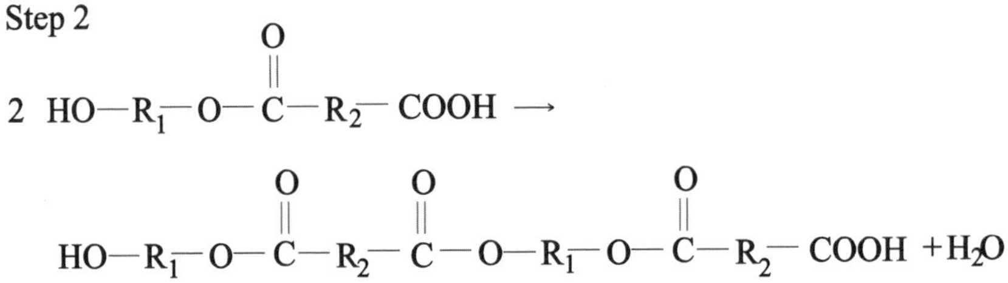

with subsequent reactions analogous to Step 2 combining intermediate esters of varying sizes to produce a polyester.

Example 1:

The reaction between n molecules of hexamethylene diamine

$$H_2N\text{-}(CH_2)_6\text{-}NH_2$$

and n molecules of adipic acid

$$HOOC\text{-}(CH_2)_4\text{-}COOH$$

yields the polyamide Nylon 6,6

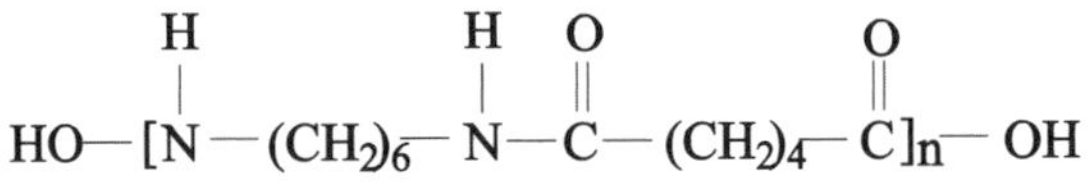

$$HO{-}[\overset{\overset{\displaystyle H}{|}}{N}{-}(CH_2)_6{-}\overset{\overset{\displaystyle H}{|}}{N}{-}\overset{\overset{\displaystyle O}{\|}}{C}{-}(CH_2)_4{-}\overset{\overset{\displaystyle O}{\|}}{C}]_n{-}OH$$

plus (2n-1) molecules of water, H_2O.

Example 2:

The reaction of n molecules of ethylene glycol

$$HO - CH_2 - CH_2 - OH$$

and n molecules of terephthalic acid

yields the polyester polyethylene terephthalate (PET)

plus (2n-1) molecules of water.

Example 3:

Polycarbonate is also produced by reactions involving the formation of ester groups, when carbonic acid

(a derivative of carbonic acid, $O=C-Cl_2$ phosgene, is actually used) reacts with bisphenol A

to yield poly(bisphenol A carbonate) or polycarbonate, PC

As illustrated, the condensation polymer molecules could be viewed as formed by the alternation of the two different monomers needed for the chemical reactions.

$$- M_1 - M_2 - M_1 - M_2 - M_1 - M_2 - M_1 - M_2 - M_1 - M_2 -$$

While this alternating order in placement of the monomers could be considered to satisfy the definition of an alternating copolymer, as was said earlier, such condensation polymers are generally *not* considered copolymers, since it is the intermediates, the esters or amides, which are actually forming the polymer, and they are identical. The term copolymer is reserved for condensation polymers formed from more than one type of monomer with a given functional group, such as using two different diacids in the polymerization, which will give rise to two different intermediates.

In some cases, a single molecule contains two different types of functional groups that are able to react with each other. In that case, a condensation polymer can be produced using only one type of monomer. The most common case is nylons polymerized from amino acids:

$$n\ H_2N-R-COOH\ \longrightarrow\ H_2N-(R-\overset{\overset{\displaystyle O}{\|}}{C}-\overset{\overset{\displaystyle H}{|}}{N})_n-H\ +\ n\ H_2O$$

3.9   Molecular Weight and Molecular Weight Distribution

Besides the chemical composition of the polymer molecule, its size, i.e. the number of monomers connected together in the molecule, is the most fundamental structural characteristic determining the properties of the polymer. By chemically connecting monomers we move from simple molecules (monomers), which are mostly gases at normal pressure and temperature conditions, to liquid, then to waxy material (oligomers), and finally to hard solids (polymers), just by adding more and more monomers. For this reason, there is a practical interest in defining and determining the size of polymers, which is normally expressed as molecular mass or molecular weight. In this section we describe different ways of describing molecular size, including degree of polymerization, average molecular weight, and molecular weight distribution. We also discuss experimental methods for determining molecular weight, and the relationship between molecular size and polymer properties.

3.9.1 Degree of Polymerization

The size of a polymeric molecule can be described by its degree of polymerization, n, which is the number of monomeric units linked together in the polymer chain. The degree of polymerization relates to the molecular mass of the polymer and the monomer as follows:

$$n = \frac{M}{M_m} \tag{3.1}$$

where M is the molar mass of the polymer chain (or molecular weight) and M_m is the molar mass of the monomer.

Example 1: What is the degree of polymerization of a PE sample with a molar mass of 280,000 daltons?

Molar mass of ethylene = 28 daltons (or g/mol)

n = 280,000 daltons/28 daltons = 10,000

Typical values of n for plastics are in the range of 4,000-100,000 monomeric units.

Example 2: Calculate the molecular mass of a sample of poly(vinyl chloride) (PVC) of degree of polymerization n=1.5 x 10^4.

Molecular weight of vinyl chloride, $CH_2=CHCl$ is 62.5 daltons (or g/mol)

M = n M_m = (1.5 x 10^4)(62.5 daltons) = 9.375 x 10^5 daltons

---

3.9.2 Molecular Mass (Weight) and Molecular Weight Distribution

"Relative molecular mass" is a recent concept that is replacing the older "molecular weight" concept. Although the term relative molecular mass is more operationally correct, it has been slow to gain acceptance and we will be using both terms interchangeably to refer to the molar mass of molecules.

The molecular mass is the mass of a single molecule, and is usually experimentally determined. While low molecular weight compounds occur at only one molecular weight (atomic isotopes excepted), the processes used to synthesize polymers never produce a single value of degree of polymerization, and thus never produce polymers with a single molecular weight. A polymer sample normally contains a very large number of macromolecules of different lengths, since the polymerization process is the result of a succession of random reactions. Not only do the lengths of the chains differ, but also the number of molecules having a given value of molecular weight varies. In general, polymers contain a small proportion of molecules which are very short, a small proportion which are very large, and the largest proportion of molecules are in between.

Therefore, when we refer to the molecular weight of a polymer, unless we are talking about a single molecule, we more accurately mean the average molecular weight of the molecules in that polymer. However, polymers with the same average molecular weight can differ significantly in how many molecules are how much above and below that average.

Thus to fully characterize the polymer, we need to describe its distribution of molecular weights. The *molecular weight distribution* (MWD) of a polymer sample is a description of the number of molecules at each chain length value. Although it is almost impossible to know the exact MWD, there are many methods of determining approximate chain lengths (or molecular weights).

Consider a sample of polymer of total mass W made of a series of fractions (intervals of molecular weight values) each with a mass of w_i. Each fraction has a representative value of molecular weight M_i with a number N_i of molecules having that weight M_i. The mass of the sample is the summation of the mass of each fraction, as indicated by

$$W = \sum_i w_i \qquad (3.2)$$

and the mass of each fraction is given by

$$w_i = N_i M_i \qquad (3.3)$$

The total number of molecules, N, is the sum of the N_i values. Using this terminology, we can define several types of average molecular weight.

3.9.3   Number Average Molecular Weight

The number average molecular weight, $\overline{M}_n$, is defined as the total mass of the sample divided by the total number of molecules (or moles of molecules) in the sample.

$$\overline{M}_n = \frac{W}{N} = \frac{\sum_i w_i}{\sum_i N_i} = \frac{\sum_i N_i M_i}{\sum_i N_i} \qquad (3.4)$$

This molecular weight average is called the number average molecular weight because it is based on the total number of molecules in the sample under consideration.

While $\overline{M}_n$ is strongly related to some properties of the polymer, in particular the colligative properties (osmotic pressure, freezing point depression and boiling point elevation), these properties are not significant for packaging materials, so other measures of average molecular weight are more useful.

3.9.4 Weight Average Molecular Weight

The weight average molecular weight is defined as the average molecular weight based on the total weight of the molecules in the sample, the summation of the weight of molecules of each molecular mass in the sample times that mass, divided by the total weight of the sample.

$$\overline{M}_w = \frac{\sum_i w_i M_i}{\sum_i w_i} = \frac{\sum_i w_i M_i}{W} \tag{3.5}$$

$\overline{M}_w$ correlates better than $\overline{M}_n$ with properties of interest for plastics processing and performance, such as melting point, viscosity, tensile strength, and elasticity.

Since $w_i = N_i M_i$, equation (3.5) can be written as

$$\overline{M}_w = \frac{\sum_i N_i M_i^2}{\sum_i N_i M_i} \tag{3.6}$$

Equation (3.6) indicates that $\overline{M}_w$ is more sensitive to high values of M_i than is $\overline{M}_n$. Note that if all the molecules have the same molecular weight, $\overline{M}_w = \overline{M}_n$; in all other cases, $\overline{M}_w > \overline{M}_n$.

The ratio $Q = \overline{M}_w / \overline{M}_n$ is called the *Dispersity* or *Dispersion Index* and gives a measure of the range of the molecular sizes in the sample. Q indicates whether a polymer sample has a narrow or broad distribution of molecular weight. Commercial polymers have dispersion indices above 2. In general, when $Q < 6$ the distribution is said to be narrow.

In Figure 3.24 the mass fraction (w_i / W) of a distribution is plotted versus molecular weight. As can be seen from the figure, $\overline{M}_w > \overline{M}_n$. Four MWD combinations of low and high average molecular weight, and broad and narrow distributions of molecular weight, are shown, as well as a bimodal distribution.

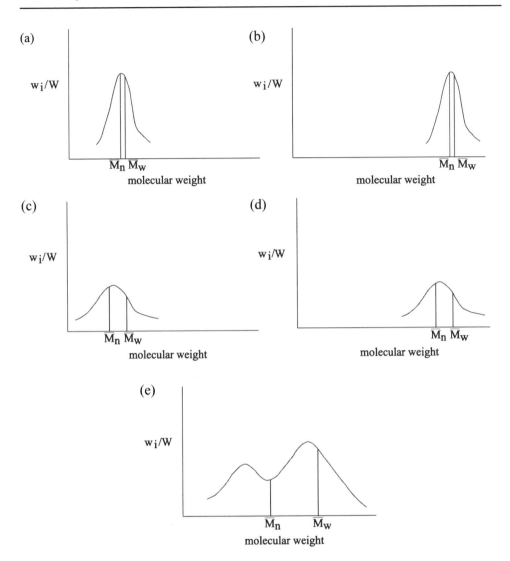

Figure 3.24 Four molecular weight distributions with different average molecular weight and shape, plus a bimodal distribution: (a) low average, narrow, (b) high average, narrow, (c) low average, wide, (d) high average, wide, (e) bimodal.

Example 1: Consider a 5.0 g polymer sample made of 2.5 g of $C_{100}H_{202}$ and 2.5 g of $C_{120}H_{242}$. Calculate $\overline{M}_n$, $\overline{M}_v$, and Q.

$M_1 = 100\,(12) + 202\,(1) = 1402$ daltons; $w_1 = 2.5$ g

$$N_1 = \frac{w_1}{M_1} = \frac{2.5\,\mathrm{g\,mol}}{1402\,\mathrm{g}} = 1.783 \times 10^{-3}\ \mathrm{mol}$$

$M_2 = 120\,(12) + 242\,(1) = 1{,}682$ daltons; $w_2 = 2.5$ g

$$N_2 = \frac{w_2}{M_2} = \frac{2.5\,\mathrm{g\,mol}}{1682\,\mathrm{g}} = 1.486 \times 10^{-3}\ \mathrm{mol}$$

$W = w_1 + w_2 = 5.0$ g; $N_1 + N_2 = 3.269 \times 10^{-3}$ mol

$$\overline{M}_n = \frac{W}{\sum N_i} = \frac{5\,\mathrm{g}}{3.269 \times 10^{-3}\ \mathrm{mol}} = 1{,}530\ \text{daltons}$$

$$\overline{M}_w = \frac{\sum w_i M_i}{W} = \frac{2.5(1402) + 2.5(1682)\ \mathrm{g\,g}}{5.0\ \mathrm{g\,mol}} = 1{,}542\ \text{daltons}$$

$$Q = \frac{\overline{M}_w}{\overline{M}_n} = \frac{1542\ \text{daltons}}{1530\ \text{daltons}} = 1.01$$

Example 2: Consider now that the sample is made of 2.5 g of $C_{100}H_{202}$ and 2.5 g of $C_{1200}H_{2402}$. Note that the second component has a much larger degree of polymerization than in the previous example.

$M_1 = 1{,}402$ daltons; $w_1 = 2.5$ g; $N_1 = 1.783 \times 10^{-3}$ mol; $M_2 = 16{,}802$ g as shown in Example 1; $w_2 = 2.5$ g

$$N_2 = \frac{w_2}{M_2} = \frac{2.5\,\mathrm{g\,mol}}{16{,}802\,\mathrm{g}} = 1.488 \times 10^{-4}\ \mathrm{mol}$$

$$\overline{M}_n = \frac{W}{\sum N_i} = \frac{5\,\mathrm{g}}{1.932 \times 10^{-3}\ \mathrm{mol}} = 2588\ \text{daltons}$$

$$\overline{M}_w = \frac{\sum w_i M_i}{W} = \frac{2.5(1402) + 2.5(16{,}802)\ \mathrm{g\,g}}{5\,\mathrm{g\,mol}} = 9{,}102\ \text{daltons}$$

$$Q = \frac{\overline{M}_w}{\overline{M}_n} = \frac{9102\ \text{daltons}}{2588\ \text{daltons}} = 3.52$$

The results of these examples agree with the following general rules:

1. $\overline{M}_n$ is always smaller than $\overline{M}_w$: $\overline{M}_n < \overline{M}_w$

2. $\overline{M}_n$ is more sensitive to low molecular mass species than $\overline{M}_w$

3. $\overline{M}_w$ is more sensitive to high molecular mass species than $\overline{M}_n$

3.9.5   Other Molecular Weight Averages

Other characterizations of average molecular weight also exist, giving increasing importance to the larger molecules. For example, $\overline{M}_z$, which is strongly related to melt elasticity, is defined as:

$$\overline{M}_z = \frac{\sum_i N_i M_i^3}{\sum_i N_i M_i^2} \tag{3.7}$$

However, for most purposes, $\overline{M}_w$ and Q adequately characterize the MWD of unblended polymers, and are commonly used in the plastics industry.

3.9.6   Determination of MWD

Common methods for the experimental determination of the molecular mass of polymeric materials include:

- Gel permeation chromatography
- Chemical analysis
- Measurement of colligative properties
    Osmometry
    Cryoscopy
- Viscosimetry of dilute solutions
- Light scattering
- Sedimentation
- Field flow fractionation

Gel permeation chromatography is a very popular method that gives a full account of the MWD of a polymer expressed in intervals of chain length values, by separating the molecules according to their chain length.

The most common measure of average molecular weight for polymers relies on viscosity measurements, since these can be done considerably more quickly and simply. The resulting viscosity average molecular weight, $\overline{M}_v$, lies between $\overline{M}_w$ and $\overline{M}_n$, but much closer to $\overline{M}_w$. Although $\overline{M}_v$ is generally determined experimentally, it can be written in the same type of formula as the other molecular weight averages:

$$\overline{M}_v = \frac{\sum_i N_i M_i^x}{\sum_i N_i M_i^{x-1}} \tag{3.8}$$

where x is between 1 and 2, but closer to 2. The exact value of x which gives the best fit of this relationship varies with the polymer.

3.9.7 Effect of Molecular Weight and Molecular Weight Distribution on Flow and Mechanical Properties

Polymer flow, as well as mechanical and thermal properties, is affected by both the average molecular weight and the molecular weight distribution. These effects are due to the manner in which the chains of the polymer interact with one another.

It seems easiest to explain by starting with the behavior of amorphous chain molecules in the melt. At low average molecular weight, polymers are viscous liquids at room temperature. As the molecular weight increases, they become solids at room temperature. Amorphous polymers do not have a melting point as do crystalline materials, but rather continue to soften over a broad temperature range until they are viscous liquids at high temperature. We still refer to the phenomenon as melting. In the melted state the molecules exist in random coil forms as shown in Figure 3.25.

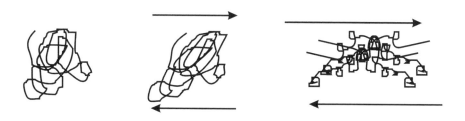

Figure 3.25 Random coil conformation of a polymer molecule and effect of shear.

At sufficiently high molecular weight, polymer molecules tend to associate mainly with themselves in the melt state, but entanglements do occur. That is, the chain of one molecule may penetrate into the coil of another molecule. It has been found that for amorphous polymers, one can define a critical molecular weight, M_c, for these entanglements to occur. This is shown in Figure 3.26. Here we see that at low molecular weights, the zero shear viscosity, η_o, of the melt varies with the molecular weight to the 1 to 1.5 power. Above the critical molecular weight, the relationship changes dramatically, becoming proportional to the molecular weight to the 3.4 power.

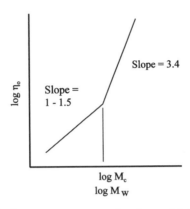

Figure 3.26 Relationship between average molecular weight and viscosity.

Some explanation of terms is appropriate here, including the concept of viscosity. As stated above, the polymers in the melt exist in coiled form. As flow occurs, a shear field is imposed on them due to the velocity gradients found in any flow channel. The shear field causes the molecules to elongate. The molecules that are entangled with one another have additional drag to dissipate energy. As the velocity increases, shear increases, the molecules become more elongated and have less interactions. Therefore the amount of energy dissipated is less than would be expected by the increase in flow rate. At low shear, approaching zero, the melt exhibits Newtonian behavior. That is, the viscosity is constant, and obeys the following equation:

$$\tau = \eta \dot{\gamma} \tag{3.9}$$

where τ = shear stress
η = viscosity
$\dot{\gamma}$ = shear rate

As the shear increases, the flow behavior becomes non-Newtonian and follows an equation of the power law form:

$$\tau = K\dot{\gamma}^{\,n} \tag{3.10}$$

where τ = shear stress
$\dot{\gamma}$ = shear rate
K = consistency index
n = power law index, usually less than 1

This behavior is shown graphically in Figure 3.27. There is even an upper Newtonian region, where the polymer molecules are completely straightened out by the flow field.

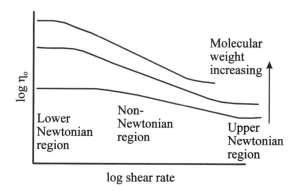

Figure 3.27 Relationship between viscosity and shear rate, as affected by molecular weight.

The molecular entanglements increase as the average molecular weight increases. The lower Newtonian region becomes smaller because the interactions occur at lower shear rates, and the non-Newtonian region persists to higher shear rates. This is illustrated in Figure 3.27.

The effect of broadening the MWD, at a given average molecular weight, is to put smaller molecules into the melt which have lower interactions and act as lubricants for the larger molecules. This means that the melt viscosity will be lowered, and the polymer will become more Newtonian as the MWD is broadened.

The solid state properties are affected similarly to the melt properties, and for similar reasons. The solid form of an amorphous polymer also consists of randomly entangled long chains, but with restricted motion because of limited thermal energy and very high viscosity. The tensile strength of a polymer is the measure of its resistance to deformation in the solid form. As force is applied to the material, the strain in the material increases until the interaction forces between the chains are overcome and slippage begins to occur. At this point we have reached the yield point. As the molecular weight increases, the tensile strength increases because the entanglements are greater. As the MWD is broadened, again the small molecules act as lubricants and the tensile strength is reduced.

From an end-use standpoint, usually one wants the highest molecular weight possible to maximize strength or minimize permeability, etc. From a processing standpoint, this may cause a problem. Commonly, the first stage of processing is to melt the polymer, usually in an extruder. The higher molecular weight raises the viscosity of the melt, which will cause the flow rate to decrease at a given energy input. One may have to put in more energy, which raises the temperature of the melt; or one may exceed the energy available from the extruder motor. Broadening the MWD is one method used to improve the processing conditions (temperature and pressure) in the extruder. As mentioned above, this will degrade some of the physical properties, so this is a balancing act. For mold filling in injection molding, one may need to lower the MW or increase the MWD to allow the part to fill. In blow molding, though, the elongational viscosity of the melt needs to be high enough to support the molten parison, so lowering the MW or broadening the MWD is detrimental to processing. Therefore, one needs to balance the needs of the packaging materials and the processing conditions when choosing the correct MW and MWD.

The molecular weight distribution is also important in determining heat seal temperatures and sealing ranges. A polymer with a narrow molecular weight distribution melts over a narrower range of temperature than a polymer with a broader distribution. Thus, a broad molecular weight distribution creates a broader range of temperature over which good seals can be made, which is generally advantageous in heat sealing.

3.10 Polymer Morphology

Polymer morphology refers to the fine structure of the plastic, including the presence, shape, arrangement, and physical state of the amorphous and crystalline regions that are often found co-existing in a polymer sample. Polymers used in packaging are either amorphous solids or, more frequently, semicrystalline solids which contain both crystalline and amorphous regions inter-mixed. A semicrystalline polymer can be thought of as containing crystalline regions embedded in an amorphous phase. LDPE, HDPE, PET, nylon 6, and EVOH are examples of semicrystalline polymers, while general purpose polystyrene, butadiene-styrene copolymers, some polyester copolymers, and nylon 6I/6T(hexamethylene diamine polymerized with terephthalic and isophthalic acid) are amorphous polymers.

The actual morphology of a polymer depends primarily on three factors: chemical composition, degree of polymerization, and chain configuration. Other variables affecting the final physical state of a polymer sample include its thermomechanical history and processing methods.

3.10.1 Crystallinity

Polymer molecules tend to move towards an arrangement which is in the lowest possible energy state (lowest Gibbs free energy). The lowest energy level that a compound can achieve is a crystal form. Crystallinity involves a regular repeating arrangement of the molecules. Although crystallization tends to occur naturally, crystallization takes place in small regions of a polymer, on the order of 1×10^{-9} m, if at all. To produce a crystal, the polymer chains must be capable of packing closely together in a regular, parallel array. The molecules must be packed side by side in extended form, either in a planar zigzag or helical format. Polymer chains must be fairly regular in structure to permit this. Thus, the ability of a polymer to crystallize is largely determined by the regular placement of atoms in the chain. Important sources of polymer irregularity include:

1. Head-to-head or tail-to-tail placement of monomer units
2. Stereochemical irregularity
3. Copolymers with random placement of comonomers
4. Branching

Linear polymers made of symmetrical unsaturated monomers, such as polyethylene and polyvinylidene chloride, crystallize easily. Asymmetric linear polymers such as polypropylene (PP) crystallize only if the configuration is regular, so isotactic and syndiotactic PP can crystallize, while atactic PP is amorphous.

Normally, step-reaction polymers which are synthesized from difunctional monomers containing alcohol, acid, or amines, can crystallize. These polymers produce highly ordered chains since difunctional monomers force the chain to grow in only one isomeric configuration. This is the case, for instance, with nylon 6 and nylon 6,6.

Absolute regularity is not always essential for crystallization. Some substituent groups are close enough to each other in size and behavior that they can substitute for each other randomly in a crystal lattice. Thus polyvinyl alcohol

$$-\left(\underset{\underset{\text{H}}{|}}{\overset{\overset{\text{H}}{|}}{\text{C}}}-\underset{\underset{\text{H}}{|}}{\overset{\overset{\text{OH}}{|}}{\text{C}}}\right)_{n}-$$

can crystallize, even though it is atactic. Similarly, polyvinyl fluoride

$$-\left(\underset{\underset{\text{H}}{|}}{\overset{\overset{\text{H}}{|}}{\text{C}}}-\underset{\underset{\text{F}}{|}}{\overset{\overset{\text{H}}{|}}{\text{C}}}\right)_{n}-$$

is highly crystalline. In these cases, the F atom, the OH group, and H atoms are all sufficiently small that they can be "accommodated" in a crystal lattice similar to that of PE to produce highly crystalline polymers. Groups which can be accommodated within the same crystal lattice sites, such as -H, -OH, and -F in the examples above, are called *isomorphous*.

On the other hand, polymers with regular structures and configurations do not crystallize if the substituent is excessively bulky, or if other characteristics of the polymer configuration do not permit it to assume the regular repeating pattern required. For example, when the difunctional monomers in condensation polymers contain aromatic and cyclohexane rings, only polymers with substituents in 1,4 positions, like PET and PC, are crystallizable. A 1,3 linkage in the ring causes a "kinking" of the chain that minimizes the ability to crystallize since the molecules are unable to pack in a parallel array. When a stereochemically regular polymer has a very bulky substituent group, this can interfere with crystallinity. While the presence of a bulky group does not usually prevent the formation of some crystallinity, it can significantly limit the amount of crystallinity which actually develops.

Branching also decreases crystallinity, since the branch points produce irregularities in the molecular packing. When the reaction proceeds to cross-linking, the result is generally an amorphous polymer. Thus for step-reaction polymerization with monomers containing three and four functional groups, the resulting polymer is generally an amorphous three dimensional network. This is the case in epoxy adhesives and polyurethanes, which are amorphous thermoset materials. Low density polyethylene is able to crystallize, despite its branching, because of the very great flexibility of the ethylene chains, which contain only small hydrogen atoms as substituents. The degree of crystallinity, however, is significantly less for LDPE than for linear HDPE.

Copolymerization also introduces structural irregularity which can interfere with crystallinity. The amount of interference depends on the amount and type of comonomer, so the resultant polymer may be partially crystalline, or may be totally amorphous. When the comonomers are isomorphous, as is the case with ethylene vinyl alcohol (EVOH), a highly crystalline polymer can result even with high degrees of copolymerization. If the comonomers are not isomorphous, introducing a small amount of a comonomer will act to decrease the crystallinity of the polymer, as well as the crystalline melting temperature, whether the homopolymer associated with the comonomer is itself amorphous or highly crystalline. In most cases, random copolymers containing a substantial amount of both comonomers will be totally amorphous.

Block and graft copolymers, if they crystallize at all, tend to form small crystalline regions containing molecular segments with only one of the included monomers. Thus a block or graft copolymer may contain small regions with totally different crystalline arrangements, one type for each of the extended segments of a single monomer type.

Woodward [1] describes seven common crystalline morphologies in polymers: faceted single lamellas, non-faceted lamellas, dendritic structures, sheaf-like lamellar ribbons, spherulite arrays, fibrous structures, and epitaxial lamellar overgrowths on microfibrils. Spherulites are complex ordered aggregations of submicroscopic crystals. PE crystal spherulites, for example, are about 10 nm thick. The spherulites are separated from one another by small amorphous regions called micelles. When such spherulites are larger than the wavelength of visible light, they produce light scattering that makes the polymer opaque. Often plastic materials with high crystallinity are opaque, while plastics with low degrees of crystallinity are transparent or clear, and amorphous materials are totally transparent. The degree of light scattering produced by the crystallites is dependent on their size. At moderately high cooling rates, the crystalline regions can sometimes be reduced in size to less than the wavelength of light, producing a combination of high crystallinity and transparency, as is the case with some polypropylene resins and film.

We will now discuss the crystallization of polyethylene in more detail to illustrate some of the details mentioned above. A linear polyethylene molecule, HDPE, will crystallize in an extended chain conformation in the crystallite if its molecular weight is below about 10,000 daltons. Above this molecular weight, the polymer forms folded chain lamellae, as shown in Figure 3.28. The crystalline regions are closely packed chains which loop back on themselves. The region above and below the crystalline region is composed of two portions. That closest to the crystallite is a boundary region where the looping occurs. Further away is an amorphous region having no order. The molecules usually exit the crystallite after making three or four folds and enter the amorphous region. The same molecules will then either re-enter the same crystallite, or perhaps another crystallite. The morphology obtained during crystallization depends on the polymer's state of stress and on the temperature conditions.

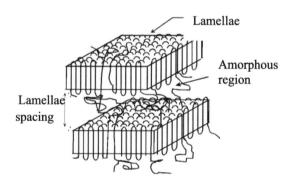

Figure 3.28 Crystallite lamellae in polyethylene.

The type of morphology obtained from a polyethylene melt differs depending on whether the melt is quiescent or under strain as crystallization occurs. In films, the melt is always under strain during the crystallization process. In an injection molded part, there will often be two regions of crystallization. The region near the walls is often under strain during crystallization, while the interior of the part may be quiescent. Another question that must be addressed about the process of crystallization to understand the probable morphology is whether or not crystallization occurred under nearly isothermal conditions. Most of the time, one can assume that nearly isothermal crystallization occurs, even in thin films. Study of non-isothermal crystallization of PE films has shown that the released heat of crystallization is enough in many cases to keep the temperature nearly constant during the crystallization process. The critical issues are the level of strain in the melt during crystallization and the temperature gradient across the material. Non-isothermal conditions affect the amount of frozen-in strain.

In any crystallization process, the polymer molecules must slip through the melt and arrange themselves in the ordered loops of the lamellae. If there is a branch point, or a side group like vinyl acetate, these will not fit into the crystal lattice and will be rejected. The time needed to pull itself from the melt relative to the time before another molecule could begin to fit itself into the structure means that most chains develop only about four continuous folds into a crystallite before being interrupted. For polymers with regular repeating small side groups, such as polypropylene, it is possible to find a way to pack the molecules uniformly close enough together to form a crystallite. However, this requires stereoregularity. A change from the isotactic conformation of the monomer to syndiotactic would cause rejection of that portion of the molecule from the crystallite. Therefore, a regular molecular structure is very important to crystallite formation. High molecular weight and long chain branching both result in molecules having more difficulty in moving through the melt to get into the ordered structure, so both decrease crystallinity. Crystallization goes on from the time the proper temperature is reached for it to commence, until the energy levels have dropped enough that molecules no longer have enough energy to rearrange themselves into crystallites. No crystallization can occur at temperatures below the glass transition temperature or above the melt temperature (see Sec. 3.11). For many polymers, the range of temperature over which crystallization can occur is much more limited than this. As crystallization proceeds, it becomes more and more difficult to rearrange the remaining amorphous fraction of the polymer into a crystalline structure. For example, the short segments of molecules in the amorphous regions between two lamellae are constrained in their mobility by the sections which are within the crystallites. Eventually, even at relatively high energy conditions, no more crystallization can occur in the polymer. At this point, it will have both crystalline and amorphous regions. The amount of crystallinity which develops in a polymer is, therefore, a function both of its structure and of the processing conditions (energy and strain) to which it has been exposed.

In a crystallization process that occurs under strain, the morphology is described as row nucleated lamellae. Transmission electron or scanning electron microscopy show a backbone of extended chains onto which epitaxial growth of lamellae has occurred. A diagram of this structure is shown in Figure 3.29. The extended chains are higher molecular weight molecules that have been oriented in the direction of flow. The lamellae are made up of lower molecular weight species that can relax more rapidly. For this ordered growth to occur on the extended chain backbone, the chains must slip through the melt, as discussed above. Therefore, the structural characteristics that retard movement through the melt, such as high molecular weight, long chain branching, and other factors, are important influences on how fast the materials can crystallize.

Lamellae are described by an a-, b-, and c-axis. The c-axis is along the molecule. The b-axis is the growth direction, and the a-axis is the remaining direction. The stress on the melt has been shown to control the orientation of the crystallites in reference to the machine direction for films and sheet.

High stress flows cause what is referred to as c-axis orientation along the machine direction. This would mean that in Figure 3.29, the machine direction was the vertical direction. Since one of the critical parameters for many films is the permeability, this orientation characteristic is important. Permeation cannot occur through the crystallites, so orientation in the thickness direction of the film would provide maximum barrier. This type of orientation cannot be achieved, but the right choice of polymer and processing conditions can provide a random orientation of the lamellae.

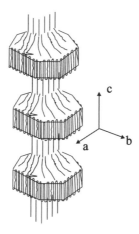

Figure 3.29 Row nucleated structure or shish-kebab, obtained from crystallization under strain.

The critical measure for characterizing a polymer is not the crystallization rate, but the elastic relaxation rate as measured by rheological means. The reason crystallization rates are not always an accurate predictor should be obvious from the preceding discussion. The rate is affected by molecular weight, molecular weight distribution, strains on the melt, and the degree and rate of cooling. Normal crystallization measurements are accomplished at a nearly quiescent state. Not all rheological measures of relaxation are necessarily accurate enough, either. Therefore, careful measurement techniques are needed to determine these parameters correctly.

As can be seen, the morphology obtained during processing can be very different depending on the conditions. The morphology that is present in the packaging material will have a pronounced effect on the properties and on the ability to further process the materials, such as material orientation, which will be discussed in the next section.

3.10.2   Polymer Orientation

Polymers exhibit anisotropic behavior, meaning that the properties of the material are dependent on the direction in which they are measured (as opposed to isotropic materials, where the properties do not depend on direction). Low molecular weight compounds are usually isotropic, but polymers typically show anisotropism in both the amorphous and crystalline phases. When a polymer flows, the molecules have some tendency to line up in the direction of the flow. On cooling, some of this molecular orientation is usually preserved. Therefore, stress in one direction will meet resistance from a preponderance of backbone chains with their strong covalent bonds, while stress in a perpendicular direction will be resisted by a larger percentage of much weaker intermolecular forces.

Processing conditions associated with the cooling rate of polymer melts, stretching in film blowing, and film orientation can enhance a polymer's natural anisotropy. Molecules in polymer film or sheet can be oriented into a more orderly morphology in response to external stress at temperatures above the glass transition temperature (see Sec. 3.11.2). As shown in Figure 3.30, crystallites are unraveled and reordered into fibrils. The amorphous regions are stretched also, and the increased order in the amorphous regions increases the intermolecular forces. One result is a decrease in permeability because of the effect that decreased molecular mobility has on the diffusion of the permeant through the film. The result of uniaxial molecular orientation (orientation in one direction) is a substantial increase in strength and toughness in the direction of the stretch (but a decrease in strength in the perpendicular direction). The reason for the decrease in strength is the weakness planes that develop between the fibrils shown in Figure 3.30. Unless there are tie molecules between the crystallites in the fibrils, there is very little strength in the perpendicular direction. Polymers can be unoriented, uniaxially oriented, or biaxially oriented (oriented in two directions perpendicular to each other).

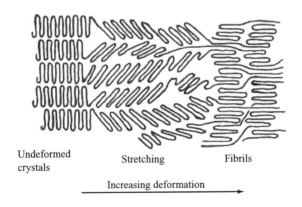

Undeformed crystals   Stretching   Fibrils

Increasing deformation

Figure 3.30 Orientation of a crystalline material, forming fibrils.

Film used to fabricate bags is generally uniaxially oriented to improve its tensile strength, since most of the force on the bag, in use, is exerted in a vertical direction. Films used for pouches are often biaxially oriented because it is expected that tensile forces may act in both directions, and also because if shrinking occurs, the pouch should shrink in both directions.

Biaxially oriented materials can be balanced (stretched the same amount in both directions), or unbalanced (stretched more in one direction than in the other). Unbalanced oriented films will have different values of tensile and tear strength when measured in the machine direction (the direction of film travel when it is made), than when measured in the cross-machine direction (perpendicular to film travel). Unbalanced films are the most commonly produced commercially. Stretching of films is expressed as a percent of the unstretched material dimension. For example, a 400% stretch ratio indicates that the film is four times its initial length.

In PVC and PET stretch blow molded containers, molecular orientation plays an important role in the mechanical, barrier, and optical properties of the container. Orientation improves the strength and barrier capability of the container, and enhances its clarity.

The orientation of the amorphous regions of these two polymers pulls the chains closer to one another, as well as aligning them. This results in a decrease in the entropy (chaos) of the amorphous region. The result is an increased driving force to cause the aligned chains to crystallize. These regions of straight chain crystallites are as impermeable as the folded chain regions, and they act as mini-crosslinks in the amorphous region to minimize molecular motion necessary for diffusion. This type of modification of the amorphous regions can be measured by comparing the degree of crystallization from both differential scanning calorimetry (DSC) and x-ray crystallography. Birefringence measurements can measure the differences in the degree of orientation in the various directions.

3.10.3 Degree of Crystallinity

The degree of crystallinity of a polymer reflects the relative amount of crystalline regions and of amorphous regions. This amount can be expressed on a volume or a mass basis. The degree of crystallinity is most accurately determined by x-ray scattering. In practice, this is a tedious operation and is rarely performed.

The approximate crystallinity of a polymer sample can be calculated from density measurements. Based on the two-phase model of polymer behavior, the mass or volume fraction of crystallinity can be calculated by measuring the density of a polymer sample, if the densities of the amorphous material and of pure crystals are known, as indicated by the following equations:

$$a_v = \frac{\rho - \rho_a}{\rho_c - \rho_a}$$

(3.11)

$$a_m = \frac{\rho_c}{\rho} a_v \tag{3.12}$$

where a_v and a_m are the volume and mass crystallinity respectively and ρ, ρ_c, and ρ_a are the density of the sample, pure crystalline, and pure amorphous materials respectively. The percent crystallinity is a_v or a_m multiplied by 100.

The density of a polymer sample can be determined by the density gradient method, as described by ASTM D1505. In this method two solutions, A and B, are prepared with densities in the range of interest. Solution A, with the lowest density, and solution B, with the highest density, are combined in a glass tube to form a vertical column of liquid in which the density varies linearly from the bottom to the top. The column is calibrated with glass beads of known density. Plastic samples are dropped in the column and will rest at the level corresponding to their density. The density of the plastic is calculated from the position of the sample compared to that of the calibration beads.

The approximate percent crystallinity of a polymer can also be calculated from measurements of the heat of fusion (see Sec. 3.11.5) made using differential scanning calorimetry. If the heat of fusion for a pure crystalline sample of the polymer is known, the mass percent crystallinity can be determined by dividing the heat of fusion of the sample by the heat of fusion of 100% crystalline polymer, and multiplying by 100%.

A number of important polymer properties depend on the morphology of the polymer. As crystallinity increases:

Density	increases
Permeability	decreases
Opacity	increases
Blocking	decreases
Tensile strength	increases
Compression strength	increases
Clarity	decreases
Tear resistance	decreases
Impact strength	decreases
Toughness	decreases
Ductility	decreases
Ultimate elongation	decreases
Heat sealing temperature	increases
Heat sealing range	decreases

While, properly speaking, all packaging plastics are either amorphous or partially crystalline, in practice the term crystalline polymer is generally used to refer to a polymer that has any substantial degree of crystallinity. The term semicrystalline is sometimes used for polymers that have only a slight amount of crystallinity. The term paracrystalline is used to refer to materials, such as polyacrylonitrile, that have a substantial degree of molecular order, but which fall short of true crystallinity. Polymers with only a very limited ability to crystallize are sometimes, incorrectly, referred to as amorphous polymers. Polyvinyl chloride, for example, has an extremely small ability to form crystals since it is primarily atactic but has some syndiotactic tendency. Nonetheless, it is often reported to be an amorphous polymer.

3.11 Thermal Properties

Thermal properties are the relationships between the polymer properties and temperature. We will discuss melting temperature, glass transition temperature, and other thermal transitions, as well as heat capacity, thermal conductivity, and dimensional changes due to temperature variation.

3.11.1 Melting Temperature

One important thermal property of a polymer is its melting temperature, T_m. For small molecules, the melting temperature is very well defined, marking the transition from solid to liquid, and characterized by an uptake in energy while the temperature remains constant, until all the solid material has become liquid. For crystalline polymers, the melting temperature is less sharply characterized, and instead is a range of a few degrees over which the crystallites in the polymer break up, and there is an uptake in energy accompanied by a gradual rise in temperature. This occurs because not all crystallites break up at the same temperature; some require more energy in order to be disrupted than do others. Therefore, we find a melting range rather than a definite melting temperature, and represent this range by some average value. ASTM methods D2117 and D3418 are methods for measuring T_m.

For amorphous polymers, the situation is more complex. We can find low temperatures where the polymer acts like a solid, and high temperatures where the polymer flows like a liquid. However we are not able to find any region of a few degrees where we clearly have a transition from one type of behavior to another. Instead, like glass, the polymer gradually gets less and less resistant to deformation as the temperature increases, and more and more resistant as it decreases. Therefore the concept of melting temperature is not defined for amorphous polymers.

3.11.2   Glass Transition Temperature

While amorphous polymeric materials, like inorganic glasses, do not have melting points, they do have a glass transition temperature, T_g, which is defined as the freezing in (on cooling) or the unfreezing (on heating) of micro-Brownian motion of chain segments 2-50 carbon atoms in length in the amorphous regions of a material. This motion is a semi-cooperative action involving torsional oscillation and/or rotations around backbone bonds in a given chain as well as in neighboring chains. Torsional motion of side groups around the axis connecting them to the main chain may also be involved.

This motion is often referred to as segmental mobility, and reflects the ability of a portion of a polymer molecule to change its position with respect to its neighbors. Although the glass transition temperature is defined only for amorphous regions of a polymer, since crystalline polymers also contain amorphous regions, both crystalline and amorphous polymers have defined glass transition temperatures.

At low temperature, an amorphous polymer is glassy, hard and brittle, but as the temperature increases, it becomes rubbery, soft, and elastic. There is a smooth transition in the polymer's properties from the solid to the melt, as discussed above, so no melting temperature is defined. At the glass transition temperature, marking the onset of segmental mobility, properties like specific volume, enthalpy, shear modulus, and permeability show significant changes, as illustrated in Figure 3.31.

Consider a melted polymer that is cooling down. Assume that we monitor the value of the specific volume (v) as a function of temperature (T). Different curves of v versus T can be obtained depending on the rate of cooling and the capacity of the polymer to crystallize.

First consider a polymer capable of crystallizing 100%, where the rate of cooling is slow enough to allow the chain polymers to form crystals. We can see that there is a drastic change in the specific volume once the melting temperature is passed (Fig. 3.32). The liquid melt state has a larger thermal expansion coefficient than the solid crystal state.

Now consider a completely amorphous polymer, i.e., a polymer which does not crystallize even when cooled from the melt at a very slow rate. Since one effect of cooling the melt is to decrease the degree of thermal agitation of the molecular segments, as the cooling process continues, the rate of segmental movement becomes more and more sluggish and then, on further decreasing the temperature, the segmental movement finally stops. At this point, the glassy state of the amorphous polymer is reached, and the temperature at which this glassy state occurs is called the glass transition temperature, T_g. This process is associated with the limitation of segmental chain mobility as defined above. The glassy state consists, then, of "frozen" entangled chain molecules with a complete absence of stereoregularity and coordinated motion. This absence of stereoregularity is typical of liquids.

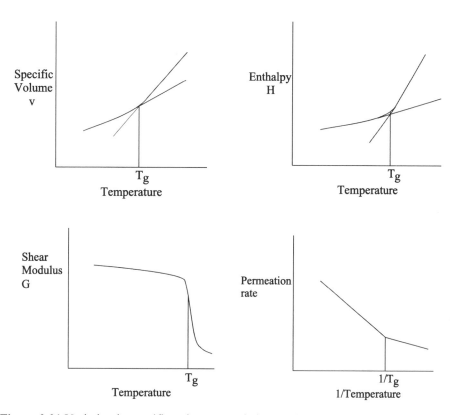

Figure 3.31 Variation in specific volume v, enthalpy H, shear modulus G′ and permeability P near the glass transition temperature.

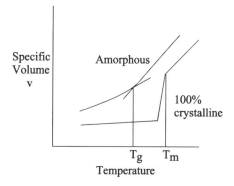

Figure 3.32 Specific volume v. temperature of crystalline and amorphous materials.

As shown in Figure 3.32, the curve of specific volume as a function of temperature shows a smooth change from the melt state to the glassy state. However, there is a change in the slope of the curve near T_g. In the glassy region, the specific volume of the amorphous phase is much larger than the value corresponding to the crystal phase of the same polymer. This is equivalent to additional space, or "free volume," trapped between the entangled and frozen polymeric chains. The reduction in specific volume as the temperature decreases below T_g is associated with the amount of molecular movement and is determined by the intermolecular forces. The presence of this additional free volume directly affects properties such as permeability; as the free volume increases, permeability increases.

For semicrystalline polymers, the curve of specific volume versus temperature follows an intermediate path between the ones for pure amorphous and pure crystalline polymers.

Whether the polymer is totally amorphous or partially crystalline, the material will be glassy (brittle) or rubber-like (soft) depending on its temperature with respect to T_g. If an amorphous polymer is at a temperature below T_g, it will be brittle and will show properties of a glassy material; for example, it will fracture more easily. As the temperature of the sample increases and approaches T_g, it adopts a leathery behavior and its elastic modulus decreases. When the sample has reached several degrees above T_g, it shows a clear rubbery behavior and is easily deformable. If the temperature is increased even more, the polymer reaches liquid flow behavior. If the polymer is semicrystalline, it exhibits similar behavior, but when it reaches the melting temperature the crystals will break up, and the polymer will then reach the melted liquid state. This behavior is illustrated in Figure 3.33 where the elastic modulus is plotted versus temperature. Table 3.2 shows values for the glass transition temperature of common plastics.

As indicated, the thermal expansion coefficient, permeability, elastic modulus and heat capacity of plastics may have very different values, depending on whether the polymer sample is above or below its T_g.

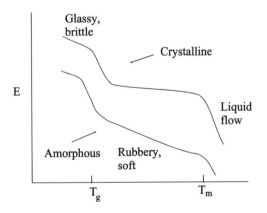

Figure 3.33 Elastic modulus as a function of temperature for amorphous and partially crystalline polymers.

3.11.2.1 Measuring T_g

Any polymer property that changes with temperature and has different values above and below T_g can be used, in principle, to determine T_g. For example, the change in specific volume, heat capacity or elastic modulus may be used to measure T_g. Differential Scanning Calorimetry (DSC) and Dynamic Mechanical Analysis (DMA) are two common methods for such determinations. An example of the results of DSC analysis is presented in Figure 3.34. It is common for different methods to yield slightly different values for T_g.

Table 3.2 Selected Values of Glass Transition Temperature of Polymers and Their Mechanical Characteristics at Room Temperature

Polymer	T_g, °C	Characteristics at room temperature
PE	-110	Soft, flexible
PP, atactic	-19	More rigid than PE
PP, isotactic	-8	Can be brittle at frozen temperature
PVDC	-19	Flexible but more rigid than PE
PVC, unplasticized	80	Rigid (without plasticizer)
PET	80	Stiff
PVOH	85	Stiff and rigid
PS	100	Brittle

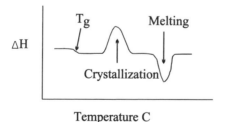

Figure 3.34 Differential scanning calorimetry trace, obtained by cooling a crystalline polymer from the melt.

An empirical relationship has been determined between T_g and T_m, the melting temperature, with both temperatures in Kelvin. This allows either T_g or T_m to be estimated, if the other is known.

$$\frac{T_g}{T_m} \approx 0.6 \tag{3.13}$$

Example: For poly(propylene), $T_g = -19°C$ and $T_m = 176°C$

$$\frac{T_g}{T_m} = \frac{273 - 19}{273 + 176} = 0.57$$

3.11.2.2 Variables Affecting T_g

The chemical structure of a polymer profoundly affects its glass transition temperature. Some of the most significant factors are the presence of bulky groups, polar groups, and the strength of intermolecular forces.

Bulky side groups increase T_g by increasing the energy required for bond rotation. For example, the $-CH_3$ group in PP and benzyl group in PS increase T_g compared to PE ($T_g = -110°C$). The bulkier the group, the larger the effect.

PP $-(CH_2-CH)_n-$ $T_g = -19°C$
                          |
                        CH_3

PS $-(CH_2-CH)_n-$ $T_g = 100°C$

Rigid groups within the main chain, such as ring structures, can also significantly raise T_g by decreasing segmental mobility. The benzene ring present in the main chain of PET is one of the major reasons why its glass transition temperature is 80°C. This stiffness in the main chain also contributes to PET's strength. In general, the presence of a ring within the main chain is considerably more effective in raising the T_g than the presence of the same ring in a side group. Cross-linking also increases T_g by creating a more rigid structure, interfering with segmental mobility.

The presence of double bonds in the main chain creates somewhat of an anomaly. The double bonds themselves are, like ring structures, rigid. There is no rotation around double bonds. However, the presence of a double bond greatly facilitates rotation around the single bonds adjacent to the double bond, in part because of decreased interference with the rotation of groups attached to that next carbon. Therefore, the presence of a double bond actually acts to increase chain flexibility, and hence decreases T_g. Synthetic rubbers such as polybutadiene contain double bonds in the main polymer chain.

Highly polar groups increase T_g by increasing intermolecular forces, which must be disrupted for rotation to occur. We can compare the effect of CH_3- in PP and $Cl-$ in PVC, and can see that T_g not only increases with respect to PE, but that the more polar Cl has a more pronounced effect.

PP $\qquad$ $-(CH_2-CH)_n-$ $\qquad\qquad$ $T_g = -19°C$
$\qquad\qquad\qquad\quad |$
$\qquad\qquad\qquad\ CH_3$

PVC $\qquad$ $-(CH_2-CH)_n-$ $\qquad\qquad$ $T_g = 80°C$
$\qquad\qquad\qquad\quad |$
$\qquad\qquad\qquad\ Cl$

Hydrogen bonds provide particularly strong intermolecular forces, and therefore have a large effect on T_g. The -OH group is not only polar, but also produces hydrogen bonds that result in an increase of 185 °C in the glass transition temperature when PE is compared to PVOH.

PE $\qquad$ $-(CH_2-CH_2)_n-$ $\qquad\qquad$ $T_g = -100°C$

PVOH $\qquad$ $-(CH_2-CH)_n-$ $\qquad\qquad$ $T_g = 85°C$
$\qquad\qquad\qquad\quad |$
$\qquad\qquad\qquad\ OH$

Table 3.3 shows the effects of intermolecular forces (expressed as cohesive energy density) on the value of T_g for three vinyl polymers.

Table 3.3 Effect of Cohesive Energy Density on the Glass Transition Temperature

Polymer	CED, J/cm^3	T_g, °C
Polyethylene	259	-100
Polyvinyl alcohol	381	85
Polyacrylonitrile	992	108

3.11.3 Other Thermal Transitions

In addition to the glass transition temperature T_g, amorphous polymers have other relaxation temperatures like T_β, T_{ll} and $T_{l\rho}$. As previously indicated, the molecular motions associated with T_g result from semi-cooperative actions involving torsional segmental oscillations and/or rotations around backbone bonds in a chain. Torsional motion of side groups around the axis connecting them to the main chain may also be involved. T_g involves, then, both intrachain and interchain segmental motions.

In contrast to T_g, T_β is associated mostly with intrachain sub-group motions (2-10 consecutive chain atoms). T_{ll} is a weak transition-relaxation temperature about 1.2 x T_g associated with the thermal disruption of intermolecular segment-to-segment contacts known as segmental melting. T_{ll} marks the onset of the true liquid state of an amorphous polymer. At a higher temperature, 30 to 50 K above T_{ll}, the molecular chains contain enough energy to break the rotational barrier marking the onset of liquid behavior in an amorphous polymer. This relaxation has been identified as $T_{l\rho}$.

The relationship between relaxation temperatures and end use properties can be described as follows. According to Boyer, physical aging of polymers occurs at all temperatures between T_g and T_β (the higher the temperature, the faster the aging process). The effect of aging can be erased by heating the polymer above T_g. At a given temperature, toughness is associated with the presence of one or more relaxation temperatures below that particular temperature. Wetting, tackiness, and adhesion appear to be fully developed at T_{ll}.

3.11.4   Heat Capacity

The heat capacity, or specific heat, of a material is the amount of energy needed to change the temperature of a unit of mass of the material by one degree. The heat capacity of a plastic, which is obtained at constant pressure, is temperature dependent, especially near the glass transition temperature. In a semicrystalline polymer, the heat capacity in the amorphous phase is larger than the heat capacity in the crystalline phase. This implies that the heat capacity values depend on the percent crystallinity of the polymer. The heat capacity values of polymers at 25 °C vary from 0.9 to 1.6 J/gK for amorphous polymers, and from 0.96 to 2.3 J/gK for crystalline polymers. Reliable data regarding the heat capacity of amorphous and crystalline phases are available for only a limited number of polymers. The usual techniques for measuring specific heat are differential thermal analysis (DTA) and differential scanning calorimetry (DSC).

3.11.5   Heat of Fusion

The heat of fusion ΔH_m, is the energy involved in the formation and melting of crystalline regions. For semi-crystalline polymers, as discussed in Section 3.10.3, the energy of fusion is proportional to the percent crystallinity. Amorphous polymers, or amorphous polymer regions, do not have a heat of fusion, since amorphous structures have a smooth transition from the solid amorphous state to the liquid state. Experimental values of the crystalline heat of fusion for common packaging plastics vary from 8.2 kJ/mol for polyethylene to 43 kJ/mol for nylon 6,6. ASTM D3417 describes a method for measuring the heat of fusion and crystallization of a polymer by differential scanning calorimetry (DSC).

3.11.6 Thermal Conductivity

The thermal conductivity is the parameter in Fourier's law that relates the flow of heat to the temperature gradient. Fourier's law is

$$\frac{dq}{dt} = -kA\frac{dT}{dx} \tag{3.14}$$

where q is heat, t is time, k is thermal conductivity, A is area, T is temperature, and x is the direction of heat flow.

In specific terms, the thermal conductivity is a measure of a material's ability to conduct heat. The thermal conductivity of a polymer is the amount of heat conducted through a unit thickness of a material per unit of area and time, with a temperature difference of one degree between the surfaces. Thermal conductivity values control the heat transfer process in applications such as plastic processing, heat sealing, cooling and heating of packaging, and sterilization processes.

Plastics have values of k much lower than metals. Thermal conductivity for plastics ranges from 3×10^{-4} cal/s cm °C for PP to 12×10^{-4} cal/s cm °C for HDPE. For aluminum, k is 0.3 cal/s cm °C, and for steel it is 0.08 cal/s cm °C. Plastic foams have values of k much lower than those of unfoamed plastic. This is due to the presence of gas trapped in the cellular structure of the plastic foam. Plastics, with their low thermal conductivity values, do not conduct heat well. Consequently foams, which are enhanced by the even lower thermal conductivity of gases, are excellent insulating materials. In addition to being thermal insulators, foams are attractive cushioning materials. Fillers in plastics may increase their thermal conductivity. Methods for measuring k are given in ASTM D4351, C518, and C177.

3.11.7 Thermal Expansion Coefficient

The coefficient of linear (or volume) thermal expansion of a material is the change of length (volume) per unit of length (volume) per degree of temperature change at constant pressure. The linear thermal expansion coefficient is $\beta = (1/L)(dL/dT)_p$; and the coefficient of volume expansion is $\alpha = (1/V)(dV/dT)_p$. Units of α and β are °K^{-1} or °R^{-1}.

For changes where the coefficients are independent of temperature, we can calculate the change in dimensions from:

$$\Delta L = \beta L_o \Delta T \tag{3.15}$$

for linear dimensions. For volume, we can calculate the change from:

$$\Delta V = \alpha V_o \Delta T \tag{3.16}$$

Compared to other materials, polymers have high thermal expansion coefficients. While metals and glass have values in the range 0.9 to 2.2 $°K^{-1}$, polymers range from 5.0 to 12.4 $°K^{-1}$.

Thermal expansion coefficients can be measured by thermomechanical analysis (TMA). ASTM D696 describes a method using a quartz dilatometer, while ASTM E831 describes the determination of the linear thermal expansion of solid materials. Volume contraction of a container from the molding operation temperature down to room temperature is called shrinkage, and its measurement is described in ASTM D955, D702, and D1299.

3.11.8   Other Dimensional Changes

Not all dimensional changes during heating and cooling are due to thermal expansion. Partially crystalline materials and oriented materials can undergo dimensional changes due to changes in morphology.

A partially crystalline polymer that is not fully crystallized during processing will shrink when heated to a certain temperature. When the chain mobility is increased enough for the chain to begin to form into crystallites, the polymer will shrink, since the crystallites occupy less volume than the amorphous regions. This behavior can be seen if one heats a non-oriented PET or nylon film. Most often it is associated with non-packaging applications like the shrinkage of polyester fabrics when washed in water that is too hot.

Oriented materials (see Sec. 7.2.6) can be caused to shrink by increasing the temperature, even if they are totally amorphous polymers like polystyrene. When orientation occurs, the amorphous regions are stretched into more close packing than is energetically at equilibrium. For partially crystalline polymers, the crystalline regions can be disrupted by orientation forces also. Therefore, when heat is applied and chain motion increases, the chains try to return to their low energy state, which is random amorphous regions and larger crystals. The decrease in dimension in the direction(s) of orientation is coupled with an increase in dimension(s) in the unoriented direction(s). An example of this is shrink film, or what happens when boiling water is poured into a PET beverage bottle.

3.11.9   Dimensional Stability

The dimensional stability of a structure refers to how well it maintains its dimensions under changing temperature and humidity conditions. ASTM D1204 describes a standard method for measurement of linear dimensional changes of flexible thermoplastic films and sheets at elevated temperatures. Dimensional stability is an important property in any flexible material converting process. During printing, for example, even a small change in dimensions may lead to serious problems in holding a print pattern. In a flexible structure, dimensional stability may produce different changes in the machine and transverse (cross) directions.

Change in relative humidity does not affect the dimensions of most polymers to any substantial degree. However, the dimensions of some water sensitive polymers can be affected by high relative humidity, especially at elevated temperatures. As discussed, plastics are relatively sensitive to temperature-induced dimensional changes.

3.12 Mechanical Properties

Polymers show an array of mechanical behaviors whose values change greatly with temperature and with speed of testing, as well as with the intrinsic structure of the polymer. Tensile strength, tear strength, impact and bursting strength, folding endurance, and pinhole and abrasion resistance are all of practical importance in plastic packaging.

3.12.1 Tensile Properties

The mechanical behavior of a polymer can be evaluated by its stress-strain characteristics under tensile deformation, as illustrated in Figures 3.35 and 3.36. The stress is measured in force per unit area, expressed in pascal or psi. The strain is the dimensionless fractional length increase. The behavior of a polymer is influenced by the rate of deformation, and stress-strain curves are typically obtained at low strain rates.

The *modulus of elasticity*, E, is the ratio between the stress applied and the strain produced, in the elastic portion of material behavior, indicating the material's resistance to elastic deformation. Thus it equals the slope of the first linear portion of the stress-strain curve. The tensile modulus also gives a measure of the material's stiffness; the larger the modulus, the stiffer the material. For example, E of LDPE is 250 mPa, while for "crystal" PS, it is 2500 mPa. Values of tensile modulus in polymers (for example, 1.9×10^3 mPa for nylon) are much lower than for glass (55×10^3 mPa), or mild steel (210×10^3 mPa). The *elastic elongation* is the maximum strain under which elastic behavior is maintained - in other words, the farthest the material can be stretched without undergoing permanent deformation. The *ultimate elongation* is the strain at which the sample ruptures, the farthest the material can be stretched without breaking.

The *elastic limit* marks the maximum stress the material can experience without undergoing permanent deformation, thus the elastic elongation is the strain corresponding to the stress at the elastic limit. The *yield point* is the point at which the material begins to undergo an increase in strain without requiring an increase in stress. The yield point, if it exists, always lies after the elastic limit, and is a sign that plastic (permanent) deformation is occurring.

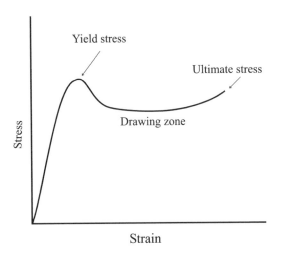

Figure 3.35 Tensile stress-strain curve, where tensile strength is exhibited at yield.

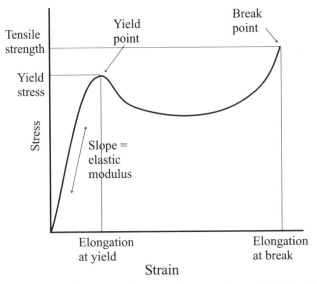

Figure 3.36 Tensile stress-strain curve, where tensile strength is exhibited at break.

Resilience is the ability of a material to absorb energy and then return to its original dimensions. Resilience is a property largely of the amorphous regions of a polymer. Increasing crystallinity and orientation tend to decrease resilience. Cross-linking (unless too extensive) increases resilience by preventing plastic deformation and thus enhancing the ability of the material to return to its original dimensions

The *ultimate strength* or *tensile strength* is the maximum tensile stress the material can sustain. Generally this occurs either at the yield point or at the break point of the material.

Toughness can be defined as the energy a material can absorb before rupturing, and is measured by the area under the stress-strain curve. A significantly different measure of toughness results when polymers are tested at high strain rates, such as in impact tests.

Brittleness is defined as inability to experience a significant amount of strain without rupture. Thus it is often associated with a lack of toughness, though it is not identical to this. Amorphous and semi-crystalline polymers become brittle when cooled below their glass transition temperature.

Creep refers to the tendency of a material to continue to deform over time when subjected to a constant load. Because of the viscoelastic nature of plastics, the molecules tend to slowly change their positions when subject to such extended stress, which can result in significant deformation over time. This is important in such diverse areas as plastic pallets and plastic bottles containing contents under pressure.

Stress relaxation is the tendency of a material, when exposed to a constant deformation, to experience a reduction in the amount of stress required to sustain that deformation. This too is a consequence of molecular rearrangement in the polymer that occurs slowly over an extended time. Stress relaxation is particularly important for stretch film, since the "holding power" of the film decreases over time.

Tests for tensile properties are described in ASTM D882; tests for flexural properties are described in ASTM D790 and ASTM D6272. Table 3.4 indicates how the softness or toughness of any material relates to the values of elastic modulus, yield stress, strength and elongation. The behavior observed in the stress-strain curve corresponds to viscoelastic behavior that is typical of polymeric materials. The viscoelastic behavior is highly dependent on the temperature at which the test is performed and its relationship to the T_g of the sample, and is also dependent on the rate of deformation, as mentioned above. In general, very rapid deformation does not allow time for molecular rearrangement to occur and results in behavior characteristic of a more brittle material.

Table 3.4 Materials Can Be Soft or Tough Depending on the Values of Tensile Properties

Type of Polymer	Elastic Modulus	Yield Stress	Ultimate Strength	Elongation at Break
Soft and weak	Low	Low	Low	Moderate
Soft and tough	Low	Low	Moderate	High
Brittle and hard	High		Moderate	Low
Hard and strong	High	High	High	Moderate
Hard and tough	High	High	High	High

3.12.2   Tear Strength

The measurement of the tear strength of a material evaluates the energy absorbed by the sample during tear initiation and/or propagation. Two standard methods are available: ASTM methods D1004 and D1922. The former is designed to measure the force required to initiate tearing. It covers the determination of the tear resistance of a flexible plastic sample at very low rates of loading, 51 mm/min. ASTM D1922 measures the energy required for propagation of a tear after it is initiated by a small hole in the sample. The value of tear strength of a film depends on the orientation stretching ratio, and whether the measurement is performed along or across the machine direction. Tear strength of films can also be evaluated along the machine direction and cross direction using a tensile apparatus; ASTM D1938 describes this standard procedure.

3.12.3   Impact and Bursting Strength

The ability of a polymer to resist fracture when exposed to a sudden impact is the basis of tests for impact strength. These are loosely related to the toughness of the polymer, and fall into two main categories, those suited for film and those suited for sheet.

The dart drop impact test, ASTM D1709 and ASTM D4272, is often used for film samples. In D1709, a sample of the film is subjected to an impact by a round-headed dart falling from a designated height. The weight of the dart which will rupture the film is determined. D4272 determines the kinetic energy lost by a dart as it passes through the film. ASTM D3420, the pendulum impact test, is also used for films. It measures the resistance of the film to impact puncture simulating high speed end use applications.

For polymer sheet, the notched Izod impact strength is often reported (ASTM D256). In this test, as well as the Charpy impact test in the same ASTM standard, the polymer sample is subjected to an impact from a falling hammer, and the energy absorbed is measured.

The *bursting strength* of a material is the hydrostatic pressure, given in psi (or pascal), required to produce rupture of a flat material (film or sheet) when the pressure is applied at a controlled increasing rate through a circular rubber diaphragm, 30.48 mm (1.2 in) in diameter, as described in ASTM method D774. This pressure, expressed in psi, is reported as "points bursting strength."

3.12.4   Other Mechanical Properties

Folding endurance is a measure of the material's resistance to flexure or creasing. Folding endurance is greatly influenced by the polymer's glass transition temperature and the presence of plasticizers. ASTM D2176 describes a standard procedure for determination of the number of folds necessary to break a film.

Pinhole flex resistance is the ability of a plastic film to resist the formation of pinholes (tiny holes) during repeated folding. Films having a low value of pinhole flex resistance will tend to generate pinholes at the fold line during repeated flexing. The test is described in ASTM standard F456. Pinhole flex resistance is related to folding endurance.

Abrasion resistance is a difficult property to define as well as to measure. It is normally accepted that abrasion resistance depends on the polymer's hardness and resilience, frictional forces, load, and actual area of contact. ASTM D1044 evaluates the resistance of transparent plastics to one kind of surface abrasion by measuring its effects on the transmission of light. Another test method to evaluate abrasion, ASTM D1242, measures the volume lost using two different types of abrasion machines: loose abrasion and bonded abrasion.

The mechanical properties of a polymer can be strongly influenced by its degree of crystallinity. When polymer crystallinity increases, the properties which depend primarily on the strength of the attractions between neighboring molecules, such as tensile strength and modulus of elasticity, increase. However, properties which depend primarily on the ability of molecules to absorb energy by temporarily or permanently changing their positions, such as impact strength and tear strength, decrease.

Stress crack resistance is another measure of plastic strength. When a solid polymer is under a stress, and it is subjected to attack by a material that acts as a weak solvent, cracks will occur in the material to relieve the stresses. Stress cracking is most often thought of as occurring in blow molded or injection molded parts. Here the cooling is rapid enough to freeze in strains in the part. When a material that interacts strongly with the polymer is put in the package, the polymer attempts to swell as it is being solvated. This swelling introduces some strains, but mainly produces increased molecular motion. When the solvated polymer zone reaches the molded-in strain, a crack usually results. Similar stress cracking can occur in films, also. In this case one is usually dealing with stress applied externally by folding the film in a package. Tight folds cause the outer layer of the fold to be in tension. If the product is a detergent or a fatty food, stress cracking can occur, resulting in small pin-holes.

3.13   Barrier Properties

Barrier properties of materials indicate their resistance to diffusion and sorption of substances. A good barrier polymer has low values of both diffusion (D) and solubility (S) coefficients. Since the permeability coefficient, P, is a derived function of D and S, a polymer with high barrier has low permeability. The diffusion coefficient is a measure of how fast a penetrant will move within the polymer, while the solubility coefficient gives the amount of the penetrant taken in (or sorbed) by the polymer from a contacting phase. Both diffusion and solubility can be applied to the reverse of sorption, that is, the migration of compounds from the polymer to a surrounding medium. Several factors influence the effective values of diffusion and solubility coefficients in polymers: (1) chemical compositions of the polymer and permeant; (2) polymer morphology (since diffusion and sorption occur mainly through the amorphous phase and not through crystals); (3) temperature (as temperature increases, diffusion increases while solubility decreases); (4) glass transition temperature, and (5) the presence of plasticizers and fillers. A detailed discussion of barrier properties is presented in Chapter 14.

3.13.1   Diffusion Coefficient

As expressed in Fick's law (see Sec. 14.5), the diffusion coefficient, D, is a parameter that relates the flux of a penetrant in a medium to its concentration gradient. A diffusion coefficient value for a polymer is always specific to a particular penetrant-polymer pair. For solid polymers, the diffusion coefficients of a large number of low molecular mass substances are in the range 1×10^{-8} to 1×10^{-13} cm^2/s. Diffusion theory states that diffusion is an activated phenomenon that follows Arrhenius' law. In addition to temperature, penetrant concentration and plasticizers also affect the value of the diffusion coefficient. Methods for the determination of D are discussed in Chapter 14.

3.13.2   Solubility Coefficient

The solubility coefficient indicates the sorption capacity of a polymer with respect to a particular sorbate. The simplest solubility coefficient is defined by Henry's law of solubility, which is valid at low concentration values for most substances. For other combinations, such as CO_2 in PET at high pressure, Henry's and Langmuir's laws must be combined. These are also discussed in Chapter 14.

3.13.3   Permeability Coefficient

The permeability coefficient, P, combines the effects of the diffusion and solubility coefficients. The barrier characteristics of a polymer are commonly associated with its permeability coefficient values. The well known relationship P = DS holds when D is concentration independent and S follows Henry's law. Standard methods for measuring the permeability of organic compounds are not yet available. ASTM E96 describes a method for measuring the water vapor transmission rate. ASTM D1434 describes a method for the determination of oxygen permeability.

3.14   Surfaces and Adhesion

A more thorough discussion of the relationship between surface properties and adhesion is presented in Chapter 6. Here we present some basic concepts.

3.14.1   Surface Tension

Surface tension is an important determinant of the surface and adhesion properties of polymers. In both solids and liquids, the forces associated with molecules inside the material are balanced because each molecule is surrounded on all sides by like molecules. On the other hand, molecules at the surface are not completely surrounded by the same type of molecules, generating unbalanced forces. Therefore, at the surface these molecules show additional free energy. The intensity of the free energy is proportional to the intermolecular forces of the material. The free surface energy of liquids and solids is called the surface tension. It can be expressed in mJ/m^2 or dyne/cm. Values of surface tension in polymers range from 20 dyne/cm for PTFE to 46 dyne/cm for nylon 6,6. The determination of surface tension values by contact angle measurement is covered by ASTM D2578. Several independent methods are available for estimation of the surface tension of liquids and solid polymers including the parachor, which is a parameter that relates surface tension and molar volume.

When two condensed phases are in close contact, the free energy at the interface is called the interfacial energy. Interfacial energy and surface energy in polymeric materials control adhesion, wetting, printing, surface treatment, and fogging.

3.14.2   Wettability

Adhesion and printing operations on a plastic surface depend on the substrate's wettability, which indicates the ability of a liquid to spread on the surface. Wettability is a function of surface tension, and ASTM D2578 describes a standard procedure for measuring the wettability of a surface by determining the surface tension required for a liquid to wet the film surface.

3.14.3   Adhesive Bond Strength

The adhesive bond strength between two surfaces, which most often involves an adhesive and a solid substrate, is a complex phenomenon. It is controlled (at least in part) by the values of surface tension and solubility parameters of the materials, and the viscosity of the adhesive. To obtain good wettability and adhesion between a polymeric substrate and an adhesive, the surface tension of the adhesive must be lower than that of the substrate. Usually, the difference between the two values must be at least 10 dynes/cm. Similarity in solubility parameters between the two phases indicates similarity of the intermolecular forces between the two phases. For good compatibility, the values of the solubility parameters must be very close. Low viscosity in the adhesive is necessary for good spreadability and wettability of the substrate. Adhesive bond strength is addressed in more detail in Section 6.3.1.

3.14.4   Cohesive Bond Strength

In contrast to adhesive bond strength, cohesive bond strength is the force within a material, reflecting the strength of the secondary forces between molecules. In a typical adhesive joint, the overall strength will be determined by the adhesive bond strength between the adhesive and each substrate, and by the cohesive bond strength within each substrate and within the adhesive itself. Cohesive bond strength is addressed in more detail in Section 6.3.2.

3.14.5   Blocking

Blocking is the tendency of a polymer film to stick to itself upon physical contact. This effect is controlled by the adhesion characteristics of the polymer. The polymer in the film has a great affinity for itself, and only characteristics of the surface can keep the molecules from attracting one another. Blocking can be measured by the perpendicular force needed to separate two sheets, and it can be minimized by incorporating additives such as silica in the polymer film. Silica roughens the film surface and forms microscopic air gaps which keep the surfaces separated. Crystallites on the film surface can have the same effect. ASTM D1893, D3354, and Packaging Institute Procedure T3629 present methods to evaluate blocking.

3.14.6 Friction

The coefficient of friction (COF) is a measure of the friction forces between two surfaces. The COF of a surface is determined by the surface adhesivity (surface tension and crystallinity), additives (slip, pigment and antiblock agents), and surface finish. It characterizes a film's frictional behavior, and is particularly important in cases such as film passing over free-running rolls, bag forming, the wrapping of film around a product, and the stacking of bags and other containers. In addition to the intrinsic variables affecting a material's COF, environmental factors such as machine speed, temperature, electrostatic buildup, and humidity also have considerable influence on its final value. The static COF is associated with the force needed to start moving an object. It is usually higher than the kinetic COF, which is the force needed to sustain movement. Determination of static COF is described in TAPPI standard T503 and ASTM D1894.

The COF for polymer films passing over surfaces harder than the film is a function of the shear strength of the material. If the second sliding surface has roughness that can gouge into the film, even on a microscopic scale, polymer will be torn away from the film and deposited on the second surface.

In the real world, the COF of a film freshly unwound from a roll is typically higher than that which would be measured 15-30 minutes later, due to blooming of the slip additive in the film. Blooming refers to the migration of the additive from the bulk of the film to the surface, and may not occur fully in the roll form. Therefore, one needs to make measurements of COF on film samples quickly after unwinding to simulate what occurs in the equipment, where the film is moving rapidly.

3.14.7 Heat Sealing

The heat sealability of a material is an important property for wrapping, bag making, or sealing a flexible structure. At a given thickness, heat sealing characteristics of flexible web materials are determined by the material composition (which controls strength), average molecular mass (controlling temperature and strength), molecular mass distribution (controlling temperature range and molecular entanglement), and thermal conductivity (controlling dwell time). Tests normally conducted to evaluate the heat sealability of a polymeric material are the cold peel strength (ASTM F88), and the hot tack strength. Hot tack is the melt strength of a heat seal without mechanical support when the seal interface is still liquid. This adhesivity while liquid is associated with the molecular entanglement of the polymer chains, viscosity, and intermolecular forces of the material. Heat sealing is addressed in more detail in Section 6.6.

3.15   Optical Characteristics

Among the most important optical properties of polymers are absorption, reflection, scattering, and refraction. Absorption of light takes place at the molecular level, when the electromagnetic energy is absorbed by groups of atoms. If visible light is absorbed, a color will appear; however, most polymers show no specific absorption of visible light, and are, therefore, colorless. Reflection is the light that is returned from the surface. It depends on the refractive indices of air and of the polymer. Scattering of light is caused by optical inhomogeneities which reflect the light in all directions. Refraction is the change in direction of light due to the difference between the refractive indices of the polymer and air. Transparency, opacity, and gloss of a polymer are not directly related to the chemical structure or molecular mass, but are mainly determined by the polymer morphology. Optical appearance properties are of two types: optical morphological properties, which correlate with transparency and opacity; and optical surface properties which produce specular reflectance and attenuated reflectance.

3.15.1   Gloss

Gloss is the percentage of incident light that is reflected at an angle equal to the angle of the incident rays (specular reflectance). It is a measure of the ability of a surface to reflect the incident light. Figure 3.37 illustrates a common procedure for measuring gloss.

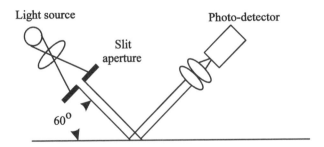

Figure 3.37 Measurement of gloss.

For films, gloss is usually measured at an angle of 45°. If the specular reflectance is near zero, the surface is said to be matte. A surface with high reflectance has a high gloss which produces a sharp image of any light source and gives a pleasing sparkle. Surface roughness, irregularities, and scratches all decrease gloss. Test method ASTM D2457 describes the determination of gloss.

3.15.2 Haze

Haze is the percentage of transmitted light that, in passing through the sample, deviates by more than 2.5° from an incident parallel beam (Fig. 3.38). Haze is caused by light being scattered by surface imperfections, non-homogeneity, and internal scattering from crystallites. The measurement of haze is described in ASTM D1003.

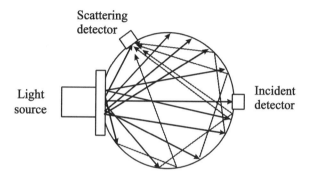

Figure 3.38 Measurement of haze.

3.15.3 Transparency and Opacity

Transmittance is the percent of incident light that passes through the sample. It is determined by the intensity of the absorption and scattering effects (Fig. 3.38). The absorption in polymers is generally insignificant, so, if the scattering is zero, the sample will be transparent. An opaque material has low transmittance and, therefore, large scattering power. The scattering power of a polymer results from morphological inhomogeneities and/or the presence of crystals. An amorphous homogeneous polymer such as "crystal" polystyrene will have little or no scattering power and, therefore, will be transparent. A highly crystalline polymer such as HDPE will be mostly opaque. Transmittance can be determined according to standard ASTM D1003. A transparent material is defined as having a transmittance value above 90%.

3.16   Electrical Properties

Electrical properties of polymers subject to low electric field strengths can be described by their electrical conductivity, dielectric constant, dissipation factor, and triboelectric behavior. Materials can be classified as a function of their conductivity (κ) in $(\Omega/cm)^{-1}$ as follows: conductors, $0\text{-}10^{-5}$; dissipatives, $10^{-5}\text{-}10^{-12}$; and insulators, 10^{-12} or lower. Plastics are considered non-conductive materials (if the newly developed conducting plastics are not included). The relative *dielectric constant* of insulating materials (ε) is the ratio of the capacities of a parallel plate condenser with and without the material between the plates. A correlation between the dielectric constant and the solubility parameter (δ) is given by $\delta \approx 7.0\varepsilon$. There is also a relation between *resistivity* R (the inverse of conductivity) and the dielectric constant at 298 K: log R=23-2ε.

When two polymers are rubbed against each other, one becomes positively charged and the other negatively charged. Whether a polymer becomes positive or negative depends on the electron donor-acceptor characteristics of the polymer. A *triboelectric series* is a listing of materials according to their charge intensity. Polymers can be ordered from more negatively charged polymers (electron-acceptors), through neutral polymers, and finally to more positively charged polymers (electron-donors). Charging of polymer films also takes place by friction during industrial operations such as form-fill-seal packaging. A brief list of the triboelectric series for polymers as a function of polymer's dielectric constant is presented in Table 3.5. Hydrophilic polymers absorb water and became more conductive as the dielectric constant increases.

Table 3.5 Brief Triboelectric Series

Polymer	Dielectric Constant
Negative:	
Polypropylene	2.2
Polyethylene	2.3
Polystyrene	2.6
Neutral:	
Polyvinyl chloride	2.8
PVDC	2.9
Polyacrylonitrile	3.1
Positive:	
Cellulose	3.7
Nylon 6,6	4.0

Adapted from Van Krevelen [2]

3.17   Plastics Identification Using IR Spectrophotometry

The radiation absorption and transmission characteristics of polymers can be used to identify polymer types. The basic principle of spectrophotometric methods of analysis is to provide an energy source (such as infrared radiation, IR) to which a sample is exposed, and to convert to some measure a response that is characteristic of the molecules. This response is, most often, absorption of a fraction of the incident light.

The frequency of an absorption band in a spectrum is usually expressed in terms of wave numbers, v, which are commonly given in reciprocal centimeters (cm^{-1}). The frequency is defined as the number of waves passing a fixed point in a unit time, and is equal to the inverse of the wavelength, which is the distance between adjacent waves. Thus:

$$\text{wave number in } cm^{-1} = 1/(\text{wavelength in microns x } 10^{-4})$$

Infrared radiation refers to the electromagnetic spectrum between the visible and microwave regions. The area of greatest interest for polymer identification is between $4000 \ cm^{-1}$ and $600 \ cm^{-1}$ (corresponding to wavelengths of 2.5 μm to 15.0 μm).

Incident radiation is only partly transmitted when it strikes a layer of a chemical substance. The remaining radiant energy is reflected or absorbed to a varying degree, depending upon the substance and the frequency of the radiation. Absorption may occur only if the energy of the incident photon coincides with an allowed energy transition within the molecule. The energies associated with molecular vibrational and rotational transitions are comparable to those of infrared radiation, thus permitting photons of certain energies to be absorbed by certain species within the molecule. Infrared spectroscopy measures the absorption of the incident radiation as a function of the radiation frequency.

In practice, the absorption spectrum in the infrared region is shown as a plot of absorption (or percent transmission) versus the wavelength in microns (or the frequency). The presence or absence of characteristic absorption bands permits the species within the molecules, and thus the type of polymer, to be identified.

Absorption of infrared radiation occurs only when a molecule, or segment of a molecule, is composed of asymmetrical atoms, which is the case in all polymers. In a symmetrical molecule (i.e., hydrogen, oxygen, nitrogen), the center of positive charge arising from the constituent nuclei coincides with the center of negative charge arising from the constituent electrons. They remain in conjunction regardless of the vibrational positions of the atoms. In contrast, with dissimilar atoms, the centers do not coincide and change their relative position during vibration, giving rise to an oscillating dipole. If radiation of a frequency equivalent to that of the molecular vibration strikes the molecule, a molecular resonance condition is established between the oscillating dipole and the electrical component in the radiation.

The absorption of infrared radiation can be expressed either as transmittance (T) or absorbance (A). Transmittance is the ratio of the intensity of radiation transmitted (I), leaving the polymer sample, to the incident intensity (I_o), hitting the sample surface.

$$T = \frac{I}{I_o} \quad \text{or} \quad T(\%) = \frac{I}{I_o} \times 100 \qquad (3.17)$$

Absorbance is related to transmittance as follows:

$$A = \log_{10} \frac{I_o}{I} = \log_{10} \frac{1}{T} \qquad (3.18)$$

If transmittance is expressed as percent,

$$A = \log_{10} \frac{100}{T(\%)} \qquad (3.19)$$

Infrared absorption can occur as a result of both vibrational and rotational transitions within the molecule. However, vibrational absorption bands are of greater practical interest as they are more easily measured, and because only relatively few compounds exhibit pure rotational transitions. Two types of molecular vibrations are of importance: stretching and bending. A stretching vibration is a rhythmical movement along the bond axis, such that the interatomic distance is increased and decreased.

References

1.   Woodward, A. E., *Understanding Polymer Morphology*, Carl Hanser Verlag, Munich, 1995.
2.   Van Krevelen, D. W., *Properties of Polymers*, 3d ed., Elsevier, New York, 1990.

Study Questions

1.   What are the four major types of copolymers, in terms of structure?

2.   Explain how addition polymerization works. What characteristic is shared by the monomers used in addition polymerization?

3.   Explain the difference between conformation and configuration. Which do you think will affect polymer properties the most? Why?

4.   What is the degree of polymerization of a PP sample with a molecular weight of 100,000? (Answer: 2381)

5. Give an example of a polymer that has more than one stereochemical configuration. Sketch the configurations, and name them. How do you expect the configurations to affect the properties of the polymer?

6. Why do we not usually consider condensation polymers made from two different monomers to be copolymers, but addition polymers made from two different monomers <u>are</u> considered copolymers?

7. For the following polymer sample, calculate the number average molecular weight, weight average molecular weight, and dispersion index. Construct a graph of the molecular weight distribution and indicate the values for the average molecular weight. (Answer: M_n =4775, M_w=4840, Q=101)

   Polymer: PVC
   20% of molecules (by count, not weight) have n=4000
   30% have n = 4500
   30% have n = 5000
   15% have n = 5500
   5% have n = 6000

8. Polymer X has a T_g of 30°C. Knowing nothing else, what can you hypothesize about its properties? Estimate its melting temperature, assuming it has one. What must you know to know whether the melt temperature exists?

9. Transparency is a highly valued characteristic of packaging materials in a variety of applications. Why do you think this is so?

10. What properties of polymers are, in general, determined by (a) the monomer chemical composition, and (b) the architectural arrangement of the monomers in the polymeric chain?

11. Explain why bifunctional monomers yield linear polymers. Also give an explanation of how a "normal" bifunctional monomer may yield a branched polymer. Discuss examples.

12. You are working with two polyethylene samples, A and B. Sample A has a melting temperature of about 104°C and a heat sealing range of 14°C. Sample B melts at about 121°C and it can be heat-sealed in the 122-119°C range. Indicate which sample probably has the greater average molecular weight, and which has the broader molecular weight distribution.

13. In Eq. 3.9 when n = 1 the K (consistency index) becomes the viscosity, η. Write the equation for n = 1. Calculate the viscosity for a shear stress of 617 psi and a shear rate value of 850 s^{-1}. (Answer: 5000 Ns/m^2)

14. Discuss the effect of MWD and chain branching on the viscous behavior of a polymer melt.

15. For each of the following polymers, indicate whether they are addition or condensation polymers: LLDPE, Nylon 66, PVC, PC, PAN, PP, HDPE, PET, LDPE, and PVDC. Write their chemical structures.

16. Indicate which of the polymers listed in Problem 15 is capable of having isotactic or syndiotactic structures.

17. Estimate the density of a PE sample having 60% crystallinity, on a volume basis. The density of totally amorphous PE is 0.855 g/cm^3, and 100% crystalline PE has a density of 1.00 g/cm^3. What is the crystallinity of the sample on a mass basis? (Answer: 0.942 g/cm^3, 63.7%)

18. How would you expect the properties of a copolymer made of ethylene and acrylonitrile ($H_2C=CHCN$) to differ from homopolymer PE?

19. Discuss potential problems that might arise from the friction of plastic against parts of a filling machine. What would be the most convenient plastic to use in order to minimize these problems? How would you proceed to evaluate a new plastic?

20. How do transparency, opacity and gloss correlate with polymer morphology? Explain.

4   Major Plastics in Packaging

4.1   Branched Polyethylenes

Low density polyethylene is the most widely used packaging plastic. It is a member of the *polyolefin* family. Olefin, which means oil-forming, is an old synonym for alkene, and was, originally, the name given to ethylene. Alkenes are hydrocarbons containing carbon-carbon double bonds, such as ethylene and propylene. In the plastic industry, olefin is a common term that refers to the family of plastics based on ethylene and propylene. The term polyolefin strictly applies to polymers made of alkenes, whether homopolymers or copolymers. It includes the family of polyethylene, and the family of polypropylene.

Polyethylene (PE) is a family of addition polymers based on ethylene. Polyethylene can be linear or branched, homopolymer or copolymer. In the case of a copolymer, the other comonomer can be an alkene such as propene, butene, hexene or octene; or a compound having a polar functional group such as vinyl acetate (VA), acrylic acid (AA), ethyl acrylate (EA), or methyl acrylate (MA). If the molar percent of the comonomer is less than 10%, the polymer can be classified as either a copolymer or homopolymer. Figure 4.1 presents a diagram of the family of polymers based on ethylene monomer.

Polyethylene was the first olefinic polymer to find use in food packaging. Introduced in the 1950s, it became a common material by 1960, used in film, molded containers, and closures. Since low density polyethylene was first introduced in 1940, strength, toughness, thermal and heat sealing properties, optical transparency, and processing conditions have been much improved. Today there are a number of polyethylene grades of relevance to packaging, as shown in Figure 4.1.

Low density polyethylene has a branched structure. The family of branched polyethylenes includes homopolymers and copolymers of ethylene that are non-linear, thermoplastic, and partially crystalline. They are fabricated under high pressure and temperature conditions by a free radical polymerization process. The random polymerization of ethylene under these conditions produces a branched polymer that is actually a mixture of large molecules with different backbone lengths, various side chain lengths and with various degrees of side-chain branching.

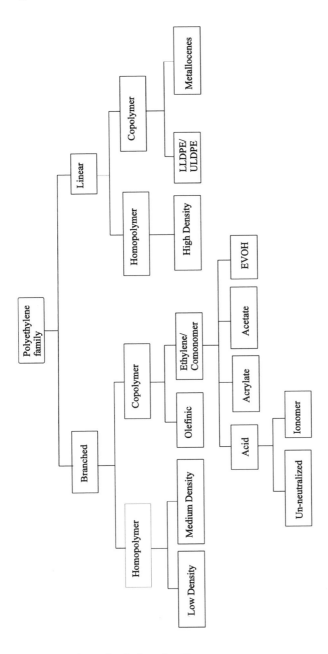

Figure 4.1 The polyethylene family.

While linear PE, because of its stereoregularity and the small size of its pendant groups, has a high percent crystallinity, from 70 to 90%, the presence of branches in the backbone chain acts to limit the formation of polyethylene crystals by introducing irregularities in the structure. Therefore, linear PE produces a highly crystalline polymer that has a relatively high density and is known as high density polyethylene (HDPE). Branched PE has lower crystallinity and consequently lower density, so is known as low density PE (LDPE). LDPE typically has a crystallinity of 40 to 60%, with a density of 0.910 to 0.940 g/cm^3; in contrast, HDPE has a density of about 0.940-0.970 g/cm^3. Comonomers such as propylene and hexene are commonly used in the reaction to help control molecular weight. A wide variety of branched polyethylenes are commercially available, with properties dependent on the reaction conditions and on the type and amount of comonomer.

4.1.1   Low Density Polyethylene

The chain branching in homopolymer LDPE gives this polymer a number of desirable characteristics such as clarity, flexibility, heat sealability and ease of processing. The actual values of these properties depend on the balance between the molecular weight, molecular weight distribution, and branching.

LDPE is also versatile with respect to processing mode, and is adaptable to blown film, cast film, extrusion coating, injection molding, and blow molding. Film is the single largest form of LDPE produced. In the U.S., more than half of total LDPE production is made into films with thickness less than 300 microns (12 mils). Products made of LDPE include containers and bags for food and clothing, industrial liners, vapor barriers, agricultural films, household products, and shrink and stretch wrap films. LDPE can be used alone or in combination with other members of the PE resin family, and is the most widely used plastic in packaging, with film the largest use.

LDPE is characterized by its excellent flexibility, good impact strength, fair machinability, good oil resistance, fair chemical resistance, good heat sealing characteristics, and low cost (about $1.60/kg). Its transparency is better than HDPE because of its lower percent crystallinity. For the same reason, while it is a good water vapor barrier, it is inferior to HDPE. Similarly, it is an even poorer gas barrier than HDPE. A summary of the properties of LDPE is presented in Table 4.1.

Medium density polyethylene (MDPE), 0.925-0.940 g/cm^3, is sometimes listed as a separate category, but usually is regarded as the high density end of LDPE. It is somewhat stronger, stiffer and less permeable than lower density LDPE. MDPE processes similarly to LDPE, though usually at slightly higher temperatures.

The major competitor to LDPE is LLDPE (discussed in Section 4.2.1), which provides superior strength at equivalent densities. However, LDPE is still preferred in applications demanding high clarity or for extrusion coating a substrate.

Ethylene can be copolymerized with alkene compounds or monomers containing polar functional groups, such as vinyl acetate and acrylic acid. Branched ethylene/alkene copolymers are essentially the same as LDPE, since in commercial practice a certain amount of propylene or hexene is always added to aid in the control of molecular weight.

Table 4.1 LDPE Properties

Density	0.910 to 0.925 g/cm^3
T_g	-120°C
T_m	105-115°C
Tensile strength	8.2-31.4 MPa (1,200-4,550 psi)
Tensile modulus	172-517 MPa (24,900-75,000 psi)
Elongation at break	100-965%
Tear strength	200-300 g/25 μm
WVTR	375-500 g μm/m^2 d at 37.8°C, 90% RH
	(0.95-1.3 g mil/100 in^2 d at 95°F, 90% RH)
O_2 permeability, 25°C	163,000-213,000 cm^3 μm/m^2 d atm
	(400-540 cm^3 mil/100 in^2 d atm)
CO_2 permeability, 25°C	750,000-1,060,000 cm^3 μm/m^2 day atm
	(1900-2700 cm^3 mil/100 in^2 d atm)
Water absorption	<0.01%

4.1.2   Ethylene Vinyl Acetate (EVA)

Ethylene vinyl acetate copolymers (EVA) are produced by copolymerizing ethylene and vinyl acetate monomers

$$
\begin{array}{cc}
\text{H} & \text{H} \\
| & | \\
\text{C} & = \text{C} \\
| & | \\
\text{H} & \text{O} \\
& | \\
& \text{C} = \text{O} \\
& | \\
& \text{CH}_3
\end{array}
$$

The result is a random copolymer, where

$$
\begin{array}{c}
-\text{O}-\text{C}-\text{CH}_3 \\
\parallel \\
\text{O}
\end{array}
$$

groups appear as side groups at random locations on the carbon chain, replacing H atoms.

EVA copolymers with vinyl acetate (VA) contents ranging from 5% to 50% are commercially available. For most food applications, VA ranging from 5% to 20% is recommended. EVA resins are mainly recognized for their flexibility, toughness, and heat sealability.

Vinyl acetate is a polar molecule. The inclusion of polar monomers in the main chain during production of branched ethylene copolymers will lower crystallinity, improve flexibility, yield a wider range of heat sealing temperature, and result in better barrier properties, as well as increasing density. These changes in properties result from the interference with crystallinity caused by the presence of random irregularities produced by the relatively bulky side groups from the comonomer, plus an increase in intermolecular forces resulting from the presence of polar groups in the comonomer. The increase in density is attributable to the presence of oxygen atoms with their higher mass, which more than compensates for the decreased crystallinity.

EVA is a random copolymer whose properties depend on the content of vinyl acetate and the molecular weight. As the VA content increases, the crystallinity decreases, but the density increases. Other properties are also affected, resulting in improvement in clarity, better flexibility at low temperature, and an increase in the impact strength. At 50% VA, EVA is totally amorphous. The increased polarity with increasing VA content results in an increase in adhesion strength and hot tack. An increase in average molecular weight of the resin increases the viscosity, toughness, heat seal strength, hot tack and flexibility.

Because of its excellent adhesion and ease of processing, EVA is often used in extrusion coating and as a coextruded heat seal layer. Examples include functioning as a heat sealing layer with PET, cellophane and biaxially oriented PP packaging films (20% VA) for cheese wrap and medical films. Because EVA has limited thermal stability and low melting temperature, it has to be processed at relatively low temperatures. However, this also results in toughness at low temperatures, which is a significant asset for packages such as ice bags and stretch wrap for meat and poultry.

4.1.3   Ethylene Acrylic Acid (EAA)

The copolymerization of ethylene with acrylic acid (AA)

$$
\begin{array}{cc}
\text{H} & \text{H} \\
| & | \\
\text{C} = & \text{C} \\
| & | \\
\text{H} & \text{C} = \text{O} \\
& | \\
& \text{OH}
\end{array}
$$

produces copolymers containing carboxyl groups (HO-C=O) in the side chains of the molecule. These copolymers are known as ethylene acrylic acid, EAA. They are flexible thermoplastics with chemical resistance and barrier properties similar to LDPE. EAA, however, is superior to LDPE in strength, toughness, hot tack and adhesion, because of the increased intermolecular interactions provided by the hydrogen bonds. Major uses include blister packaging and as an extruded tie layer between aluminum foil and other polymers.

As the content of AA increases, the crystallinity decreases, which implies that clarity also increases. Similarly, adhesion strength increases because of the increase in polarity, and the heat seal temperature decreases due to the decrease in crystallinity.

Films of EAA are also used in flexible packaging of meat, cheese, snack foods, and medical products; in skin packaging; and in adhesive lamination. Extrusion coating applications include condiment and food packages, coated paperboard, aseptic cartons, composite cans and toothpaste tubes. FDA regulations permit use of up to 25% acrylic acid for copolymers of ethylene in direct food contact.

4.1.4   Ionomers

Neutralization of EAA or a similar copolymer, for example EMAA (ethylene methacrylic acid), with cations such as Na^+, Zn^{++}, Li^+, etc., produces a material that has better transparency and toughness, and higher melt strength than the un-neutralized copolymer. These materials are called ionomers because they combine covalent and ionic bonds in the polymer chain. The structure of an ionomer of the ethylene sodium acrylate type is:

$$-CH_2-CH-CH_2-CH_2-CH_2-CH-CH_2-CH_2-$$
$$\begin{array}{ccc} | & & | \\ C=O & & C=O \\ | & & | \\ O^-\,Na^+ & & O^-\,Na^+ \end{array}$$

Ionomers were developed in 1965 by R.W. Rees and D. Vaughan, while working for DuPont, which uses the trade name Surlyn for these materials.

The ionic bonds produce random crosslink-like ionic bonds between the chains, yielding solid-state properties usually associated with very high molecular weight materials. However, ionomers behave as normal thermoplastic materials because the ionic bonds are much more readily disrupted than covalent bonds, allowing processing in conventional equipment. Normal processing temperatures are between 175 and 290 °C. The presence of ionic bonds decreases the ability of the molecules to rearrange into spherulites, thus decreasing crystallinity. The high elongational viscosity caused by the ionic bonds imparts excellent pinhole resistance.

Barrier properties of ionomers alone are relatively poor, but combined with PVDC, HDPE, or foil they produce composite materials that are excellent barriers.

Ionomers are frequently used in critical coating applications, films, and laminations. Applications include heat seal layers in a variety of multi-layer and composite structures. They are used in combination with nylon, PET, LDPE, and PVDC. Coextrusion lamination and extrusion coating are the most common processing techniques.

Ionomers are used in packaging where formability, toughness, and visual appearance are important. Food packaging films are the largest single market. They are highly resistant to oils and aggressive products, and provide reliable seals over a broad range of temperatures. Ionomers stick very well to aluminum foil. They are also used extensively as a heat-sealing layer in composite films for fresh and processed meats, such as hotdogs. Other applications of ionomers include frozen food (fish and poultry), cheese, snack foods, fruit juice (Tetra Pak™ type container), wine, water, oil, margarine, nuts, and pharmaceuticals. Heavy gauge ionomer films are used in skin packaging for hardware and electronic products due to their excellent adhesion to corrugated board and excellent puncture resistance. Ionomers can resist impacts at temperatures as low as -90°C (lower than LDPE).

Sodium can be replaced by other metals such as zinc, and other comonomers such as methacrylic acid

$$
\begin{array}{cc}
H & CH_3 \\
| & | \\
C = & C \\
| & | \\
H & C=O \\
& | \\
& OH
\end{array}
$$

can be used. There are more than fifty commercial grades of ionomer with a wide range of properties. In general, sodium ion types have better optical properties, hot tack and oil resistance. Zinc ionomers are more inert to water, and have better adhesion properties in coextrusion and for extrusion coating of foil.

4.2   Linear Polyethylenes

High density polyethylene is the second most widely used packaging plastic. As already discussed, it is produced by polymerization of ethylene, but it has a nearly totally linear structure, in contrast to the highly branched structure of low density polyethylene. This results in a greater ability to crystallize, producing a tighter packing of molecules and consequently a higher density.

Linear polyethylenes have traditionally been produced using a Ziegler-Natta stereospecific catalyst, named after the two chemists who invented them in the 1950s. These catalysts are made from a combination of a transition metal halide, such as titanium tetrachloride, with a reducing agent, such as an aluminum alkyl. An example of the catalyst structure is shown in Figure 4.2. The first application of these catalysts was production of HDPE. In a high pressure reactor, ethylene polymerization could not achieve densities much over 0.94 g/cm³. These new catalysts, which only allow the combination of free radicals in a specific orientation, allowed the density to be increased to 0.97 g/cm³. Later, in the 1970's, the use of Ziegler-Natta catalyst technology was extended by Dow and Union Carbide to the production of lower density polyethylenes by incorporating comonomers such as 1-butene, 1-hexene, and 1-octene into the polymer backbone. The resulting randomly placed side groups decreased the crystallinity of the material, consequently lowering density. Ziegler-Natta catalyst technology allows control over the density through the percentage of comonomer incorporated into the backbone. Since the reactions are run at relatively low temperatures and pressures, the hydrogen abstraction reactions responsible for the branching in LDPE do not occur to any significant extent. The milder reaction conditions also facilitate improved control over the average molecular weight and the molecular weight distribution.

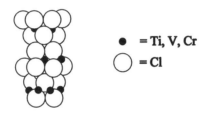

δ -form MX₃ catalyst crystal

Figure 4.2 A typical Ziegler-Natta catalyst.

Linear polyethylenes can be divided into the following groups, with density given in parentheses:

ULDPE - ultra low density PE (0.89-0.915 g/cm³)
LLDPE - linear low density PE (0.916-0.940 g/cm³)
HDPE - high density PE (0.940-0.965 g/cm³)
HMW-HDPE - high molecular weight HDPE (0.940-0.965 g/cm³)

The molecular weight of linear polyethylene ranges from medium to ultra high molecular weight, as shown in Table 4.2.

4.2.1   High Density Polyethylene (HDPE)

High density PE is a milky-white, nonpolar, linear thermoplastic. Its density ranges from 0.940 to 0.965 g/cm³, and it has a melting temperature of about 128-138°C. It is one of the most versatile polymers, and is the second most commonly used plastic in the packaging industry. Typical applications include:

- Containers for milk, detergent, bleach, juice, shampoo, water, and industrial chemical drums made by extrusion blow molding.
- Buckets, thin walled dairy containers, and closures made by injection molding.
- Cosmetic containers, pharmaceutical bottles, and deodorant containers made by injection blow molding.
- Blown and cast films utilized in flexible packaging applications such as cereal, cracker and snack food packaging, wrap for delicatessen products, and produce bags.

Table 4.2 Molecular Weight Range of Linear Polyethylene

Type	Relative Molecular Mass (Molecular Weight)
Medium Molecular Weight	<110,000
High Molecular Weight	110,000 - 250,000
Very High Molecular Weight	250,000 - 3,500,000
Ultra High	>3,500,000

The molecular chains of HDPE homopolymers are long and straight with little branching. HDPE forms a large fraction of ordered, crystalline regions as it cools below its T_m. This close packing produces HDPE with a crystallinity of 65-90% and contributes to HDPE's good moisture-barrier properties and non-transparency. HDPE has excellent resistance to a wide range of chemical compounds: water based products, medium molecular weight aliphatic hydrocarbons, alcohols, acetone, ketones, dilute acids and base. Not acceptable for aromatic hydrocarbons such as benzene. Tensile strength of HDPE can be as high as 45 MPa. Impact strength is primarily controlled by MW although it is affected by MWD. HDPE is also characterized by good moisture barrier properties, and poor oxygen and organic compound barrier characteristics. It has good machinability characteristics. Table 4.3 summarizes HDPE properties.

As is the case for other polymers, the properties are affected by the average molecular weight and molecular weight distribution. As relative molecular mass (or molecular weight) increases, tensile strength, impact strength and resistance to stress cracking increase.

Environmental stress cracking is a common problem for high density polyethylene. It can be defined as the failure of a material that is under stress and exposed to a chemical substance, in conditions where exposure to either the stress alone or the chemical alone does not cause failure. Products such as laundry detergent are known stress-crack agents. In food products, fatty acids can also lead to stress cracking, but it is not as severe as with detergents. A high degree of crystallinity increases the tendency for environmental stress cracking. Therefore, copolymer HDPE is employed for applications such as detergent bottles. The comonomer reduces crystallinity by disrupting the close packaging of the linear molecular chains, and thereby gives greater resistance to crack propagation.

Table 4.3 Properties of HDPE

Density	0.94 to 0.965 g/cm^3
T_g	-120°C
T_m	128-138°C
Tensile strength	17.3-44.8 MPa (2,500-6,500 psi)
Tensile modulus	620-1.089 MPa (89,900-158,000 psi)
Elongation at break	10-1200%
Tear strength	20-60 g/25 μm
WVTR	125 g μm/m^2 d at 37.8°C, 90% RH
	(0.32 g mil/100 in^2 d at 95°F, 90% RH)
O_2 permeability, 25°C	40,000-73,000 cm^3 μm/m^2 d atm
	(100-185 cm^3 mil/100 in^2 d atm)
CO_2 permeability, 25°C	200,000-250,000 cm^3 μm/m^2 day atm
	(500-640 cm^3 mil/100 in^2 d atm)
Water absorption	<0.01%

4.2.2   Linear Low Density Polyethylene (LLDPE)

As mentioned above, linear polyethylene can be produced as a homopolymer, resulting in high density polyethylene, HDPE, or as a copolymer having as comonomer alkenes such as butene, hexene and octene.

butene	$H_2C = CH\ CH_2\ CH_3$
hexene	$H_2C = CH(CH_2)_3\ CH_3$
octene	$H_2C = CH(CH_2)_5\ CH_3$

The presence of a comonomer in the polymerization process, when a stereo-specific catalyst is used, results in the production of a rather linear polymer with very short branch-like pendant groups. This polymer is called linear low density polyethylene (LLDPE) or ultra low density polyethylene (ULDPE), depending on the density achieved by the addition of the comonomer. The larger the amount of comonomer added, the lower is the density of the copolymer. For example, if hexene is used, the pendant groups are as follows:

$$-CH_2-CH\ -CH_2-CH_2-CH_2-CH\ -CH_2-CH_2-$$

$$\underset{\displaystyle CH_3}{\overset{\displaystyle (CH_2)_3}{|}} \qquad\qquad \underset{\displaystyle CH_3}{\overset{\displaystyle (CH_2)_3}{|}}$$

Normally the amount of comonomer ranges from 1 to 10% on a molar basis.

The physical properties of LLDPE are controlled by its molecular weight (MW) and density (0.916-0.940 g/cm^3). Due to the increased regularity of the structure and narrower molecular weight distribution, LLDPE tends to have improved mechanical properties, compared to LDPE at the same density. The greater stiffness results in an increase of 10-15°C in the melting point of LLDPE compared to LDPE. LLDPE has higher tensile strength, puncture resistance, tear properties and elongation than LDPE. However, LDPE has better clarity and gloss than LLDPE. LDPE also has better heat seal properties.

Typical uses for LLDPE include stretch/cling film, grocery sacks, and heavy duty shipping sacks. LDPE and LLDPE are often blended to optimize the benefit obtained from both materials, with LLDPE adding strength and LDPE adding heat seal and processability.

It has been found that as the density is pushed below 0.91 g/cm^3 by the incorporation of higher levels of comonomer, the level of hexane extractables increases to a level beyond that sanctioned by the FDA. These extractables also can oxidize, resulting in off odors and off flavors.

Polyethylenes with larger amounts of comonomer and consequently density below the normal LLDPE range are called very low density polyethylene, VLDPE, or ultra low density polyethylene, ULDPE. While these can be produced using Ziegler-Natta catalysts, often they are made using metallocene catalysts, as described next.

4.2.3   Metallocene Polymers

In the 1990s, a new family of polyethylenes, based on metallocene catalysts, emerged. These catalysts offered significant new ability to tailor the properties of linear polyethylenes and other polyolefins. In particular, they have the ability to provide more uniform incorporation of comonomers.

Metallocene catalysts (Fig. 4.3) were first discovered in the early 1950s by Natta and Breslow, and were first used to make polyethylene in 1957. They were used to produce syndiotactic polystyrene in 1984 and syndiotactic polypropylene (FINA) in 1986. However, commercialization for polyethylene did not come until the mid-1990's, since until that time the advantages the new catalyst systems offered were not fully appreciated. Metallocene catalysts employed today commonly contain a co-catalyst to increase the catalyst activity.

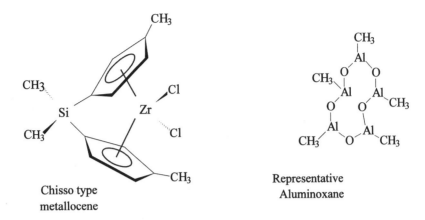

Chisso type
metallocene

Representative
Aluminoxane

Figure 4.3   Single-site metallocene catalyst with aluminoxane co-catalyst.

The first metallocene catalysts were biscyclopentadienyl titanium complexes and dialkylaluminum chloride. These catalysts were not stable and produced very low yields. However, they were the first catalyst systems to produce copolymers of polypropylene and 1-butene with very high comonomer uniformity, due to the fact that they had only one type of active site.

In the 1980s and 1990s, improved polymer characterization techniques were used to explain some of the characteristics, particularly higher haze and higher extractables, of LLDPE. Traditional Ziegler-Natta catalysts were found to have three different types of sites on the catalyst particles. As shown in Figure 4.4, one type of site produced a low MW species with a high proportion of comonomer. Another site produced a high MW species with very little comonomer, and the third type of site produced the predominant medium MW species with a medium amount of comonomer, which was the desired polymer. When the comonomer content was pushed up to produce densities below 0.91 g/cm^3, the percentage of low molecular weight material with a high concentration of comonomer increased. The extractables and off odors are due to this low MW species. The haze in LLDPE is primarily due to the high MW, linear fraction that crystallizes to a high degree.

Metallocene catalysts, on the other hand, contain only one type of site geometry, so are often referred to as single site catalysts (Fig. 4.5). They produce the desired copolymer, incorporating the comonomer in proportion to the amount added to the reactor. This results in improved properties. Compared to Ziegler-Natta catalysts, metallocene catalysts, by providing greater control over comonomer content, produce more uniform incorporation and improved MWD control. This results in improved clarity and lower extractables, permitting a higher level of incorporation of comonomer. Tensile strength and tear strength are both improved, and the polymer has a softer feel.

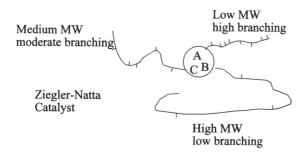

Figure 4.4 Ziegler-Natta catalyst sites. (Note: "branching" stems from incorporation of comonomer, so the side groups are not true branches.)

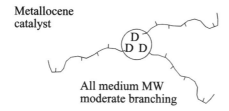

Figure 4.5 Metallocene catalyst - single site. (Note: "branching" stems from incorporation of comonomer, so the side groups are not true branches.)

The main class of metallocene catalysts used today is Kaminsky-Sinn catalysts. They are based on titanium, zirconium, or hafnium, and use methylaluminoxane as a cocatalyst. These catalysts produce very uniform comonomer incorporation and very narrow molecular weight distributions.

Figure 4.6 shows the results for hexane extractables on conventional LLDPE's and on metallocene polymers of lower density, but similar comonomer. Figure 4.7 shows the effect of the catalyst change on the haze in films.

Metallocene catalysts also permit the incorporation of novel comonomers that cannot be used with older Ziegler-Natta catalysts. Long alpha olefins can be incorporated, giving the effect of controlled long-chain branching, and offering some of the benefits of LDPE such as improved heat sealing, along with the benefits provided by control over MW and MWD. Constrained geometry catalysts (Fig. 4.8) are used to produce LLDPE with controlled "long chain branching" (LCB). These so-called long chain branches arise from incorporation of higher α-olefins, long alkenes (longer than octene), with a double bond at one end.

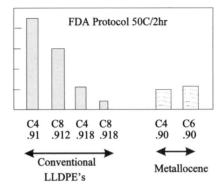

Figure 4.6 Hexane extractables, at densities indicated.

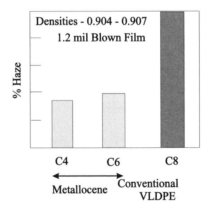

Figure 4.7 Haze.

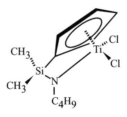

Figure 4.8 Constrained geometry catalyst

Processing is similar to LLDPE. The narrower MWD of the metallocenes results in higher viscosity at high shear rates, and therefore higher horsepower requirements for the extruder.

The improved control over the polymer structure offered by these catalysts offers the polymer producer a significantly greater ability to tailor the polymer to the end-user requirements. Polymer research with metallocene catalysts continues, so more advances can be expected for polyethylene, polypropylene, and other polyolefins.

4.2.4   Property Trends in the Polyethylene Family

The family of polyethylenes has many properties in common. Tables 4.4 and 4.5 show the relationship of these properties to molecular weight, MWD, and density.

Table 4.4 Effect of Density on the Permeability of Oxygen and Water in Polyethylene

Density of polyethylene g/cm^3	WVTR $g\ \mu m/m^2$ day	Oxygen Permeability $cm^3\ \mu m/m^2$ day atm
0.910	0.866	275
0.915	0.779	256
0.920	0.685	225
0.925	0.579	201
0.930	0.465	165
0.935	0.366	137
0.940	0.276	104
0.945	0.244	91.3
0.950	0.208	76.4
0.955	0.185	70.1
0.960	0.145	61.0

4.3   Polypropylene (PP)

Polypropylene is a thermoplastic produced by addition polymerization of propylene, and has the following structure:

$$- (\underset{\underset{H}{|}}{\overset{\overset{H}{|}}{C}} - \underset{\underset{H}{|}}{\overset{\overset{CH_3}{|}}{C}})_n -$$

Polypropylene is commercially available as both PP homopolymer, and PP random copolymer. The latter is produced by the addition of a small amount of ethylene (2%-5%) during the polymerization process. Thermoplastic PP polymers are characterized by low density (0.89-0.92 g/cm^3) and good resistance to chemicals and to mechanical fatigue, including environmental stress cracking. There are a wide variety of applications for PP, from automobile parts to packaging film and containers. Manufacturers of PP continuously are offering PP grades with improved or modified properties.

Table 4.5 Property Trends in Polyethylene

As molecular weight increases:	
Tensile Strength	increases
Impact Strength	increases
Clarity	increases
Ultimate Elongation	increases
Melt Strength	increases
Tear Resistance	decreases

As molecular weight distribution widens:	
Ultimate elongation	decreases
Tear Impact	decreases
Impact Strength	decreases
Melt Pressure	decreases
Melt Strength	increases

As density increases:	
Tensile Strength	increases
Melting Temperature	increases
Clarity	decreases
Ultimate elongation	decreases
Tear Resistance	decreases
Impact Strength	decreases
Blocking	decreases
Gas Permeability	decreases

4.3.1   PP Homopolymer

Depending on the type of catalyst and other polymerization conditions, the molecular structure of PP homopolymer can consist of any of the three different types of stereochemical configurations for vinyl polymers: isotactic, syndiotactic and atactic.

Isotactic PP (iso-PP), the most common commercial form of PP, is synthesized using Ziegler-Natta catalysts under controlled conditions of temperature and pressure. The placement of the methyl groups all on the same side of polymer backbone provides a structure that readily yields a highly crystalline material, resulting in good solvent and heat resistance. Industrial processes are designed to minimize the production of atactic PP, a lower-value, noncrystalline, tacky byproduct that is used mainly in adhesives.

PP has lower density, higher melting temperature, and higher stiffness (higher modulus) than LDPE and HDPE. These properties determine the applications for PP homopolymer. For example, its higher stiffness and ease of orientation makes PP homopolymer suitable for stretch applications, while its higher heat resistance allows containers made of this material to be sterilizable in an autoclave.

The molecular weight of PP typically averages between 200,000 and 600,000. Broad MWD materials are easy to process in injection molding applications. Compared with PE, isotactic PP is more sensitive to oxidative degradation due to heat and light. Oxidative degradation may produce chain scission that reduces the molecular weight and increases the flow rate. To control this process, antioxidants are added to the resin. Another common additive is an antistatic agent, used to dissipate static charges.

Isotactic PP has extremely good flow properties at a wide range of flow rates, and therefore good processing behavior. The melt flow index typically ranges from 0.5-50 g/10 min. Films, which can be produced by both blown and cast methods, can be oriented to provide improved optical characteristics and better strength. Because of the rapid crystallization of PP, blown films must be produced by either water quench or mandrel quench processes, unlike PE, which is cooled by air.

PP is an excellent moisture barrier, and has medium transparency. The melting temperature T_m is between 160 and 170°C. PP is commonly oriented to improve optical, mechanical, and barrier properties. Table 4.6 compares selected properties of oriented and non-oriented PP.

4.3.2   Random Copolymer Polypropylene

Random copolymer PP typically contains 1.5 to 7% ethylene, by weight, as a comonomer. The polymer structure is similar to that of isotactic PP with the addition of random insertion of ethylene groups.

The addition of ethylene, placed randomly in the backbone, prevents the high crystallinity obtained with isotactic PP. Lower crystallinity results in improved clarity and flexibility, and a lower melting point (152°C with 7% ethylene). The density is also slightly lower, 0.89-0.90 g/cm^3. This polymer has better toughness and low temperature impact strength than homopolymer PP.

Random copolymers show good chemical resistance to acids, alkalis, alcohols, and to low-boiling hydrocarbons, but are not suitable for packaging aromatic hydrocarbons. Moisture-barrier properties are good. For example at 38°C (100°F) and 90% RH, permeance to water vapor for one such copolymer is 235 g µm/m^2 24 h.

PP random copolymers are processed mainly by blow molding, extrusion, and injection molding. Applications include medical packaging, packaging for clothes, and packaging for bakery products, produce, and other foods.

Table 4.6 Effect of Orientation on Typical PP Properties

Property	Non-Oriented PP	Oriented PP
T_g, °C	-10	
T_m, °C	160-175	
WVTR g µm/m^2 day	590	240
at 90% RH and 38°C		100-300
Tensile strength, MPa	31-42	50-165
Tensile modulus, MPa	1,140-1,550	High, similar to cellophane
Tear strength, g/25 µm, film	50	Very low CD, very high MD
Heat sealability	Yes, T 177-232°C	No, film distorts
Density, g/cm^3	0.902	0.902
Transparency	Very good	Excellent
Surface adhesivity to inks, etc.	Low	Low
WVTR, g µm/m^2 d at 25°C	100-300	
O_2 permeability, cm^3 µm/m^2 day atm	146,000	98,000 50,000-94,000
CO_2 permeability, cm^3 µm/m^2 day atm		200,000-320,000
Water absorption, %	0.01-0.03	

Unoriented random PP copolymer films are soft and are easy to heat seal. Oriented PP films (both homopolymer and copolymer) have improved strength, clarity and gloss (brightness) compared to unoriented films. Applications for oriented PP include shrink wrap (records, toys, games, hardware items, frozen foods and cigarette wrap). A 7% ethylene copolymer is often used as a heat-seal layer in food packaging.

4.4   Polyvinyl Chloride (PVC)

Polyvinyl chloride is a thermoplastic homopolymer of vinyl chloride monomer, with the following structure:

$$-(CH_2-CH)_n-$$
$$\underset{Cl}{|}$$

80% of commercial PVC in packaging is produced by addition polymerization using a suspension method; other polymerization methods include emulsion and solution. The predominant configuration of the monomer in the chain is a head-to-tail alignment. The polymer produced has a slight degree of syndiotacticity, just enough to permit a very small degree of crystallinity. PVC has strong attractions between neighboring molecules, because of the polarity of the C-Cl bond, and in its normal state is stiff and rigid at room temperature.

The melting temperature, about 212°C, and decomposition temperature of PVC are very close together, rendering unmodified PVC very difficult to process. Decomposition produces HCl, which is highly corrosive, especially in the presence of water. To reduce problems associated with decomposition, stabilizers are added to PVC; octyl tins are most often used in rigid PVC for food and pharmaceutical packaging. The first indications that PVC is decomposing occur at temperatures as low as 100°C (212°F).

PVC without plasticizers (rigid PVC) has a T_g of 82°C (180°F) and is very difficult to process into a useful product. Plasticizers are additives that, by virtue of a "lubricating" action at the molecular level, increase the flexibility of the material. They also produce flow at lower temperatures, thereby decreasing processing temperatures. The polar nature of PVC gives it a strong affinity for plasticizers as well as other additives. As a result, PVC can be produced with a wide stiffness range, from rigid containers to very soft and flexible films. Virtually all properties of PVC are strongly affected by the amounts and types of additives incorporated in its formulation. Relatively unplasticized PVC resin has reasonably good barrier properties, for example, while highly plasticized films provide poor barriers. The addition of a liquid plasticizer permits the production of a flexible film with a moderate oxygen permeability that can be used in meat packaging. With addition of different plasticizers, blow molded bottles, blown films, and flexible materials are possible. Therefore, the manufacture of a wide variety of products from PVC is possible because of the miscibility of the polymer with a range of plasticizers. Formulations to produce specific products made of PVC are mostly proprietary. Many types of plasticizers are available, some suitable for food contact applications. Impact modifiers, such as methacrylate-butadiene-styrene (MBS), acrylonitrile-butadiene-styrene (ABS), chlorinated polyethylene, and acrylics are also blended with PVC. Resin characteristics can also be modified by copolymerization. Table 4.7 illustrates the properties of a typical flexible PVC film.

Table 4.7 Properties of a Typical Flexible PVC Film

Property	Flexible PVC, meat packaging type
Density	1.23 g/cm^3
Tensile Strength, MD	34.5 MPa (5000 psi)
Tear Strength	116 N/mm
Service Temperature	-29 to 66 °C
Heat Seal Temperature	143 to 160 °C
O$_2$ Permeability	3,342 cm^3 μm/m^2 day kPa (12,800 cm^3 mil/m^2 day atm)
WVTR at 38°C and 90% RH	6096 g μm/m^2 day (240 g mil/m^2 day)

Processing of PVC is carried out by conventional methods such as injection molding, extrusion, blown film and blow molding. Additives used in PVC (as well as other plastics) must be FDA-approved if the resin is used in food contact applications

Flexible PVC film is often used for packaging food products, particularly fresh red meat. The oxygen permeability of PVC film is well suited to maintain the necessary oxygen requirements of the meat, to preserve its red color and appearance of freshness. PVC films also provide good toughness and resilience. PVC is also used to wrap fresh fruits and vegetables. Almost all poultry producers in the U.S. once used PVC stretch films for chilled, tray-packed poultry parts. However, multilayer barrier films have made considerable inroads in this market. PVC is also used for a variety of other food and non-food packaging applications, as illustrated in Table 4.8.

Table 4.8 Typical Uses of PVC Packaging

Food packaging	
Bottles	Milk, dairy products, edible oil, liquor
Food wraps	Butter, fresh meat, frozen meat, cured meat, fish, poultry, fresh produce
Medical packaging	
Blisters	Medical devices, pharmaceutical products
Bags	Blood, intravenous solutions
Non-food packaging	
Blisters	Hardware, toys
Bottles	Toiletries, cosmetics, shampoos, household products
Wraps	Cartons and bottles for various products

PVC, over the years, has been subjected to scrutiny on health and environmental grounds. One issue is the level of residual vinyl chloride monomer in the material, which may migrate into food. Vinyl chloride monomer has been determined to be a carcinogen, at least under some conditions. In the polymerization process for PVC, less than 100% of the vinyl chloride monomer (VCM) is converted to polymer. This means that relatively high values of VCM may remain unreacted and trapped in the resin. To remove this residual monomer, the resin is subjected to repeated applications of vacuum. In this manner, VCM concentrations in the resin are reduced substantially. PVC packaging resins currently produced have much lower levels of residual vinyl chloride (under 10 ppb) than those used in containers in the mid-1970s when this concern first surfaced.

The disposal of PVC raises other environmental questions, especially when it is incinerated. Burning PVC produces HCl, and some suspect it contributes to the formation of chlorinated dioxins. Added to more general concerns about the effects of chlorinated organics in our environment, the result is a rather negative environmental image for PVC, which has accelerated its replacement by PET and other plastics that provide some of the same functions without the perceived environmental side-effects.

4.5   Vinylidene Chloride Copolymers (PVDC)

Polyvinylidene chloride, PVDC, is an addition polymer of vinylidene chloride with the following structure:

$$- (CH_2CCl_2)_n -$$

Vinylidene chloride homopolymer and copolymers were first produced as Saran, a registered trademark of Dow Chemical. The polymers are based on vinylidene chloride (VDC) and comonomers such as vinyl chloride (VC), acrylates (methyl acrylate) and vinyl nitrile. PVDC homopolymer has a melting point of 388- 401°C, but it decomposes at 205°C, producing HCl in a manner similar to PVC. These conditions make PVDC homopolymer impossible to melt-process. By adding comonomers, the melting point is decreased to about 140-175°C, making melt processing feasible. Polymers may also contain 2-10% plasticizer (for example, dibutyl sebacate or diisobutyl adipate), and will incorporate heat stabilizers as well. Molecular weight ranges from 65,000-150,000. The most notable attributes of PVDC copolymers are their chemical resistance and extremely low permeability to gases and liquids, comparable to EVOH resins. Oxygen permeability values range from 7.9 to 2700 cm^3 $\mu m/m^2$ day atm. The modification of the structure required to decrease crystallinity and improve processability also somewhat increases the permeability of the material.

Incorporation of a comonomer reduces crystallinity and the crystalline melting point, permitting processing at lower temperatures, or imparting solubility in organic solvents. Vinyl chloride and methyl acrylate are commonly used as comonomers for extrudable resins, typically in amounts from 6 to 28%. Vinylidene chloride copolymers with methyl acrylate and methyl methacrylate are commonly used for latex (water-based) coatings. Copolymers with acrylonitrile, methacrylonitrile, and methyl methacrylate are common for solvent-based coatings. All commercially available PVDC resins are copolymers.

PVDC resins can be processed in a variety of ways, including extrusion, coextrusion, laminating, latex coating, and injection molding, to meet specific packaging requirements. Both blown and cast films are also produced.

The main applications of PVDC resins are in food packaging as barrier materials to moisture, gases, flavors and odors. Monolayer films are used in household wrap. Industrial applications of monolayer films include laminations, unit dose packaging, and drum and pack liners for moisture-, oxygen- and solvent-sensitive products. PVDC is also used in food, pharmaceutical and cosmetic packaging. Multilayer films, generally coextrusions with polyolefins, are used to package meat, cheese, and other moisture- or gas-sensitive foods. The structures, usually containing 10-20% of PVDC copolymer, are commonly used as shrinkable films to provide a tight barrier around the food product. PVDC copolymers are also used as barrier layers in semirigid thermoformed containers.

PVDC coatings are commonly used on paper, cellophane, plastic films and rigid containers. Paper and paperboard can be coated with PVDC to give grease, oxygen and water barrier; cellophane is coated for oxygen and water barrier, along with heat-sealability. PVDC-coated plastic films have improved barrier to flavor and odor compounds. Similar benefits are obtained in semirigid containers such as latex-coated PET bottles. Barrier to oxygen and carbon dioxide is also improved. On average, the resins used for coating have less modification (copolymer or plasticizer) than those which are melt processable, so PVDC coating resins typically have better barrier characteristics than extrusion resins.

Typical properties of PVDC resins are shown in Table 4.9.

4.6   Polystyrene (PS)

Polystyrene (PS) is an addition polymer of styrene, and has the structure:

$$-(\underset{\underset{H}{|}}{\overset{\overset{H}{|}}{C}}-\underset{\underset{\bigcirc}{|}}{\overset{\overset{H}{|}}{C}})_n-$$

Table 4.9 Properties of PVDC

Density	1.60-1.75 g/cm^3
T_g	-15 to +2°C
T_m	160-172°C
Tensile strength	19.3-34.5 MPa (2,800-5,000 psi)
Tensile modulus	344-551 MPa (50,000-80,000 psi)
Elongation at break	160-400%
Tear strength	10-30 g/25 μm
WVTR	7.9-240 g μm/m^2 d at 37.8°C, 90% RH
	(0.02-0.61 g mil/100 in^2 d at 95°F, 90% RH)
O$_2$ permeability, 25°C	7.9-2,700 cm^3 μm/m^2 d atm
	(0.02-6.9 cm^3 mil/100 in^2 d atm)
CO$_2$ permeability, 25°C	1,250-17,300 cm^3 μm/m^2 day atm
	(3.2-44 cm^3 mil/100 in^2 d atm)
Water absorption	0.1%

The PS used in packaging is atactic, so cannot crystallize, and therefore is an amorphous polymer. The bulkiness of the benzene ring substituent results in considerable resistance to rotation of the chain, so PS is a highly stiff, brittle material. It has a density of 1.05 g/cm^3, and T_g of 74-105°C. Its lack of crystallinity makes it highly transparent. It is not suitable for use at high temperatures, as it experiences liquid flow at about 100°C (212°F). Since it is an amorphous polymer, it does not have a defined melting point, but gradually softens through a wide range of temperatures. The tendency of PS to flow under stress at moderately elevated temperatures makes it easy to extrude and thermoform. The brittleness of PS can be reduced by biaxial orientation of the polystyrene sheet or film. Polystyrene is, in general, a low cost polymer, and a relatively poor barrier to water vapor and gases. Its chemical reactivity is greater than that of PE and PP.

PS is available in the following grades:

- *Crystal* polystyrene is used when clarity is required. Products made with crystal PS are brittle and amorphous.
- *High impact polystyrene (HIPS)* is an opaque material that has added butadiene rubber, partially as a blend and partially as a graft copolymer, to improve impact resistance.
- *PS foam* is a form of crystal PS which has been treated with a blowing agent, typically a hydrocarbon or carbon dioxide, to produce a cellular structure in the material which reduces brittleness and makes it an excellent cushioning and insulating material. (Note: Styrofoam is a Dow Chemical Co. trademark for building insulation, not a packaging material.)

PS is one of the most versatile packaging resins. The amorphous grade, crystal PS, is used to make bottles for pills, tablets, and capsules. High impact PS (HIPS) is commonly used as thermoformed containers for dairy products. PS foam has good shock absorbing and heat insulation characteristics. Applications in food packaging include egg cartons and meat trays. Varieties of styrene-based copolymers have also been developed to exhibit special combinations of properties. Typical PS properties are shown in Table 4.10.

Table 4.10 Properties of PS

Density	1.04-1.05 g/cm^3
T$_g$	74-105°C
Tensile strength	35.8-51.7 MPa (5,200-7,500 psi)
Tensile modulus	2,270-3,270 MPa (330,000-475,000 psi)
Elongation at break	1.2-2.5%
Tear strength	4-20 g/25 μm
WVTR	1,750-3,900 g μm/m^2 d at 37.8°C, 90% RH
	(4.4-10 g mil/100 in^2 d at 95°F, 90% RH
O$_2$ permeability, 25°C	98,000-150,000 cm^3 μm/m^2 d atm
	(250-380 cm^3 mil/100 in^2 d atm)
CO$_2$ permeability, 25°C	350,000 cm^3 μm/m^2 day atm
	(900 cm^3 mil/100 in^2 d atm)
Water absorption	0.01-0.03%

4.7   Polyvinyl Alcohol (PVOH) and Ethylene Vinyl Alcohol (EVOH)

4.7.1   Polyvinyl Alcohol

Polyvinyl alcohol (PVOH) has the following structure:

$$-(CH_2-CH)_n-$$
$$|$$
$$OH$$

One might think that PVOH is an addition polymer of vinyl alcohol. However, it cannot be produced in that manner since the vinyl alcohol monomer is unstable. Consequently, PVOH is produced by hydrolysis of polyvinyl acetate, PVA. The resulting material is amorphous at first, since PVA is amorphous, but it tends to crystallize when oriented. PVOH is atactic, but since the -H and -OH are isomorphous groups, this does not interfere with crystallization. Very strong intermolecular forces are imparted by the OH groups with their ability to hydrogen bond. The forces are so strong that PVOH cannot be melt processed, since its decomposition temperature is below its melting temperature. Without stress, the molecules, though they have a strong tendency to crystallize, are not able to do so because they lack the mobility to rearrange into a crystalline structure.

PVOH has excellent barrier properties to oxygen and many other substances, because of its crystallinity and strong intermolecular forces. In its pure form, however, it is water soluble. Since the water solubility is imparted by the OH groups, the degree of solubility can be modified by controlling the amount of hydrolysis of the PVA. The more acetate groups that are retained, the lower is the water solubility of the PVOH, as well as the less effective the barrier properties.

Because of its extreme water sensitivity and difficulty in processing, PVOH has few packaging applications. Some important markets include packaging for toxic chemicals, such as pesticides, which can be encapsulated in PVOH pouches and placed, package and all, in the mixing tank. The PVOH dissolves, freeing the chemical, which can then be sprayed as usual. PVOH biodegrades in the environment, and does not present problems such as clogging spray nozzles or otherwise interfering with the application of the chemicals. In a similar approach, PVOH is sometimes used for bags for soiled linen in hospitals. The bags and linen can be placed directly in the washer, where the bags dissolve, and thus hospital personnel are able to avoid handling contaminated items.

4.7.2   Ethylene Vinyl Alcohol

Ethylene vinyl alcohol (EVOH) is produced by a controlled hydrolysis of ethylene vinyl acetate copolymer. The hydrolytic process transforms the vinyl acetate group to vinyl alcohol, in a manner analogous to production of polyvinyl alcohol. The highly polar OH groups increase the intermolecular forces, while the ethylene groups maintain molecular mobility. Since, as discussed above, the OH group is isomorphous with H, the polymer can achieve a high percent of crystallinity, even though ethylene and vinyl alcohol units are randomly distributed in the chain. The OH groups also make the polymer more compatible with water molecules, so it is much more hydrophilic than PE. The magnitude of all of these effects depends on the percentage of vinyl alcohol. The polymer will have properties similar to PE at a small amount of vinyl alcohol (VOH) content, or will be more like polyvinyl alcohol at a very high VOH content.

For packaging applications, the most important characteristic of EVOH is its outstanding O_2 and odor barrier properties. Packaging structures with EVOH provide high retention of flavors, and prevent quality loss associated with reaction of oxygen with the product. EVOH also provides a very high resistance to oils and organic vapors. This resistance decreases somewhat as the polarity of the penetrating compound increases. For example, the resistance to linear and aromatic hydrocarbons is outstanding, but for ethanol and methanol it is low; it may absorb up to about 12% of ethanol.

EVOH resins were introduced commercially in 1970 in Japan, and their use has expanded rapidly, especially where excellent oxygen barrier is required. Formulations currently in use typically contain 27 to 48 mole % ethylene. The lower the amount of ethylene, the better is the barrier - when the polymer is dry. The lower the percentage of ethylene, the greater is the moisture sensitivity, and the difficulty of processing. EVOH resins are melt processable, and exhibit high strength, toughness, and clarity. For these reasons, EVOH resins are often the best choice for applications where excellent barrier is required, as long as exposure to moisture can be controlled.

As indicated above, the hydroxyl group, -OH, makes the polymer hydrophilic, attracting water molecules. The presence of water decreases the oxygen barrier properties of the material. Because of its moisture sensitivity, EVOH is usually incorporated into package structures as a buried inner layer in a coextrusion, surrounded by polyolefins or other good water vapor barrier polymers. These structures typically contain an adhesive, or tie layer, between the EVOH and the polyolefin to provide adequate adhesion between the polar EVOH and the nonpolar polyolefin. For retorted products, this level of moisture protection may not be enough to prevent unacceptable oxygen permeation during and shortly after retorting. In that case, it is possible to incorporate a desiccant in the tie layer between the EVOH and the polyolefin. The desiccant absorbs moisture that penetrates the polyolefin during retorting, preserving the dryness, and hence the oxygen barrier, of the EVOH. EVOH properties are summarized in Table 4.11.

Table 4.11 Selected Properties of EVOH Copolymers

Property	EVOH 32% Ethylene	EVOH 44% Ethylene
Density, g/cm^3	1.19	1.14
Tensile Strength, MPa	88	68
Tear Strength, N/mm	154	193
T_m, °C	181	164
T_g, °C	70	55
Heat Seal Temperature, °C	179-238	177-238
Oxygen Permeability, cm^3 μm/m^2 day atm		
0% RH	4	2.4
65% RH	13	45
WVTR, g μm/m^2 day at 38°C 90% RH	2500	800

EVOH can be extruded in films, blow molded, or injection molded. It can also be processed by coextrusion or lamination, in combination with PE, PP, PET, nylons, and other resins. Applications in packaging include flexible structures and rigid containers, as shown in Table 4.12. Typical applications are ketchup and barbecue sauce bottles; jelly, preserves, vegetable juice, and mayonnaise containers; and meat packages. Non-food applications include packaging of solvents and chemicals.

Table 4.12 Applications of EVOH

Processing Method	Sample application	Sample structure
Cast coextrusion	Processed meats, cheese	PET/EVOH/EVA
Blown coextrusion	Red meat	LLDPE/EVOH/LLDPE
Lamination	Condiments	OPP/EVOH/LDPE
Coextrusion coating	Aseptic packaging	LDPE/paperboard/EVOH/Ionomer
Thermoforming	Yogurt	PP/EVOH/PP
Coextrusion blow molding	Ketchup	PET/EVOH/PET

4.8   Nylon

Nylons are condensation polymers, linear thermoplastic polyamides that contain the amide group

$$-\underset{\underset{O}{\|}}{C}-\underset{\underset{H}{|}}{N}-$$

as a recurring part of the chain. In general they are clear; thermoformable, strong and tough over a broad range of temperatures, and have good chemical resistance and good barrier to gas, oil and aromas. They are moisture sensitive or hydrophilic; left in normal environmental conditions, nylon can easily absorb 6-8% of its weight of water. The amount of water in a nylon sample can be described as a function of relative humidity by a sorption isotherm curve. For most packaging applications, nylons are used in film form, as a single component or in multi-layer structures.

Nylons can be made from the condensation of diamines and dibasic acids, or from the condensation of amino acids, which contain both amine and acid functional groups in a single molecule. For example:

(A) $H_2N\text{-}(CH_2)_6\text{-}NH_s$ + $HOOC\text{-}(CH_2)_4\text{-}COOH \rightleftharpoons$

      Hexamethylene diamine   Adipic acid

$$H-[\overset{\overset{H}{|}}{N}-(CH_2)_6-\overset{\overset{H}{|}}{N}-\overset{\overset{O}{\|}}{C}-(CH_2)_4-\overset{\overset{O}{\|}}{C}]_n-OH$$

Nylon 6,6 (or Nylon 66)

(B)

$$n\ H_2N-(CH_2)_{10}-COOH \rightleftharpoons H-[\overset{\overset{\displaystyle H}{|}}{N}-(CH_2)_{10}-\overset{\overset{\displaystyle O}{||}}{C}]_n-OH$$

11 - amino undecanoic acid   Nylon 11

Nylons are identified by numbers corresponding to the number of carbon atoms in the monomers (diamine first, as in case A). Two numbers are needed in case A and only one in the case of condensation of amino acids (case B), where the number indicates how many C-atoms are in the amino acid.

Examples:			
Case (A)	Number of C-Atoms in the diamine	Number of Carbons in the diacid	Name of the Nylon
	6	6	Nylon 6,6 (Nylon 66)
	6	12	Nylon 6,12 (Nylon 612)
Case (B)	Number of C-atoms in the amino acid		Name of the Nylon
	6		Nylon 6
	11		Nylon 11

Nylons are polymers with strong intermolecular forces because of the presence of H-bonding between the -C=O and HN- groups of adjacent molecules. These strong intermolecular forces are combined with crystallinity to yield tough, high melting thermoplastic materials. Nylon 6,6 has a melting point of 269°C (516°F). Beside toughness, nylons have good puncture resistance, impact strength and temperature stability. The flexibility of the aliphatic portion of the chain permits film orientation that enhances strength. By using copolymers (for example, two different di-acids), amorphous nylon can be produced. Oxygen permeability of nylons generally increases with increasing moisture content.

Nylons are melt-processable using conventional extrusion. Film can be produced by either the cast film process or the blown film process. During film production, different degrees of crystallinity are obtained depending on the temperature and rate of quenching. When the cooling rate is increased, a less crystalline nylon is obtained since the polymer was given less time to form crystals. The decrease in crystallinity produces a more transparent and more thermoformable film. Biaxial orientation of nylon films provides better crack resistance, mechanical properties and barrier characteristics. Blow molding processes are used with nylon resins to produce industrial containers, moped fuel tanks and oil reservoirs, as well as some containers. Thermoformed nylons are employed for disposable medical devices, meat and cheese packaging and in thermoform/fill/seal packaging.

Nylons are often used in coextrusion with other plastic materials, providing both strength and toughness to the structure. Polyolefins are commonly used in the coextrusions to provide heat sealability and moisture barrier, and to reduce cost. Ionomers or EVA are also often used. Multilayer films containing a nylon layer are used in vacuum-packing bacon, cheese, bologna, hot dogs and other processed meats. Polyvinylidene chloride copolymer coating on nylons is available for improved oxygen, moisture vapor, and grease barrier properties. Nylon is used to extrusion coat paperboard to obtain heavy duty paperboard.

Table 4.13 shows selected properties of Nylon 6, Nylon 11, and MXD6.

Table 4.13 Selected Properties of Nylon 6, Nylon 11, and MXD6 Nylon

Property	Nylon 6	Nylon 11	MXD6
Density, g/cm^3	1.13-1.16	1.03-1.05	1.20-1.25
Tensile Strength, MPa	41.3-165	55.1-65.4	220-230
Tear Strength, g/25 μm		400-500	
T_m, °C	210-220	180-190	64
T_g, °C	60		243
Elongation at break, %	300	300-400	72-76
O_2 permeability, 25°C, cm³ μm/m² day atm	470-1,020	12,500	60-260
CO_2 permeability, 25°C, cm³ μm/m² day atm	3,900-4,700	47,500	
WVTR, g μm/m² day at 38 °C 90% RH	3,900-4,300	1,000-2,000	
Water absorption, %, 0.32 cm thick, 24 h	1.3-1.9	0.4	

One of the specialty nylons is MXD6, which is a semicrystalline nylon produced from polymerization of adipic acid and meta-xylene diamine, so has the structure:

$$\text{H}-(\text{NHCH}_2 \quad \text{CH}_2\text{NHC}-(\text{CH}_2)_4-\text{C})_n-\text{OH}$$

The rigidity imparted by the ring in the main chain results in better gas barrier and thermal properties than nylon 6. MXD6 also is reported to have better moisture resistance than EVOH. At 100% RH its oxygen barrier is superior to EVOH. At low relative humidities and low temperatures, its barrier is inferior to PVDC, but it is a better barrier than PVDC at high temperatures and low RH. It also has good odor and flavor barrier properties.

4.9   Polyester

Polyesters are a class of polymers containing ester linkages, with the general formula:

$$H\text{-}O\text{-}(CO\text{-}R\text{-}CO\text{-}O\text{-}R'\text{-}O)_n\text{-}H$$

Polyesters can be either thermoplastics or thermosets, depending on their chemical composition. Polyethylene terephthalate (PET) is, by far, the most commonly used polyester. It has wide uses outside packaging as well, including polyester clothing and carpets.

4.9.1   Polyethylene Terephthalate (PET)

PET is a condensation polymer with the following structure:

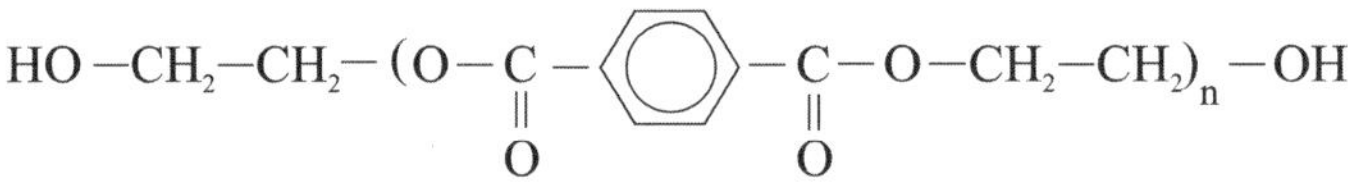

$$HO\text{-}CH_2\text{-}CH_2\text{-}(O\text{-}\underset{O}{\overset{\|}{C}}\text{-}\langle\bigcirc\rangle\text{-}\underset{O}{\overset{\|}{C}}\text{-}O\text{-}CH_2\text{-}CH_2)_n\text{-}OH$$

Polyethylene terephthalate

It is produced from para-xylene and ethylene. The p-xylene is converted into either dimethyl terephthalate or terephthalic acid, and the ethylene into ethylene glycol. These monomers are then polymerized by a condensation process, producing water as the byproduct molecule if terephthalic acid is used, and methanol if dimethyl terephthalate is used. Following the condensation polymerization, the molecular weight is increased by "solid-stating," in which the dried and crystallized resin chips from the original polymerization are subjected to high temperature and vacuum.

PET provides reasonably good oxygen and carbon dioxide barrier, which is improved by biaxial orientation. Properties of PET are summarized in Table 4.14. The largest single application of PET is in soft drink bottles, but the use of PET in non-soft-drink "custom" bottles has increased rapidly in the last few years, and continues to grow. In 2001, PET use in soft drink bottles totaled over 790 thousand tonnes, and PET use in other types of bottles and containers totaled over 827 thousand tonnes [1]. Biaxially oriented PET film has excellent odor barrier properties, which can be further improved, if desired, by coating with polyvinylidene chloride, by metallizing, or by coating with silicon oxide.

Table 4.14 Typical Properties of Polyethylene Terephthalate

T_g	73- 80°C (163-176°F)
T_m	245-265°C (473-509°F)
Density	1.29-1.40 g/cm^3
Typical yield, 25 μm (1 mil) film	30 m^2/kg (21,100 in^2/lb)
Tensile strength	48.2-72.3 mPa (7.0-10.5 x 10^3 psi)
Tensile modulus	2,756-4,135 mPa (4-6 x 10^5 psi)
Elongation at break	30-3,000 %
Tear strength, film	30 g/25 μm (0.066 lb/mil)
WVTR	390-510 g μm/m^2 day at 37.8 °C, 90% RH (1.0-1.3 g mil/100 in^2 24 h at 100 °F, 90% RH)
O_2 permeability, 25°C	1.2-2.4 x 10^3 cm^3 μm/m^2 d atm (3.0-6.1 cm^3 mil/100 in^2 24 h atm)
CO_2 permeability, 25°C	5.9-9.8 x 10^3 cm^3 μm/m^2 d atm (15-25 cm^3 mil/100 in^2 24 h atm)
Water absorption, 0.32 cm thick, 24 h	0.1-0.2 %

While PET is able to crystallize to a high extent, this can occur only over a limited temperature range. Therefore, the degree of crystallinity of PET is strongly influenced by processing conditions. PET films and bottles typically have a limited degree of crystallinity, with small crystallites and excellent transparency. Amorphous grades of PET (APET) are available which have been modified by copolymerization to remain amorphous. Crystallized PET (CPET) grades have had nucleating agents added to speed up and maximize crystallization. These containers, with their higher degree of crystallinity and larger crystallites, are an opaque white in appearance. Crystalline PET is much less subject to deformation under stress, especially at elevated temperatures, than is amorphous PET. It is also significantly more brittle at cold temperatures.

One of the disadvantages of PET is its low melt strength, which makes standard grades difficult or impossible to process by extrusion blow molding. Melt strength is dependent on the presence of long chain branches or higher molecular weight species to provide entanglements in the melt. PET has a narrow molecular weight distribution and no long chain branching. Specialty grades can be produced, by copolymerization to give long chain branching, or by increasing the molecular weight of the material; these grades have improved melt strength and can be extrusion blow molded. PET is also subject to hydrolysis at elevated temperatures, so pelletized PET must be dry before processing. Moisture content should be less than 0.005%.

Films can be produced using chill roll casting. Injection blow molding and stretch blow molding are used to produce bottles. PET is also used in extrusion coating, and PET sheet is often thermoformed.

Since oriented PET tends to deform when it is subjected to high temperatures, both film and bottles can be heat-set for improved stability. This allows use of PET for hot fill applications, for example.

PET is used for packaging food, distilled spirits, carbonated soft drinks, noncarbonated beverages, and toiletries. Typical food products include, for example, mustard, peanut butter, spices, edible oil, syrups, and cocktail mixers. Its crystallized form (CPET) is the basic material for microwavable containers for frozen meals. Biaxially oriented PET is used in meat and cheese packaging and as a base for snack food laminations. PET coating on paperboard produces ovenable board for use in applications such as frozen dinners. PET pouches are used for boil-in-bag frozen foods, and for sterilizable pouches for medical applications.

4.9.2   Glycol Modified PET, Other PET Copolymers, and PET Blends

PET copolymers can be produced by introducing an additional glycol, an additional diacid, or both, into the polymerization. These comonomers generally reduce the polyester's crystallinity and increase its melt strength, permitting easier formability and improved impact resistance. Glycol modified PET (PETG) is the copolyester with the greatest packaging use.

PETG is a copolymer of cyclohexane dimethanol with ethylene glycol and terephthalic acid. It is amorphous, clear, and colorless and has better melt strength than PET, so it can be processed by blow molding, extrusion, and injection molding. It has high stiffness and hardness, and good toughness even at low temperatures. Uses of PETG include bottles for household and food products and blister packages. It is very useful for medical device packaging, since it can be sterilized by both ethylene oxide and gamma radiation.

Blends of PET and polyarylates are finding uses in cosmetics packaging due to their high tensile strength (71 MPa compared to 56 MPa for PET) and heat resistance. Containers left in automobiles or transported by trucks in high temperatures will not deform at temperatures up to 80°C.

In many cases blends of PET with other polymers are only partially blends and partially copolymers, due to transesterification reactions which take place during extrusion. This is the case, for example, for the PET/PEN blends which will be discussed in Section 4.9.3.

4.9.3   Polyethylene Naphthalate (PEN)

Polyethylene naphthalate is a condensation polymer of ethylene glycol and naphthalate dicarboxylate (NDC), with the structure:

$$HO-(\overset{\overset{\displaystyle O}{\|}}{C} \cdots \overset{\overset{\displaystyle O}{\|}}{C}-O-CH_2-CH_2-O)_n-H$$

Because it is a homopolymer with a regular structure, it can crystallize, and, as with PET the amount of crystallinity developed depends on its processing history. PEN , compared to PET, has 400-500% better oxygen and water vapor barrier, 35% higher tensile strength, and 50% higher flexural modulus. It also has better chemical resistance than PET, including greater resistance to hydrolysis. PEN is able to block passage of UV light, and has high resistance to UV-induced degradation. It is suitable for hot-fill applications without requiring the heat-setting process needed for PET. In addition, molding and blowing cycles are shorter than for PET, allowing for increased productivity. However, PEN prices are three to four times those of PET, because of the high price of the NDC monomer.

The FDA approved PEN for food contact applications in April, 1996. PEN and PET can be blended together to make useful materials, and also PEN/PET copolymers can be produced. When PEN/PET blends are used, one is really making copolymers in the extruder. Analysis of the material that is extruded shows that transesterification reactions occur in the extruder, forming molecular bonds between PET and PEN molecules. Therefore, processing conditions are important to the quality of product one makes when using blends. Blends, as well as copolymers manufactured by polymer suppliers, can open up markets for this polyester that are not accessible to PEN alone because of its high price. While these materials have not yet been approved for food contact applications in the U.S., there appear to be no technical barriers to their approval.

PEN and PET copolymers fall into two groups. Low-NDC copolymers contain less than 15% NDC, and high NDC copolymers have 85% or more NDC. Copolymers with intermediate ranges of NDC are not used because they cannot crystallize and therefore have inferior properties. Because homopolymer PET and PEN are immiscible, blends require special mixing techniques to cause sufficient transesterification to occur. This amounts to the production of a copolymer during the extrusion process, as mentioned earlier. Blends of homopolymers with copolymers are easier to process than blends of the homopolymers themselves. Usually low-NDC copolymers are blended with PET, and high-NDC copolymers with PEN.

4.10 Polycarbonate (PC)

Polycarbonate is actually poly(bisphenol-A carbonate) and has the structure:

$$\text{H}-(\text{O}-\langle\bigcirc\rangle-\underset{\underset{\text{CH}_3}{|}}{\overset{\overset{\text{CH}_3}{|}}{\text{C}}}-\langle\bigcirc\rangle-\text{O}-\overset{\overset{\text{O}}{\parallel}}{\text{C}})_n-\text{OH}$$

It is a mostly amorphous polymer that has excellent clarity with a very slight yellowish tinge. PC is very tough and rigid, with good impact strength, dimensional stability, heat resistance, and low temperature performance. Its barrier to water and gases is relatively poor. Some PC properties are summarized in Table 4.15.

Table 4.15 Selected Properties of Polycarbonate

Density	1.20 g/cm^3
T_g	150°C
T_m	265°C
Heat deflection temperature	127-132°C at 1.8 MPa
Tensile strength	63-72 MPa
Tensile modulus	2,380 MPa
Elongation at break	110-150%
Tear strength, g/25 μm	10-16
WVTR	1,900-2,300 g μm/m^2 d at 37.8°C, 90% RH
	(4.9-5.9 g mil/100 in^2 d at 100°F, 90% RH)
O$_2$ permeability, 25°C	110,000 cm^3 μm/m^2 d atm
	(300 cm^3 mil/100 in^2 d atm)
CO$_2$ permeability, 25°C	675,000 cm^3 μm/m^2 d atm
	(1,700 cm^3 mil/100 in^2 d atm)
Light transmittance	88-91%
Water absorption, 0.32 cm thick, 24 h	0.15%

PC has good resistance to water, oil, alcohols, fruit juices, aliphatic hydrocarbons, and aqueous solutions of ethanol, but it is attacked by some solvents such as acetone and dimethyl ethyl ketone, as well as by alkalis. PC is FDA approved, and food-contact applications include microwave, ovenware and food storage containers.

PC thermoforms well, and has major uses in medical packaging, where it can be sterilized by autoclaving or by electron beam or gamma radiation. Its toughness is PC's most impressive property. For instance, PC is the material of choice for school windows and sports equipment. Being tough and clear make PC a material well suited for reusable bottles, particularly 22.7 L (5 gallon) water bottles or 3.8 L (1 gallon) milk bottles. Systems with washing stations have been developed for reuse of PC bottles. Polycarbonate films are odorless, have no taste, and do not become stained through normal contact with natural or synthetic coloring agents. In Europe, food applications include pre-baked bread, biscuits, confectionery, meat and processed cheese. Other emerging applications include hot filling, modified atmospheric packaging, rigid packaging as a substitute for PVC, high gloss coatings for paper, and barrier layers for fruit juice cartons. In addition to packaging applications, PC has major applications in the automobile industry and appliance markets.

PC can be processed by injection molding, extrusion, coextrusion and blow molding. Coextrusions with EVOH or polyamides are carried out with the help of adhesives. PC can be laminated or coextruded to PP, PE, PET, PVC and PVDC. PC is a hydrophilic polymer, and at ambient conditions can reach moisture levels of 0.35%.

4.11 Fluoropolymers

Fluoropolymers are a family of polymers containing C-F bonds. Polytetrafluoroethylene (PTFE) has the following structure:

$$-(\underset{\underset{F}{|}}{\overset{\overset{F}{|}}{C}} - \underset{\underset{F}{|}}{\overset{\overset{F}{|}}{C}})_n -$$

PTFE was first commercialized by DuPont, under the brand name Teflon. PTFE is a very highly crystalline polymer, extremely inert, an excellent barrier, and exhibits a very low coefficient of friction. Its glass transition temperature (T_g) is about -100°C, and its melt temperature is about 327°C (621°F). However, its very high viscosity makes it very difficult to process. PTFE is used most often as a component in packaging equipment, such as providing a non-stick surface on heat sealers, rather than in packages themselves.

PCTFE is polychlorotrifluoroethylene:

$$-(\underset{\underset{F}{|}}{\overset{\overset{F}{|}}{C}} - \underset{\underset{Cl}{|}}{\overset{\overset{F}{|}}{C}})_n -$$

The stereochemically random presence of the Cl atom limits the crystallinity. In commercial resins, it is generally further modified by copolymerization, resulting in a semicrystalline material, with a glass transition temperature about 45°C and a melt temperature of about 190°C. It is sold by Allied Corp. under the name Aclar. PCTFE can be melt processed, though not easily. Its main advantage is its extremely good water vapor barrier, which is the best of any plastic film available. The WVTR at 38°C and 90% RH is 9.8-17.7 g $\mu m/m^2$ d (0.025-0.045 g mil/100 in^2 24 h).

PCTFE also provides a good gas barrier, with an oxygen permeability constant of about 2800-5900 cm^3 $\mu m/m^2$ d atm (7-15 cm^3 mil/100 in^2 24 h atm), and carbon dioxide permeability 6300-15,700 cm^3 $\mu m/m^2$ d atm (16-40 cm^3 mil/100 in^2 24 h atm). It is highly inert, although it is not particularly strong or tough. Most often, PCTFE is used as a component in a laminated structure, especially for packaging moisture-sensitive drugs.

4.12   Styrene-Butadiene Copolymers

Styrene butadiene copolymers come in a wide variety of types, with a similar wide variety of properties. As discussed in Section 4.6, HIPS (High Impact PS), is partially a graft copolymer and partially a physical blend of polystyrene and polybutadiene. HIPS, which is opaque, typically contains 2 to 15 weight % polybutadiene. In addition to significantly decreased brittleness, it has a broad processing window and is easy to thermoform, either as sheet or as extruded foam.

Block copolymers of styrene and butadiene can also be used to add impact resistance to polystyrene. These materials have good transparency and toughness, excellent shatter resistance, and are easily fabricated, with the properties strongly dependent on the ratio of the comonomers and the length of the blocks. They are widely used in medical packaging applications because they can be sterilized by both gamma irradiation and ethylene oxide, and are also used in single-service food packaging, bottles, blister packs, overcaps, and film applications.

Styrene-butadiene copolymers are often blended with other polymers. Transparent blends can be made with styrene, styrene-acrylonitrile copolymers, or styrene-methyl methacrylate copolymers. Blends with styrene have low impact strength even at low styrene levels, while blends with styrene-methyl methacrylate copolymers can have greatly improved impact strength. Blends with high impact polystyrene, polypropylene, and polycarbonate are opaque.

4.13 Acrylonitrile Copolymers

Polyacrylonitrile (PAN) has the following structure:

$$-(\underset{\underset{\text{H}}{|}}{\overset{\overset{\text{H}}{|}}{\text{C}}} - \underset{\underset{\text{C}\equiv\text{N}}{|}}{\overset{\overset{\text{H}}{|}}{\text{C}}})_n-$$

PAN has very strong intermolecular forces resulting from the strong polarity of the carbon-nitrogen bond. This results in a polymer that is an excellent gas barrier, but is very stiff and brittle. It cannot be melt-processed as it degrades at 220°C, which is too low for adequate flow. Therefore a copolymer is used to somewhat reduce the intermolecular forces and make the polymer melt-processable.

High-nitrile resins (HNR) are copolymers with high acrylonitrile content. They are generally very tough materials, with excellent barrier and good transparency.

A ratio of about 3:1 styrene to acrylonitrile by weight results in styrene-acrylonitrile copolymer (SAN), which is amorphous and transparent, with excellent chemical resistance, heat resistance, and gloss, and good rigidity and tensile and flexural strength. SAN is often used for cosmetic packaging, bottles, overcaps, closures, sprays, and nozzles. Since it does not contain much acrylonitrile, its gas barrier is poor. A ratio of about 7:3 acrylonitrile to styrene results in acrylonitrile-styrene copolymers (ANS) which have very good gas barrier properties.

Acrylonitrile-butadiene-styrene copolymers (ABS) are random styrene-acrylonitrile copolymers grafted to butadiene, which are amorphous, opaque, and process easily. The properties depend on the ratios of the comonomers used. ABS is used in cosmetics packaging, and has been used in margarine tubs.

Terpolymers of acrylonitrile, methyl methacrylate, and butadiene are also produced. They generally have only moderate gas barrier properties unless the acrylonitrile content is high. Barex, produced by BP Amoco Chemicals, is an HNR with a 75:25 ratio of acrylonitrile to methacrylate polymerized onto a nitrile rubber backbone, which has excellent barrier properties. These resins can be used in the production of blow molded and injection-molded containers, film, and sheet, and are cleared by the FDA for direct food contact applications. In addition to excellent gas barrier, they have good chemical resistance and sealability. Applications include rigid containers for spices, household chemicals, cosmetics, pesticides, agricultural chemicals, and fuel additives. They are also used in medical packaging, and can be sterilized by either ethylene oxide or gamma radiation. Thermoformed blisters are used for meat and cheese packaging. Oxygen permeability for the best barrier high nitrile resins is about 310-630 cm^3 $\mu m/m^2$ d atm (0.8-1.6 cm^3 mil/100 in^2 d atm). Water vapor barrier is not as good as polyolefins, at about 2000-2900 g $\mu m/m^2$ d (5.0-7.5 g mil/100 in^2 d) at 38°C, 90% RH.

HNR can also used in coextruded structures, particularly with polyolefins, in sheet, film, or bottle form. The HNR provides gas barrier and chemical resistance, while the polyolefin provides water vapor barrier.

4.14   Cyclic Olefin Copolymers

Cyclic olefin copolymers (COC) are copolymers of ethylene and norbornene (2,2,1 bicycloheptane), made using metallocene catalysts. They are amorphous polymers, with excellent clarify, low density, and high strength and stiffness. Currently the major manufacturer is Ticona, which sells them under the trade name Topas.

Uses for COCs include blister packaging for pharmaceuticals, food packaging, and non-packaging applications. They are reported to be easily metallized, and processable by injection molding, film extrusion, blow molding, and thermoforming. Blends and coextrusions of COCs with other materials, including LLDPE and EVOH, are also of interest.

While these materials are still relatively new in terms of commercial availability, they may provide advantages in downgauging due to their excellent stiffness and water vapor barrier characteristics. Their glass-like transparency is also a significant advantage. A broad range of glass transition temperatures can be provided, depending on the norbornene content. Increasing the norbornene levels increases stiffness, strength, T_g, and heat deflection temperature.

4.15   Liquid Crystal Polymers

Liquid crystal polymers (LCP) are polymers which retain some alignment and crystal-type organization in the liquid state. They contain rigid segments which tend to self-align during shear flow, much like logs in a river, rather than the random coil conformations assumed by most polymer melts. The locally-oriented domains tend to create macroscopic oriented regions. While uniaxially oriented materials tend to split under transverse load, biaxially or multiaxially oriented materials can have very favorable properties, including strength-to-weight and strength-to-cost ratios.

Several types of LCPs are available commercially, all consisting of copolymers composed of molecules with rigid and flexible monomers. The rigid part, called the mesogenic monomer, imparts high temperature capability and good mechanical properties, while the flexible monomer provides processability. A typical biaxially oriented LCP film has a tensile strength about 40% higher than biaxially oriented PET, and a tensile modulus more than twice as high. Density is approximately the same. Tear initiation strength is about 595 kN/m, compared to about 35 for PET, and tear propagation strength is 175-525 kN/m, compared to 9-53 for PET. LCPs have a melting point of about 300°C, and can be used at temperatures over 200°C. Very significantly, LCPs are excellent barriers. A typical LCP has an oxygen permeability constant of 0.0092 cc μm/m^2 d atm and a water vapor transmission rate of 0.0068 g μm/m^2 d at standard ASTM conditions. The oxygen barrier of these materials is not affected by moisture, and is 6 to 8 times higher than EVOH at relative humidities over 85%. Barrier to carbon dioxide, nitrogen, and other gases is also excellent.

The major drawback of these materials is their high price, currently $15-22/kg. However, to maximize performance at acceptable cost, thin LCP layers can be coextruded with less costly plastics. The price is expected to drop as use and production increases, perhaps to as low as $11/kg. Blends and alloys with other thermoplastics can also be used.

A key to processing LCPs into useful materials is to control the melt flow process to achieve the desired orientation. One process that was developed uses a counter-rotating circular die in the production of blown film to produce good biaxial orientation. In conventional blown film processes, the shear forces act in only one direction, so the orientation is almost entirely in that direction. The long relaxation time of the melt phase (over a minute) means that even large transverse extensional strains have little effectiveness in producing orientation. However, once the shear produces some biaxial orientation, biaxial extensional forces are quite effective at increasing the orientation. Stretch blow-molding is the typical process used for producing containers.

The cosmetics industry uses some liquid crystal polymers in packaging, and efforts to obtain FDA approval for food and beverage packaging are underway. The metallic look of LCP containers and the processing ease of the material are two features which the cosmetics industry finds attractive.

4.16   Conductive Polymers

For some applications, particularly in packaging of sensitive electronic components, it is desirable to have a packaging material that is able to conduct electricity sufficiently to dissipate static charges which could otherwise accumulate. As will be discussed in Section 5.10, one way to accomplish this is by using additives. However, inherently conductive polymers for such applications are now becoming available. These are known as inherently dissipative polymers (IDPs), and are able to dissipate static charges without the use of conductive fillers or chemical antistats.

BFGoodrich Specialty Chemicals produces a family of thermoplastic alloys which are inherently conductive. Alloys with 15% of the active conductive polymer differ little in surface resistivity from alloys with 25% active polymer. Further, the conductive polymer content remains relatively constant through processes such as injection molding, extrusion, and thermoforming, unlike resins containing conductive fillers. In contrast to chemical antistats, the conductive polymers are active at all humidity levels, do not lose their potency over time, and add no ionic contaminants to the surrounding atmosphere.

The polymers are elastomeric polyethers, sold under the Stat-Rite S-Series in alloys with base polymers such as acetal, ABS, PP, PETG, and others. The conductive polymer reportedly has minimal effect on the mechanical properties of the base resin. Prices are higher than carbon-black filled systems, but lower than carbon-fiber materials.

Inherently conductive polymers are also being used as coatings on other polymers, as well as on electronics themselves. One such material, produced by Ormecon Chemie GmbH & Co. of Ammersbek, Germany, uses a dispersion of polyaniline. Another material, being developed by Bayer AG of Leverkusen, Germany, is based on polythiophene. A polythiophene coating for sheet and film applies the conductive polymer in a dispersion in polyurethane [2].

4.17   Thermoplastic Elastomers

Elastomers are, by definition, materials with very good elasticity, good ability to return to their original shape after deformation. Rubber, either natural or synthetic, is the major example. However, rubber, because it is a thermoset, is more difficult to process than the thermoplastics that we generally use in packaging. Thermoplastic elastomers combine the processability of thermoplastics with the functional performance and properties of a conventional thermoset rubber, greatly increasing the speed, efficiency, and economy of processing. For this reason, use of thermoplastic elastomers (TPEs) is growing rapidly, expecially in the rubber and automobile industries. Packaging use of TPEs began, for the most part, in the late 1990s, when a variety of thermoplastic elastomers and plastomers based on olefin and styrene monomers were introduced to provide high-clarity tough film and sheet, particularly as alternatives to PVC.

Most TPEs fall in one of six categories, listed in order of increasing cost and performance: styrenic block copolymers, polyolefin blends (TPOs), elastomeric alloys, thermoplastic polyurethanes (TPUs), thermoplastic copolyesters, and thermoplastic polyamides [3].

Styrenic TPEs are the most widely used. One such material, commercialized by BASF in 1999, is a styrene-butadiene block copolymer with a styrene content of about 70%, intended for thin film for food packaging. It has high oxygen and water permeability, and excellent toughness and optical properties. Cling films with EVA layers on the outside are also available, which provide complete recovery of deformation at elongations up to 400%, and elongation at break of over 650%.

Shell is developing styrene-ethylene/butylene-styrene copolymers, some marketed as an alternative to silicone rubber. Other members of the family are intended for production of high-clarity film and sheet, alone or in blends with crystal polystyrene. Dow is producing ethylene-styrene "pseudo-random" copolymers, called interpolymers, with unique properties. Fina Chemicals is producing high-styrene content elastomers for blends with polystyrene and polyethylene to improve impact resistance while maintaining clarity and adding gloss, as well as imparting crinkly sound similar to cellophane. These polymers can also be used as a tie layer in coextruded thermoformable sheet. Other companies are producing similar materials.

4.18   Thermosets

Polymers that crosslink during polymerization are known as thermosetting polymers. They form a three-dimensional structure through the crosslinking reaction, and they can not be reshaped once they are set.

Phenol-formaldehyde polymers, phenolics, were not only one of the first commercially available materials, but they were the first commercially used thermosets. Normally one does not think of thermosetting polymers as useful for packaging applications because of the need to reshape the material. Phenolics were commonly used for closure manufacture in the past, because of their excellent chemical resistance. Phenolic closures have nearly disappeared from the market today.

Another major application for thermosets is can linings. In this application, acrylics, alkyds, and epoxies are used to prevent interaction between the tin-free steel, tin plate, or aluminum and the product. While acrylic polymers can be thermoplastic, coatings from acrylics that are used widely in packaging are usually thermoset. By changing one of the monomers to a difunctional or a multifunctional species, acrylic polymers that crosslink can be produced. The exact formulation of these types of polymers varies widely depending on the properties such as toughness, chemical resistance, and flexibility desired for the end-use application. Acrylics are also often used as coatings on films to give scuff resistance. These types of coatings are often applied as a 'lacquer' over printing.

Acrylics and urethanes are used as adhesives for laminations, also. These adhesives are known as *two-part* systems, and they require time to cure after the lamination is formed before they reach their final bond strengths. The chemical cross-linking reactions take place during the curing process.

Phenolics are made by the polymerization of phenol and formaldehyde, as shown in Figure 4.9. The reaction is carried out to a low degree of polymerization, making what is called a *prepolymer*. This material may be diluted with a solvent or more monomer to make a workable liquid. The crosslinking reaction is carried out by the addition of a catalyst and heat.

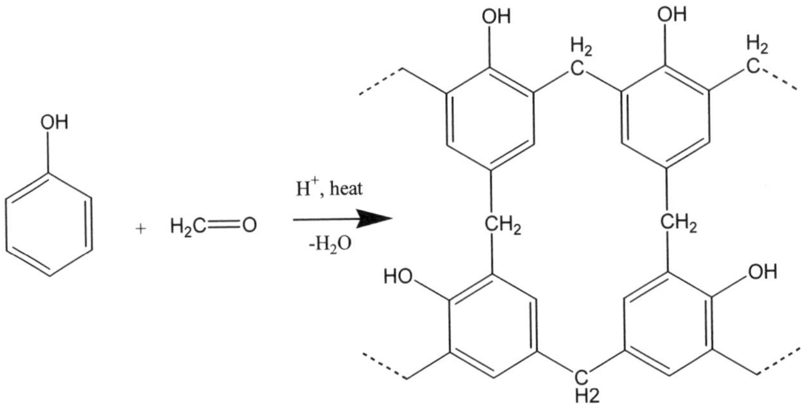

Figure 4.9 Phenol-formaldehyde plastics.

Epoxies can be formed from several types of monomers, but often start with epichlorohydrin and bisphenol-A, as shown in Figure 4.10, because the costs of the ingredients are low. Again the initial reaction is carried out to a low degree of polymerization. The system is then stabilized and diluted with monomer to achieve a workable consistency. Further reaction is carried out with the addition of an amine, amide, urea-formaldehyde, or phenol-formaldehyde.

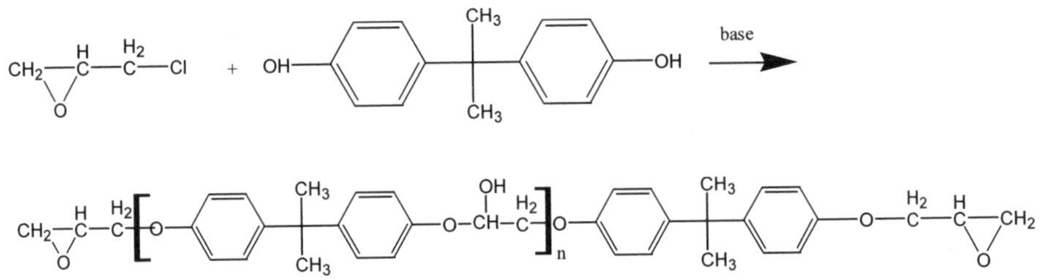

Figure 4.10 Epoxy.

In both of these systems, various monomer systems can be combined to add toughness, flexibility, or other desired properties. The resulting polymer is a three-dimensional network with short links between chains. The crosslink density, how close together the crosslinks occur in the backbone of the polymer, significantly affects polymer properties. Increasing the crosslink density causes thermoset polymers to become stiffer and more brittle. For example, the chemical composition of rubber bowling balls and rubber tires is very similar, but they differ substantially in crosslink density, and therefore in behavior.

4.19   Cellophane and Cellulosic Plastics

As mentioned, cellophane, though a polymer formed by condensation reactions of glucose, is not a plastic, since it will not melt and cannot be formed by heat and pressure. It was, however, the first transparent packaging film, and as such had extensive use. Some thermoplastics derived from cellulose, the cellulosic plastics, are still used today, though not in large quantities.

Cellophane and cellulosic plastics are derived from the cellulose in biomass materials, usually wood. The general reaction involved in forming cellulose is:

$$n\ C_6H_{12}O_6 \rightarrow (C_6H_{12}O_5)_n + (n-1)\ H_2O$$

The glucose residues are linked together by a series of β-(1-4) bonds, producing a linear polymer which is structurally and stereochemically regular. Cellulose is typically about 70% crystalline. The degree of polymerization is not known precisely, but is believed to be in excess of 10,000. Chemists have not been able to duplicate this reaction in the laboratory, so all cellulose is produced in natural sources by reactions involving enzyme systems. Because of the abundance of hydroxyl (-OH) groups, cellulose has strong inter- and intra-chain hydrogen bonds. The energies required to disrupt this bonding are so high that such disruption cannot be achieved without breaking main chain bonds. Thus cellulose, like polyvinyl alcohol, cannot be melted. In fact, the glass transition temperature of cellulose, like the melting temperature, is above the decomposition temperature.

4.19.1   Cellophane

Cellophane consists of much smaller molecules containing the basic cellulose chemical structure, but lacking its crystallinity. The lack of crystallinity, achieved by dissolving the cellulose since it cannot be melted, makes it transparent. Since cellulose is insoluble if chemically unchanged, the manufacture of cellophane involves chemical modification of the cellulose to decrease the molecular weight and render it soluble. In subsequent processing, the modified cellulose is deposited in the desired shape and then regenerated, meaning that the modifying groups are removed.

The first step in manufacture of cellulose is to treat the cellulose with caustic soda (NaOH) and allow it to oxidize until the degree of polymerization is down to 200 to 400. Treatment with CS_2 then yields sodium cellulose xanthate, which is dissolved in aqueous NaOH. This material, called *viscose*, is extruded through a small slit onto a roller immersed in a tank of weak sulfuric acid and sodium sulfate, which hydrolyzes the xanthate groups, regenerating the cellulose and yielding CS_2 and H_2S as byproducts.

After additional washing, bleaching, and other treatments, the water-swollen cellulose is dried, becoming a transparent film.

Since cellophane cannot melt, it is not heat-sealable. The many hydroxyl groups which it contains make it sensitive to water. It will not dissolve, but its properties can change markedly on exposure to moisture. For both these reasons, cellophane used in packaging is generally coated. Common types of coatings include vinylidene chloride-acrylonitrile copolymers, polyolefins, and mixtures of cellulose nitrate, wax, and resin. It may also have plasticizers added to improve its flexibility.

In addition to coatings and additives, the properties of cellophane are affected by the molecular weight of the cellulose, the degree of stretch imparted during the regeneration process, the rate and degree of drying, and other process variables. Typical properties are shown in Table 4.16.

In addition to the need to protect cellophane from moisture, its other major disadvantage is its high cost, and the environmental issues associated with its production. Cellophane has largely been replaced in packaging by polypropylene and other synthetic plastics, which often have superior properties at a lower cost

Table 4.16 Cellulose Properties

Density, g/cm^3	1.5
Elongation, %	
MD	16
CD	60
Tensile Strength, MPa	
MD	124
CD	55
Tensile Modulus, MPa	
MD	5,500
CD	2,800
Oxygen permeability, 25 °C, cm^{3}μm/m^2 d atm	
50% RH	1,200-2,000
90% RH	20,000-31,000
Water vapor transmission, 38 °C, 90% RH, g μm/m^2 d	43,000

4.19.2 Cellulosic Plastics

Cellulose esters, thermoplastics derived from cellulose, were, until 1950, the most important group of thermoplastic materials. Like cellophane, they have been, for the most part, displaced by synthetic plastics with their superior performance and lower cost. They still maintain some niche markets, however.

Cellulose nitrate is the oldest plastic, as was mentioned in Chapter 1. Cellulose trinitrate is formed by replacing all three hydroxyl groups on the glucose unit with nitrates. Since it is explosive, it has no use in packaging and is not made commercially. The nitrocellulose plastics have a degree of substitution of about 1.9, meaning, on average, slightly less than 2 of the hydroxyl groups per glucose have been replaced. However, this material is highly flammable and has poor chemical resistance, so has no significant use in packaging.

Cellulose acetate was first produced in about 1865, by acetylating cellulose with acetic acid and acetic anhydride, and hydrolyzing it to a DP of about 175-360. The acetyl groups reduce the hydrogen bonding, increasing the inter-chain separation. However, it will still decompose below its softening point, so requires the addition of plasticizers, usually dimethyl phthalate, to make it melt-processable. The result if a greaseproof, transparent plastic, which is FDA approved. It has high permeability, and can be sealed using either heat or solvents, though solvents are used most often. Unlike cellophane, it is not susceptible to softening on exposure to moisture.

Cellulose butyrate and cellulose propionate are also available. They tend to be tougher than cellulose acetate, easier to process, and more resistant to water, though less resistant to organic compounds. The butyrate polymers sometimes have an undesirable odor.

What can be regarded as copolymers, cellulose acetate propionate and cellulose acetate butyrate, can be produced by using a mixture of either propionic acid and propionic anhydride, or butyric acid and butyric anhydride, with the acetic acid and acetic anhydride. These polymers thermoform and vacuum metallize very well.

4.20   Polymer Blends

Because the physical, rheological, or chemical properties required for some processing or packaging applications cannot always be achieved using a single commercially available polymer, polymers are often blended. Polymers may also be blended in a processing operation to economically use the in-process generated scrap. Polymer property modification might also be achieved through a copolymerization reaction, but creating such specialized polymers for low volume usage would not be economical.

Because the knowledge of how to achieve certain polymer characteristics by blending is often considered a "trade secret" by processors, it is not often discussed or even mentioned by converters. There has been considerable research about the properties of blends, but often there is little information available about the composition of commercially available blends, making it difficult to tie theory to practice.

The blending of polymers is more complex than the blending of low molecular weight liquids. For low molecular weight materials, the compatibility of a mixture is usually controlled by the similarity of the cohesive energy density and the polarity of the materials being mixed. That is, if the molecules of both materials in a binary mixture have a similar affinity for the other material as for themselves, a compatible mixture will result. Similar factors affect the compatibility of polymer mixtures, but additional factors must be considered because of the large molecules. The effect of molecular weight cannot be ignored. Large differences in molecular weight can cause separation of molecules even in the same polymer. Small differences in molecular structure can also result in separation of molecules. These effects may be helpful or harmful in achieving the desired properties.

The simplest types of blends are blends of those of like polymers. For example, two different polyethylenes, an LLDPE and an LDPE, might be blended to give a better stretch film. The LLDPE has excellent drawdown characteristics, but it will continue to yield under stress, so the LDPE is added to give some strain-hardening characteristics. Strain hardening means that the stress in the polymer rises as the material is stretched. This characteristic is necessary if the stretch film is to hold the load together. Such blends are as close to homogeneous as one can find in polymers. Even in such blends, issues of compatibility can arise. While LDPE and LLDPE are both polyethylene, the molecular structure of the two is very different. Therefore, if one measures the melting characteristics of the blended polymers, one will see a broadening of the melting peak. An example is shown in Figure 4.11, where one can see that the melting curve is broadened and two peaks occur close to one another. This is still considered a compatible blend by industry standards because the changes in the DSC are due to changes in the lamellar thickness of the crystallites caused by cocrystallization.

A major application of blends is combining clean trim scrap and defective materials produced during the manufacture of packaging with new plastic for production of new containers. The ability to reuse these materials can significantly reduce production costs. If the used material, called regrind, is uniform and clean, it can generally be incorporated in substantial amounts without adversely affecting the product quality. If the material is dirty or contaminated in other ways, it may be impossible to use without further processing. Some companies specialize in handling contaminated scrap such as multilayer materials and labeled bottles, producing resin formulations for a variety of applications.

Difficulty may also arise if the film or rigid container is a multi-layer structure. The polymers to be blended may be quite dissimilar, particularly when barrier polymers such as EVOH and nylon are used in polyolefin-based structures. These barrier polymers are very polar compared to polyethylene and polypropylene, and do not blend well with the polyolefins. However, usually an adhesive polymer is included in such multilayer structures to bond the dissimilar polymers together. This adhesive polymer will usually act as a compatibilizer in the mixture, permitting more uniform blending. Most often, such multicomponent regrind is fed into a special regrind layer in the packaging material, rather than blended with one of the other layers. This maintains the functionality of the individual layers, while allowing the economics of the process to be more favorable, thus lowering the the cost of the packaging material.

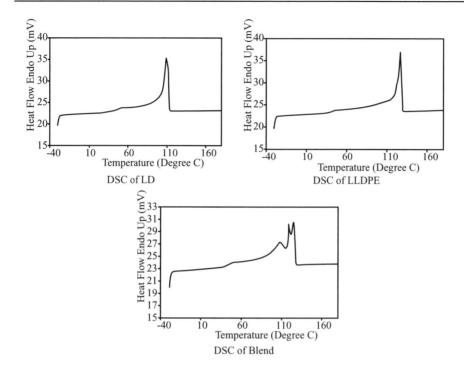

Figure 4.11 DSC of LDPE, LLDPE, and blend of LDPE and LLDPE.

Incompatible mixtures are deliberately made for certain applications. HIPS is an example. HIPS contains a rubber component, polybutadiene, that is partially blended with the polystyrene and partially copolymerized as a graft copolymer. The rubbery phase segregates into its own domains when the molten polymer cools. These domains act as stress absorbing sites during impact stressing of the material. Another example is the making of a easy-peel heat seal layer by blending polybutene with LDPE or LLDPE. The polybutene, being a very similar polyolefin, might be thought to be miscible with the LDPE or LLDPE. However, the regular CH_3CH_2- groups attached to the backbone of the polymer affect its cohesive energy density, and it prefers to associate with itself. In fact, the regularity of the branch-like structures allows it to crystallize more than LDPE, and raises its melting point. Therefore, during the heat sealing process, the polybutene generally will not melt. If it is in contact with the LDPE or LLDPE across the heat seal area, it will not bond. Either of these mechanisms creates weak spots in the heat seal, giving it an easy-open characteristic. The fact that this approach to easy-peel seals depends on a temperature effect means that the easy peel characteristic will be eliminated if excessive heat is applied.

References

1. U.S. Environmental Protection Agency, *Municipal Solid Waste in the United States, 2001 Facts and Figures*, EPA 530R-3-011, November 2003.
2. Graff, Gordon, *Modern Plastics*, March 1999, pp. 30-32.
3. Rader, Charles, *Modern Plastics Encyclopedia Handbook*, New York:McGraw-Hill, 1994, p. 64.

Study Questions

1. How do high density, low density, and linear low density polyethylene differ in structure? How do these structural differences affect the properties of the polymers? Why?

2. Why is PP stiffer than HDPE? Why does it have a higher melt temperature? How does this affect packaging uses for these materials?

3. Ionomers are known for their excellent toughness and excellent heat seal characteristics. Relate these characteristics to the chemical structure of the polymer, to explain why they perform so well in these areas.

4. How is the dependence of permeability on density in polyethylene, as illustrated in Table 4.4, related to the structure (chemical or physical) of the polymer? What is the single factor most responsible for the difference in barrier ability?

5. What is the most significant reason that PVDC is a much better barrier than HDPE?

6. Draw the structures of the monomers used to form nylon 12 and nylon 6,10.

7. Explain why the oxygen barrier of EVOH is strongly affected by the amount of water present, but the oxygen barrier of PVDC is not much affected.

8. When we use PVDC and PAN, we commonly use copolymers, even though copolymerization reduces their barrier capability. Why?

9. Polyethylenes, especially low density PE, are referred to as soft and flexible, while nylons and PET are said to be stiff. What molecular feature(s) cause(s) a polymer chain to be stiff?

10. Why do we say that polyethylene is actually a family of polymers?

11. How would you design a copolymer containing ethylene that is more transparent, heat seals better, and is more permeable to water than LDPE?

12. What is the impact on polymer properties of catalysts like the Ziegler-Natta family and the newer single-site metallocenes?

13. Based on what you have learned in Chapters 2-4, explain the property trends of PE listed in Table 4.5.

14. Why are there three stereochemical configurations of PP? Explain why this affects the packaging applications of PP. What would be the effect of these configurations on the properties of a copolymer of PP?

15. Unplasticized PVC presents an important problem during processing. What is it, and why does it happen? What is the recommended solution? Explain.

16. In what aspect is PVC superior to HDPE as a packaging material? Why are the properties of PVC so different from those of PVDC?

17. Name a plastic that is completely transparent and brittle at room temperature. Give a list of uses for such a plastic. Explain.

18. Compare the properties of PVOH and EVOH. Explain the similarities and differences.

19. What family of polymers is very tough, has high melting temperatures, good impact strength, excellent temperature stability, and is moisture sensitive? Explain these properties based on the chemical structure of the polymers.

20. List the types of polyesters discussed in this chapter. Write their chemical structures, and list their major characteristics.

21. List possible packaging applications for polytetrafluoroethylene.

22. How does Barex^{TM} differ from SAN, ANS and ABS? Explain.

23. Imagine that liquid crystal polymers are as inexpensive as PET. Suggest possible applications for LCP in packaging.

24. Do you think conductive polymers have a future in packaging? Explain.

25. What are thermoplastic elastomers, and how do they apply to packaging?

26. What are acrylic, epoxy, and phenolic thermosets?

27. Compare cellophane and polypropylene films.

Table 4.14 Comparative Properties of Common Packaging Plastics

Property	HDPE	LDPE	PC	PET
Tensile strength, MPa (x 10³ psi)	22-31 (3.2-4.5)	19-44 (2.7-6.5)	66 (9.5)	48-72 (7.0-10.5)
Elongation, %	10-1200	600	110	50-300
Flexural Modulus, MPa (x 10³ psi)	1,000-1,600 (145-225)	280-410 (40-60)	2,350 (340)	2,420-3,100 (350-450)
Mold shrinkage, cm/cm	0.015-0.040		0.005-0.008	0.020-0.025
Clarity	poor	hazy	clear	clear
Impact strength	good	very good	excellent	poor
Oxygen barrier	poor	poor	poor	good
Water vapor barrier	excellent	good	poor	good
Heat distortion temp., 455 kPa, °C (°F)	62-91 (144-196)	40-44 (104-111)	138 (280)	38-129 (100-264)

Table 4.14 - continued

Property	PP	PS	PVC	SAN
Tensile strength, MPa (x 10³ psi)	31-38 (3.6-4.5)	45-83 (6.5-12.0)	41-69 (6.0-10.0)	69-82 (10.0-11.9)
Elongation, %	100-600	1-4	5-135	2-3
Flexural Modulus, MPa (x 10³ psi)	1,170-1,730 (170-250)	2,620-3,380 (380-490)	2,620-3,588 (380-520)	3,450-4,000 (500-580)
Mold shrinkage, cm/cm	0.015-0.025	0.004-0.007	0.002-0.006	0.003-0.005
Clarity	poor/good	clear	clear	clear
Impact strength	fair	poor	good	poor
Oxygen barrier	poor	poor	good	good
Water vapor barrier	excellent	poor	fair	fair
Heat distortion temp., 455 kPa, °C (°F)	107-121 (225-250)	68-96 (155-204)	57-82 (135-180)	104-107 (220-224)

5 Additives and Compounding

5.1 Introduction

Pure resins are rarely processed into final products without the addition of selected compounds, called additives, that are incorporated during the process of extrusion and molding of a plastic resin or applied externally on the formed material. Compounding refers to the process of uniformly mixing the additives into the resin. A blend, by definition, is formed from two or more polymer resins which have been mixed together.

There are a variety of reasons for the use of additives in a resin, including: (1) to improve the processing conditions, (2) to increase the resin's stability to oxidation, (3) to obtain better impact resistance, (4) to increase or decrease hardness, (5) to control surface tension, (6) to facilitate extrusion and molding, (7) to control blocking, (8) to reduce cost, and (9) to increase flame resistance.

The number and amount of additives incorporated in a resin vary with resin type and application. A polyethylene resin, for instance, may have only an antioxidant incorporated, or a colorant may also be added. A PVC resin may require several plasticizers, a filler, a heat stabilizer, and a colorant. There are many variations, and in most cases, the final resin formulation is considered by the manufacturer to be proprietary information. A confidentiality agreement may be necessary to make certain that the material meets regulatory compliance, unless the end-user can give the supplier sufficient information for him to make the determination.

Additives are incorporated by the resin manufacturer and/or by the packaging processor. The presence of additives in packaging applications raises the question of additive migration. Most additives diffuse within the polymer and often tend to migrate to the surface of the material. When a packaged product is in direct contact with a compounded polymer, there may be a transfer of the additive to the product. (Of course, it is also possible for components of the product to be transferred to the polymer.) The extent of transfer depends on a series of conditions, and is discussed in Chapter 13. We will briefly examine compounding, and then discuss some common additives used in packaging.

5.2   Compounding

The process of compounding refers to mixing additives into the plastic to create a homogeneous material. The process often starts with some type of dry mixing, and usually ends with an extruder (see Chapter 7), which completes the blending and produces pellets of plastic resin carrying the uniformly distributed additives. Depending on the type of additive and other factors, the additive may be present in the amounts desired for the final processing, or it may be more concentrated. A complete discussion of compounding is beyond the scope of this text, so only some general principles and terminology will be covered.

Concentrates of additive dispersed in a polymer are termed master batches. The user will mix a desired proportion of the master batch with the polymer being processed as it is fed into the extruder used to produce the films, containers, or other package material or components. The reason a master batch is used is that the pre-blending simplifies the task of obtaining an even distribution of additives in the desired amounts. Ideally, the polymer component of the master batch is the same type of resin as that being processed. In practice, the master batch must at least be capable of blending well with the process resin, and not degrade its performance significantly.

Resins without additives are sometimes referred to as "barefoot" resins. The process of blending the master batch with the barefoot resin to achieve the desired additive concentration is known as "letting down" the additive.

To make a highly uniform finished product, the dispersion of the additives must be very uniform; large agglomerates of materials should not be present. The mixing is characterized as to the degree of dispersive and distributive mixing. These terms mean just what one would expect. Dispersion is the measure of the breaking up of large agglomerates, whereas distributive mixing is the measure of how evenly the small domains are distributed through the whole quantity of the polymer. Figure 1 illustrates these ideas.

Mixing the additives, which are usually liquids or solids, in this uniform manner cannot be done in a single screw extruder without a mixing tip or mixing section. The reason for this situation is that the flow in a screw is laminar (Fig. 2). This means very little stretching flow and no folding of the flow to give the distributive mixing. The mixing tip on the extruder screw, or mixing section along a screw, can provide some improvement. However, the best approach is to use an intensive mixer. A sigma blade mixer or a twin-screw extruder (Figs. 5.3 and 5.4) provide the high shear and folding flows necessary to carry out the mixing. The stretching flows off the edge of the blade

or the screw flight tip do the breaking up of the agglomerates. The shape of the elements
of the mixer blade or screw creates the folding of the material.

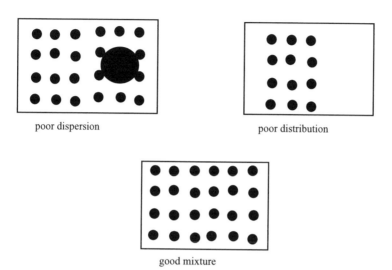

poor dispersion   poor distribution

good mixture

Figure 5.1 Dispersion and distribution of additives.

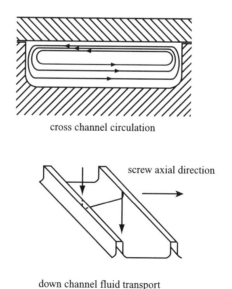

cross channel circulation

screw axial direction

down channel fluid transport

Figure 5.2 Laminar flow in a single screw extruder.

Figure 5.3 Sigma blade mixer for compounding. Courtesy of Krupp Werner & Pfleiderer.

Figure 5.4 Twin-screw extruder screws. Courtesy of Krupp Werner & Pfleiderer.

5.3   Antioxidants

Polymeric materials chemically deteriorate during fabrication, processing, and storage, due to a series of complex chemical oxidation reactions with atmospheric oxygen. The major result is chain scission, which can yield a significant decrease in molecular weight, and/or cross-linking, which increases molecular weight. Both chain scission and cross-linking affect flow properties of the polymer. Several factors promote oxidation reactions, including high temperature during processing, ultraviolet light, ionizing radiation, mechanical stress, and chemical attack. In an oxidative degradation process such as one generated by an increase in temperature, covalent bonds in the polymer chain are broken and free radicals are formed, as follows:

$$H\text{-}(P)_n\text{-}H + heat \rightarrow H\text{-}(P)_n\bullet + H$$

$$H\text{-}(P)_n\text{-}H \rightarrow H\text{-}(P)_{m1}\bullet + H\text{-}(P)_{m2}\bullet$$

where $H\text{-}(P)_n\text{-}H$ represents a polymeric chain and $H\text{-}(P)_n\bullet$ a free radical. (Recall that a free radical is a species with an unpaired electron.)

These free radicals, in turn, react with diatomic oxygen to form peroxide radicals and hydroperoxide species that propagate the formation of more free radicals.

$$H\text{-}(P)_n\bullet + O\text{-}O \rightarrow H\text{-}(P)_n\text{-}OO\bullet$$

$$H\text{-}(P)_n\text{-}OO\bullet + H\text{-}(P)_n\text{-}H \rightarrow H\text{-}(P)_n\text{-}OO\text{-}H + H\text{-}(P)_n\bullet$$

$$H\text{-}(P)_n\text{-}OO\text{-}H \rightarrow H\text{-}(P)_n\text{-}O\bullet + OH$$

$$H\text{-}(P)_n\text{-}O\bullet + H\text{-}(P)_n\text{-}H \rightarrow H\text{-}(P)_n\text{-}OH + H\text{-}(P)_n\bullet$$

The propagation mechanism continues until two free radicals are combined to form a stable molecule in the termination process.

As the free radical oxidative degradation process continues, considerable damage can result in the polymer due to chain scission and crosslinking. Usually, the mean molecular weight changes, and the molecular weight distribution broadens, and large quantities of oxygen are introduced into the molecular chain. In general, elongation, melt flow, impact strength, colors, and clarity of the polymer are affected.

To prevent damage to the polymer caused by oxidative degradation, chemical additives called antioxidants are incorporated into the polymer. Since rates of oxidation increase with temperature, and because such reactions are typically autocatalytic, as described above, stabilizing polymers during processing, where they are exposed to both oxygen and heat, is crucial in most applications of polymers. After processing, the degree of stabilization required is a function of both polymer sensitivity and conditions of use or storage.

Two main types of antioxidants are available commercially, primary antioxidants and secondary antioxidants. Primary, or chain breaking, antioxidants work by removing the free radical species, converting them to less reactive forms and thereby inhibiting the propagation process. Secondary, or preventive, antioxidants work by interfering with the formation of free radicals.

Chain-breaking donor antioxidants, often referred to as CB-D antioxidants, work by donating an H to the free radical, as in the following example, where (A)-H is the CB-D antioxidant:

$$H\text{-}(P)_n\text{-}OO\bullet + (A)\text{-}H \rightarrow H\text{-}(P)_n\text{-}OOH + (A)\bullet$$

For this to be effective, the (A)• radical must not initiate further degradation of the polymer. Typically, it is stabilized by resonance, causing it to have low reactivity. Examples of CB-D antioxidants include hindered phenols such as butylated hydroxytoluene (BHT), and secondary arylamines.

Chain breaking acceptor (CB-A) antioxidants interfere with propagation by oxidizing radical species in a stoichiometric reaction, such as:

$$H\text{-}(P)_n \bullet + (D) \rightarrow \text{ nonradical products}$$

where (D) is a chain-breaking acceptor antioxidant. Examples included stable free radicals and quinones.

As mentioned, secondary, or preventive, antioxidants work by preventing the generation of free radicals. Since decomposition of peroxides into free radicals is a key step in autocatalytic oxidation reactions, decomposing unstable peroxide molecules into stable, non-radical products is a primary mode of operation of preventive antioxidants. Phosphites and sulfur compounds are examples of peroxide-decomposer secondary antioxidants. Some act in a catalytic fashion, while others act stoichiometrically. Other categories of secondary antioxidants include UV light absorbers, which will be discussed in Section 5.5, metal deactivators, and excited state quenchers.

Excited state quenchers are agents that can absorb the excess energy from polymer species, and again dissipate it in a way that does not result in formation of active free radicals. Many of these agents also have other protective modes of behavior. For instance, benzotriazole compounds can act both as UV absorbers and as excited state quenchers.

Metal ions are catalysts for oxidation, so their inactivation helps prolong polymer life. Metal deactivators are usually chelating compounds that combine with metal ions and consequently decrease their effectiveness in catalyzing decomposition reactions.

Because primary and secondary antioxidants differ in their mechanism of attack to prevent oxidative degradation, in practice both type of antioxidants are often used together to obtain the best results. Often such combinations provide a synergistic effect, where the combined antioxidant package provides greater protection than the sum of the two alone. However, antioxidants can also interact in an antagonistic fashion, so proper pairing is needed.

Saturated polymers containing tertiary carbons (carbons bonded to three other carbons) in the backbone, such as PP, are more susceptible to oxidation than saturated linear polymers containing only secondary carbons (carbons bonded to only two other carbons) in the backbone, such as PE.

In packaging, three resins account for the majority of the market for antioxidants: PP, PE, and HIPS. For PP, a combination of hindered phenol and phosphite antioxidants is commonly used, with the total concentration normally from 0.08% to 1%, depending on formulation and end use. Ciba Specialty Chemicals has developed the phenolic antioxidant family of Irganox™ for use in PP and PE, and also the Irgafos™ family, which are phosphite stabilizers used in combination with phenolic antioxidants. For LDPE, BHT, a phenolic antioxidant, is normally incorporated at levels of 50-500 ppm;

however there is a tendency to employ less volatile additives to prevent their migration from the resin. For HDPE and LLDPE, antioxidants less volatile than BHT, such as polyphenols, at higher concentrations, are normally used in combination with phosphites. For HIPS, hindered phenols are used in combination with UV absorbers. Alpha-tocopherol (Vitamin E) is sometimes used as an antioxidant for polyolefins.

In the case of PVC, dehydrochlorination is much more important than oxidation, although the reaction is accelerated by oxygen. Typical PVC stabilizers are discussed next.

5.4   Heat Stabilizers

Heat stabilizers are used to prevent degradation of polymers when they are exposed to heat, especially during processing. For most polymers, the primary mode of reaction is oxidation, so antioxidants function effectively as heat stabilizers. For a few sensitive polymers, however, other forms of reaction are more important. In particular, a major disadvantage of PVC is its poor thermal stability. Degradation takes the form primarily of dehydrochlorination, yielding HCl and resulting in formation of a double bond in the main polymer chain.

$$-CH_2 - CHCl - CH_2- \rightarrow -CH = CH-CH_2- + HCl$$

The reaction is catalyzed by HCl, and it occurs preferentially at sites adjacent to double bonds, resulting in the formation of areas with alternating single and double bonds that can impart a visible color to the PVC - first yellow, then amber, reddish brown, and finally black. Organometallic compounds; salts derived from lead, cadmium, barium, zinc and tin; epoxides; and phosphites are the most common heat stabilizers used for PVC. Barium-cadmium, organotin, and organolead compounds account for more than 90% of the total heat stabilizers used in PVC in the US. Use of lead-containing stabilizers is decreasing due to toxicity considerations. In fact, there is no longer any significant use of lead-based stabilizers in packaging PVC. One of the most common organotin stabilizers is dibutyltin.

$$(C_4H_9)_2-Sn-(S-CH_2-COO-C_8H_{17})_2$$

While most of the degradation is encountered during processing, PVC films and bottles can degrade when they are heated to moderately high temperatures or subjected to gamma ray sterilization or UV radiation. Proper selection of additives can prevent dehydrochlorination and the development of color, as described above. The best way to control the degradation of PVC is to carefully select the correct heat stabilizer package for a specific PVC application.

The additives in PVC bottles for cooking oil and other food products must have FDA clearance. For flexible packaging materials, the most common stabilizers are mixed metals such as barium-zinc and calcium-zinc, which are replacing cadmium-zinc formulations. For rigid blow molded containers and calendered sheets, organotin formulations are the most commonly employed. Dibutyltin, calcium-zinc compounds, and methyltin have FDA clearance for use in food packaging.

PVDC copolymers also tend to be subject to thermal degradation, undergoing the same general types of reaction as PVC, so they also require the use of heat stabilizers.

5.5   UV Stabilizers

Ultraviolet (UV) radiation from outdoor light and gamma radiation used for sterilization of medical and biomedical products can cause photo-oxidation in PS, polyolefins (especially PP), PVC, and other polymers. Highly energetic UV photons can be captured by a polymeric chain, resulting in the breaking of covalent bonds and production of free radicals. Although the particular response to UV varies with each polymer due to different sensitivities, the global effect is eventually destructive to the polymer. Changes of color, loss of flexibility and gloss, and lower molecular weight are some of the effects that can be produced by photo-oxidation. (UV radiation has similar degradative effects when it is absorbed by human skin.) An example of the need for light stabilizer is polyolefin stadium seats, where extended light protection is needed for long life of the seats. To protect polymers from the destructive action of UV radiation, different approaches can be adopted.

One method is to use UV absorbers, agents that absorb the harmful UV radiation and emit harmless radiation of larger wavelength and lower energy, which is dissipated throughout the polymer matrix. An example of this type of stabilizer is hydrobenzophenone, which is commonly used with PVC, PE, PP, cellulosics, and PET. The Tinuvin™-P family, manufactured by Ciba Specialty Chemicals and based on benzotriazoles, is commonly used although it is not as effective in polyolefins as the hydrobenzophenones. One commercial example is Tinuvin™ 326, which is commonly used in concentrations of 0.25-0.30%. Another example is carbon black, which is an inexpensive material that offers UV protection for PVC, PE, and PP, even at low concentrations. UV absorbers can protect products, as well as protecting the plastic package from UV-induced degradation. Obviously, any UV light that is absorbed by the package will not reach the product inside the package.

A second group of stabilizers acts by "quenching" a polymeric chain that has been excited to a higher level of energy by the UV photon, thus bringing it back to a stable state by absorbing the energy in a way that does not promote polymer reactions. An example of this type of stabilizer is organosalts of nickel, which are often used in PP and PE.

Finally, a third mode is when the stabilizer acts as a free radical scavenger, accepting free radicals and preventing them from reacting with the polymer molecules. The highly effective family of hindered amine light stabilizers (HALS) belongs to this group. HALS are antioxidants, but in contrast to phenolic and phosphite antioxidants, they provide a regenerative radical trapping process. A single HALS molecule can eliminate several free radicals before it is converted into inert derivatives, making them two to three times as effective as other stabilizers. HALS have permitted new outdoor applications of polyolefins, extending product life and decreasing cost. Applications of HALS include their use in pigmented polymers, and as a radiation stabilizer for PP in biomedical products. The growing concern about the use of ethylene oxide in sterilization has increased the acceptance of gamma radiation in the sterilization of medical supplies. This requires the use of stabilizers to protect the polymeric materials used. HALS have become very important UV stabilizers. Many formulations of HALS are available, and the correct selection must take into account both the polymer and the intended application. Tinuvin™ 622 and Tinuvin™ 783 are examples of HALS cleared by the FDA that have low volatility and are effective in carbon black systems and films, respectively. For polypropylene, Tinuvin™ 770 is commonly used.

Combining different stabilizers acting with different mechanisms may trigger a synergistic effect that can provide excellent protection to the polymer. On the other hand, some combinations result in antagonistic interactions that reduce the efficacy of the stabilizers.

To evaluate light stabilizers for polyolefins, tests can be performed either by exposing samples to real outdoor environments that will be encountered in service, or by using accelerated test methods employing artifical light sources. The amount of incident energy is measured in cal/m^2 or angley. For example, one year of outdoor exposure in Florida is about 140 kiloangley.

5.6   Additives to Modify Surface Attractions

In many packaging applications, it is useful to be able to modify the behavior of materials that come in contact with one another. For example, blocking in a roll of film decreases our ability to unwind and use it. If a plastic part tends to stick to a mold, the molding process will be slower and less efficient. Friction between a plastic and the barrel of the extruder is desirable, but friction between the plastic and the screw inside the barrel is not. In this section, we will discuss various additives that, in one way or another, modify these behaviors.

5.6.1   Antiblocking Agents

Blocking is the tendency of two adjacent layers of a material, such as a polymer film, to stick to each other by simple physical contact. Blocking is particularly a concern for polyolefins such as PE and PP, soft PVC films, and PET. Blocking tends to increase with increasing temperature and pressure. The degree to which a film is susceptible to blocking is mainly determined by the smoothness of the surfaces; the smoother the surface, the greater is the degree of intimate physical contact, and therefore the greater the blocking. To reduce blocking, then, the smoothness of the surfaces may be reduced by incorporating tiny particulates into the polymer. These antiblocking agents produce irregularities on the film surface which reduce the amount of contact between the layers of material.

In packaging films, synthetic or natural silicas and minerals are common antiblocking agents. Performance is affected by both shape and size of the particles. As a rule, finer particle sizes are used in thin films, and larger particle sizes in thicker materials. The preferred average particle size for antiblocking agents in LDPE and PP films is 6-20% of the film thickness.

Synthetic silicas such as "micronized" silica gel, fumed silica, and zeolites are often used in high quality packaging films, while naturally occurring silicas and minerals such as clay, diatomaceous earth, talc, and quartz predominate in lower quality materials. Diatomaceous earth is a compact, granular or amorphous mineral composed of hydrated silica formed of fossil diatoms. Talc is a soft mineral of fine colloid particles with a soapy feel, made of hydrated magnesium silicate, $4SiO_2$-$3MgO$-H_2O. Talc has some advantages over diatomaceous earth as an antiblock for PE films, including platelet morphology, particle size distribution, and the possibility of being coated to make it more compatible with PE. Levels used typically range between 0.1 and 0.5% of the resin weight.

When antiblocking agents are incorporated into PE films, other important properties of the polymer are also affected. These include an increase in stiffness, a decrease in the coefficient of friction, and an increase in haze. Interaction effects with processing aids can also result, especially with fluorocarbon elastomers that are added to prevent melt fracture in the blowing process of films. Worker exposure to dust generated by these additives can be hazardous if impurities such as crystalline silica or asbestos are present.

5.6.2   Slip Agents

Slip agents are related to antiblocking agents, but instead of decreasing surface contact, they reduce blocking by introducing a thin, low-friction coating between the plastic materials. These additives are usually mixed into the polymer film, but have a strong tendency to migrate to the film surface, where they perform their function. Slip agents are often used in polyolefin films. Common choices are oleic acid amide for polyethylene, and erucamide for polypropylene.

Combinations of a slip and an antiblocking agent can improve performance through more rapid development of slip properties and a more efficient use of the antiblock; in both cases this is the result of improved dispersion of the additives. These agents are commonly added to a resin in the form of a multicomponent master batch which also includes other additives. Amounts of the antiblock in the master batch commonly range from 5 to 50%.

Both Toshiba and General Electric market solid spherical silicone powders which are reported to have combined antiblocking and slip agent properties, distributing evenly on film surfaces and reducing friction between film layers, without detracting from film clarity.

5.6.3 Antislip Agents

While blocking is a concern in many applications, at other times, there is a need to increase blocking. For example, a common problem in stacking plastic sacks is a tendency for the sacks to slide off one another. An antislip agent can be used to combat that problem by increasing friction between the surfaces, or by increasing the attractions between nearby surfaces. Common antislip agents include ethylene/maleic anhydride copolymers, colloidal silica, and finely powdered sand or other minerals. These agents may be compounded into the plastic, or sprayed on surfaces. The particle sizes are generally less than 1 micron, and concentrations of the agents are usually under one percent by weight.

5.6.4 Lubricants

Lubricants are materials that promote flow and reduce the tendency of plastics to stick to surfaces. Slip agents, discussed in Section 5.5.2, and mold release agents, discussed next, can be regarded as examples of special types of lubricants. Lubricants can be used to promote the flow of plastics over or through surfaces of dies, mold cavities, etc. Common lubricants include fatty-acid esters and amides, paraffin and polyethylene waxes, stearates, and silicones. These and other lubricants can be compounded into the plastic resin, or applied externally.

In flexible PVC film, waxes and low molecular weight polyethylene are often used as external lubricants, and fatty acids, esters, and metallic soaps are often used as internal lubricants. As the names suggest, external additives, including lubricants, are applied to the surface of the film or other plastic article, while internal lubricants are blended into the polymer before it is formed.

5.6.5   Mold Release Agents

Mold release agents are often used to facilitate the removal of plastic parts from molds by acting to decrease the adhesion between the plastic and the mold. Mold release agents may be sprayed or otherwise applied to the mold surfaces, as an alternative to, or in combination with, compounding the additives into the resin. These agents may be liquids that provide lubrication, such as those mentioned above, or may be fine solids such as dusting powder.

5.7   Colorants

The use of colorants in plastics is driven almost exclusively by marketing considerations, using product appearance to influence consumers. Colorants do not add mechanical strength nor improve mass barrier properties. However, they may give an opaque appearance that can contribute to light protection of a packaged product.

Selecting, combining and matching colors is a complicated art that only well-trained individuals are able to perform correctly. There are hundreds of different colorants used in the plastic industry, and there are as many types of colorants as different applications for plastics. Similarly to measurement of optical properties of paper and paperboard, the characterization of a color for plastics is based on the measurement of color (hue), brightness, and opacity. Other important variables to be considered in colorant selection include dispersability in the plastic, migration, toxicity, light stability, and chemical resistance.

The plastic industry continues to move away from toxic colorants, especially those based on heavy metals such as chromium, cadmium and lead. In the US, at least18 states ban the use of heavy metals, including colorants, in packaging materials. The European Union and some other countries have restrictions on their use, as well. The use of organic colorants, or heavy-metal-free (HMF) colorants, is continuously increasing, and many colorant producer companies are replacing all of their heavy-metal-containing colorants with systems that present fewer environmental problems and legal restrictions. Colorants incorporated in plastic containers in direct contact with food have to be cleared by FDA, as is the case for other additives..

The three major categories of colorants used in plastics are dyes, organic pigments, and inorganic pigments.

5.7.1 Dyes

A dye is a colorant that is soluble in the plastic. Normally, dyes are low molecular weight organic compounds. They do not interfere with transparency of the plastic, but migrate easily from it. Because of their tendency to migrate, dyes have limited use in the plastics industry. Some dyes are very toxic and their use is regulated by OSHA (Occupational Safety and Health Administration). Dyes also tend to be susceptible to light degradation, and therefore fade quickly. Dyes include azo, diazo, pirazalone, anthraquinone, quiniphthalone, and quinoline dyes.

5.7.2 Organic Pigments

Organic pigments, unlike dyes, are insoluble in the plastic matrix. They are produced in the form of very fine particles that give the plastic an opaque appearance. Organic pigments tend to migrate less than dyes, though, like dyes, some are very toxic and OSHA directives must be followed in handling them. Some organic pigments are listed in Table 5.1.

5.7.3 Inorganic Pigments

Inorganic pigments, widely used in the plastic industry, do not have the brightness or the intensity of color that characterize organic pigments. However, they are generally less expensive, more opaque, more stable to high temperatures, and have less tendency to migrate. Many of the inorganic pigments are extremely toxic since they are oxides of heavy metals such as chromium, lead, cadmium, or nickel. When handling these colorants, OSHA guidelines must be followed. As mentioned, in a number of places pigments and other additives based on lead, cadmium, mercury, or hexavalent chromium cannot be used in packaging. Some common inorganic pigments are listed in Table 5.1. As is the case with dyes, pigment formulations often contain a variety of different pigments to provide the desired colors.

5.7.4 Specialty Pigments

Lake pigments consist of a dye associated with an inorganic support such as alumina hydrate. They are used in packaging to obtain special visual effects.

Table 5.1 Common Pigments

Color	Organic Pigments	Inorganic Pigments
Black	Carbon blacks*	Carbon blacks* Iron chromite Iron oxide
Blue	Phthalocyanines	Cobalt aluminate
Brown		Iron oxide
Green	Phthalocyanines	Chromium oxide
Maroon		Cadmium sulfoselenides
Orange	Azo's Benzimidalones Pyrazalones Quinacridones	Cadmium sulfide Cadmium sulfoselenides Lead chromates Lead molybdate
Red	Benzimidalones Disazo's Quinacridones Pyrazalones	Cadmium sulfoselenides Iron oxide
Violet	Dioxazines Quinacridones	
White		Titanium oxide
Yellow	Benzimidalones Disazo's	Cadmium sulfide Chrome titanate Iron oxide Lead chromate Nickel titanate Zinc chromate

* Carbon blacks are actually organic in structure, but are commonly classified as inorganic pigments.

Pearlescent colorants are used to impart a special pearly luster and provide iridescent effects. Titanium oxide-coated mica and ferric oxide-coated mica are the major pearlescents in use. They form thin platelets of high refractive index, which both reflect and transmit the incident light. Fluorescent pigments absorb visible radiation and re-emit a narrow band of light at somewhat higher wavelengths, producing a very bright appearance. They usually consist of a fluorescent dye dissolved in a transparent resin and ground to a fine size.

Metallic pigments are generally based on fine flakes of aluminum. They impart a shiny look to the plastic.

5.7.5   Colorants and the FDA

Besides economic factors, the use of colorants in plastic packaging requires health and safety considerations. The FDA has made public a list of sanctioned colorants in Title 21 of CFR §178.3297. Colorants listed there "may be safely used as colorants in the manufacture of articles or components of articles intended for use in producing, manufacturing, packing, processing, preparing, treating, packaging, transporting, or holding food." There are also provisions related to the definition of colorants, migration to food, and conformation under section 409 of the Federal Food, Drug and Cosmetic Act that should be reviewed when dealing with this subject.

Inorganic colorants listed in 21CFR §178.3297 include aluminum, aluminum hydrate, aluminum and potassium silicate, aluminum silicate, barium sulfate, bentonite, calcium carbonate, calcium silicate, calcium sulfate, carbon black (channel process, prepared by the impingement process from stripped natural gas), chromium oxide green Cr_2O_3, cobalt aluminate (with restrictions), diatomaceous earth, iron oxides, kaolin - modified for use in olefin polymers in amounts up to 40%, magnesium oxides, magnesium silicate (talc), sienna, silica, titanium dioxide, titanium dioxide-barium sulfate, ultramarines, zinc carbonate (limited use), zinc chromate (less than 10%), zinc oxide (limited use), and zinc sulfide (less than 10%).

Organic colorants listed in 21CFR §178.3297 include all FDC certified colors, C.I. Pigment Blue 15, C.I. Pigment Violet 19, C.I. Pigment Red 38, C.I. Pigment Orange 64, C.I. Pigment Yellow 95, C.I. Pigment Yellow 138, and C.I. Pigment Red 177. In recent years, the FDA has sanctioned few new colorants for food-packaging or extended the use of others. Some organic pigments have limited thermal stability which probably makes them unsuitable for use with high-heat resins like nylons and polycarbonate. A final decision on colorant selection, as with other additives for food-contact packaging, should be done in accordance with FDA regulations.

5.8   Antifogging Agents

In many cases, rather than relying on a pretty color to give a plastic package the desired appearance, the user wishes to have the product clearly visible through the package.

Several types of plastics can provide excellent transparency. However, when condensation of water molecules take place on the surface of a package, a thin layer of small droplets of water can be formed. In transparent films and structures, the droplets of water act to scatter the incident light. This phenomenon, called fogging, makes the film or structure appear opaque. The droplets are formed when the polymer surface tension is lower than the surface tension of water, causing the water to remain in droplets rather than forming a continuous layer (see discussion of adhesion, Sec. 6.3.1).

Antifogging additives function by increasing the critical surface tension of the polymer surface, allowing the molecules of water to wet the surface, forming a continuous layer of water which does not scatter light and therefore does not interfere with transparency.

Common antifogging agents are fatty acid esters such as glycerol and sorbitol stearate, fatty alcohols, and ethyloxylates of nonyl phenols. Antifogging agents are incorporated in the resin in levels ranging from 0.5% to 4%. Factors involved in the selection and use of these additives include polymer type, thickness of the structure or film, performance life, and type of product. If the product is a food, for instance, FDA clearance is necessary for additives. Antifogging compounds can be applied on the surface of the material or compounded internally in the plastic.

As is the case with other additives, the same additive may have more than one effect. For example, in PVC films, esters of multifunctional alcohols are often used to provide both antifogging and antistatic properties.

5.9   Nucleating Agents

The use of nucleating agents is also related to obtaining a highly transparent package. As was discussed in Chapter 3, the presence of crystallinity is often associated with opacity of plastic materials, as the crystallites scatter light. However, if the crystallites are small enough, they produce much less interference with light transmission than do larger crystallites. Nucleating agents can be used to decrease the average crystal size, resulting in improvements in clarity.

If you have ever seen the result of dropping a seed crystal into a supersaturated salt solution, you have observed the importance of providing a site for crystal growth to begin. Nucleating agents provide such sites. Although one might imagine that the result of nucleating crystals would be an increase in crystallinity and consequently a decrease in transparency, that is not always the case. The nucleating agent provides a multitude of sites for crystal growth to begin. The resulting formation of a large number of crystallites causes the crystals to interfere with one another, so the crystals are much smaller. Smaller crystallites do not interfere as much with light transmission although the overall level of crystallinity may be higher. Improved transparency can be the result.

Nucleating agents are often used in highly transparent grades of polypropylene. Adipic acid, benzoic acid, and some metal salts of these acids are often used in polypropylene. For nylon, colloidal silicas are a common choice of nucleating agents. Nucleating agents are not often used with polyethylene.

PET, because of its low rate of crystallization, has only a low level of crystallinity and small crystallites when it is formed in standard processes that provide reasonably rapid cooling. Addition of nucleating agents, in this case, can increase overall crystallinity and render the formed object opaque. Thus, the effect of adding nucleating agents in PET is to decrease, rather than increase, crystallinity.

5.10 Antistatic Agents

Static electricity can be generated on a polymer surface by friction, such as by rubbing it against another surface. The surface is usually a solid, but may be just air. In packaging, the fast moving film in a continuous converting operation or in a form-fill-seal processing line, for instance, promotes the generation of static electricity on the film. PE, PS, PP, PET, PAN, PVC, and nylon are all highly susceptible to accumulation of static charges.

Static can adversely affect a manufacturing operation or process by introducing uncontrolled electrical forces that may result in, for example, materials folding and sticking together. It may also create dangerous conditions such as the formation of sparks leading to vapor explosions. Most polymers are susceptible to the accumulation of electric charges on their surface because of their high resistivity, which prevents the conduction of electrons to dissipate the charge.

Static charges on polymer surfaces can be controlled by the presence of antistatic agents that make the surface more conductive, or less resistive. For example, water within a hydrophilic polymer can act as an antistatic agent and prevent static buildup. The amount of water is important; for example, water in a polyamide in equilibrium with air at 65% RH acts as an antistatic agent, but at low RH values, water is not effective. Since most polymers used in packaging are not hydrophilic, antistatic additives may be used to control static. Generally, these agents are cationic, anionic, or nonionic surfactants.

A common group of cationic antistats is alkyl quaternary ammonium salts. These are mostly employed in polar substrates such as PVC and styrenic polymers. Other types include alkyl phosphonium and alkyl sulfonium salts. Flexible PVC may contain up to 7% of these antistatics for non-food uses, as they have not been approved by the FDA. Sodium alkyl sulfonates, similar to common detergents, have gained wide acceptance as anionic antistatic agents, and are used in PVC and styrenic polymers. Other anionic antistats include alkyl phosphonic, dithiocarbamic, and carboxylic acids.

For non-polar polyolefins, nonionic antistatics are the most commonly used. These include ethoxylated fatty amines, fatty acid esters, ethanolamides, and polyethylene glycol-esters. The amount used in LDPE is typically around 0.05%. For packaging of electronics, which can be highly sensitive to damage caused by static charges, considerably higher levels are used, up to 10% by weight.

Antistats can be applied internally or externally. Internal antistats are compounded into the resin, and act once they migrate to the surface of the polymer. External antistats are applied directly on the surface by spraying, or sometimes by dipping the polymer in a solution of the antistatic. Internal antistatic agents can often provide much longer-term protection than external agents, since additive lost from the surface can be replenished by additional migration of the additive from the bulk. One disadvantage, however, is that antistatic activity does not begin immediately, since time is required to develop a reasonably high surface concentration of the antistat.

An alternative to the use of antistatic additives is the incorporation of electrically conductive fillers or reinforcements into the polymer to make the whole structure conductive. Typical additives that are used for this purpose include aluminum, steel, or carbon powders, and metal-coated glass fibers or carbon fibers. Powdered fillers are generally less expensive than fibers. Maintaining the desired fiber distribution during processing is also problematic.

Rather than using chemical antistats or conductive fillers, it is also possible to use conductive polymers, as discussed in Section 4.16.

Testing of antistats is described in ASTM D 257, which uses measurement of the electric resistivity; and in Federal Test Method Standard 101C Method 4046 which uses measurement of the generation or decay of static electricity.

5.11   Plasticizers

A plasticizer is a substance that is incorporated into a rigid plastic to increase its flexibility, workability, and extensibility. By reducing the glass transition temperature and increasing chain lubricity, plasticizers also improve processing and extrusion characteristics, reduce the minimum required processing temperature, reduce hardness, and improve low temperature flexibility.

Not all plastics require the use of plasticizer, but for certain plastics, such as PVC, the use of an appropriate plasticizer for the desired end use is essential. Indeed, PVC applications depend on the level of plasticizer. Without plasticizer, PVC is a semicrystalline, brittle polymer that is very difficult to process. At lower concentrations, the plasticizer helps to reduce processing temperatures, and this helps minimize thermal degradation of the polymer. At higher concentrations, besides improving processing conditions, plasticizer reduces hardness and increases flexibility of the final product. The polar nature of PVC gives it a high affinity for plasticizers, and therefore PVC accounts for more than 80% of the total use of plasticizers.

In order for a plasticizer to work, it has to have a correct balance of functional groups. Normally, plasticizers have a slight to strong polar functionality for compatibility with polar polymer groups, and a non-polar (hydrocarbon) group for internal lubrication. Different ratios of polar to non-polar groups make a plasticizer more suitable for one application than for others. To improve processing conditions at high temperatures, a more polar plasticizer, such as dibutyl phthalate, is preferred. To improve low temperature performance, that is, to make the PVC more flexible by depressing T_g, a more non-polar plasticizer, such as dioctyl sebacate, is better.

Most plasticizers are liquids at room temperature, and have high boiling points and low vapor pressure, so that they are not easily lost from the plastic by vaporization. The most common plasticizers are the phthalates, and among them, diethylhexyl phthalate (DOP) is the most widely used. Safety concerns about DOP were raised in the 1980's but have still not been fully clarified. To date, no action has been taken to regulate the production or use of DOP. Both in the US and Europe it appears that DOP is making something of a comeback. Diethylhexyl pthalate (DHP) and bisphenol A have both been targeted for concern about their use in polycarbonate packaging as well as in PC baby bottles.

In films for food packaging, only FDA-approved plasticizers can be used. Most often, an adipate plasticizer, usually di(2-ethylhexyl) adipate (DEA), combined with epoxidized soybean oil is used. The oil epoxide is frequently termed a secondary plasticizer, because it is not used alone, and has heat-stabilizer properties in addition to plasticization effects.

In contrast to use of the term in other categories of additives, "external" plasticizers refer to additives blended homogeneously into the resin. The term "internal" plasticizer refers to modification of the molecular structure of the polymer by incorporating comonomers that provide greater flexibility.

5.12 Oxygen Scavengers, Desiccants and Fragrance Enhancers

Just as UV absorbers in a package can help protect the product against degradation caused by UV light, oxygen scavengers in a package can help protect the product from oxidation. This is a relatively new area, with few applications at present, but a large potential. The first oxygen absorbers used in packaging were sachets of iron oxide which were placed inside the package, and thus would not be classified as additives. For a variety of reasons, including simplifying packaging line operations and avoiding the risk of consumers inadvertently consuming the scavenger along with the product, there is a desire to replace these systems with packages with built-in oxygen scavengers. These systems represent one variety of "active packaging" - packaging which is designed to interact with the product and/or the environment to modify the conditions inside the package.

Some packaging systems have used pressure-sensitive labels, applied inside the package, which contain an oxygen scavenger. Opaque coatings of a scavenging component for the insides of packages are also available. Most of the newer systems are designed to be added to the packaging resin in the form of master batches, and be dispersed throughout the material. For oxygen, the additives generally consist of an oxidizable component, often a metal, oxidation promoters, and sometimes fillers. Metal-free absorbents commonly use mixtures of organic compounds such as phenolics, glycols, and quinones. In some cases, oxygen-absorbing monomers can be incorporated into the polymer structure during polymerization, to make an inherently oxygen-absorbing plastic. Most of the additive systems are activated by moisture.

Cryovac markets an oxygen scavenger film for modified atmosphere packaging, which has a proprietary polymer coextruded as an invisible layer of the final structure. The scavenging activity is activated just prior to sealing by a Cryovac ultraviolet light triggering unit which is installed on the processor's packaging line. The activity then continues without requiring further light exposure, until the capacity of the film is exhausted. The system reportedly can reduce residual oxygen levels from the normal 0.5-1% range down to parts per million in 4-10 days. The polymer can also be used in lidstock for thermoformed packages.

A PET bottle developed in 1989 used an oxidizable layer of MXD6 nylon, in amounts of 1 to 5%, along with 50-200 ppm of cobalt salt as a catalyst, to scavenge oxygen permeating through the PET. In 1998, both Amoco and Mitsui Chemicals were testing oxidizable copolyesters for use as oxygen scavengers in PET containers, including beer bottles. One of the PET beer bottle structures test-marketed in the U.S. in 1999 reportedly used two oxygen-scavenger nylon layers with a cobalt catalyst, one of the first uses of oxygen scavenging layers in a rigid container.

Building desiccants into packages has a longer history than incorporation of oxygen scavengers. The first applications used a sachet or capsule containing a desiccant that was placed inside the package. Such systems are still in common use for products varying from leather shoes to over-the-counter drugs.

One of the first applications of desiccant additives in packages was in multilayer coextruded containers for retorted oxygen-sensitive food products. In the retorting process, the filled and sealed package is exposed to high temperature steam heating for a sufficient time to kill microorganisms and render the product shelf-stable so that it can be stored without refrigeration, in a similar fashion to canning processes. Most such products are susceptible to oxidative degradation. While a metal can provides an essentially perfect barrier to oxygen, as discussed in Chapter 4, few plastics can provide adequate protection. One of the best oxygen barrier plastics is ethylene vinyl alcohol (EVOH). However, EVOH is water sensitive, and when it is exposed to high relative humidities, its oxygen barrier ability decreases significantly. Burying the EVOH in an inner layer, surrounded by plastics such as PP which are good moisture barriers, maintains the protection at adequate levels in most instances. During retorting, the EVOH can pick up significant quantities of moisture. This water will then leave the EVOH layer only at a slow rate after the retorting is completed, because of the good water vapor materials surrounding it. This permits gain of an undesirable amount of oxygen early in the product's shelf life, when it will do the maximum damage. An ingenious solution to this problem was the incorporation of a desiccant in the tie (adhesive) layers between the EVOH and the main structural polymer in the package. (These adhesive layers provide the adhesion between the layers necessary to maintain package integrity.) Most moisture penetrating through the structural polymer, either from the steam heating on the outside or the moisture of the product on the inside, is captured by the desiccant, and thus does not reach the EVOH. The result is a significantly higher overall level of protection against oxidation, which can mean the difference between an acceptable and an unacceptable shelf life for the product.

Built-in scavenger systems consisting of films impregnated with chemically reactive additives are also being developed for ethylene and other compounds which affect spoilage of products. E-I-A Warenhandels GmbH of Vienna, Austria, offers a master batch additive system which absorbs ethylene, as well as other undesirable compounds that can form inside packaged fruits, vegetables, and flowers, such as ethanol, ethyl acetate, ammonia, and hydrogen sulfide.

Fragrance enhancers, as their name indicates, are used either to eliminate undesirable odors or to produce desirable ones. Trash bags and kitty litter liners, for example, may contain additives which absorb undesirable odors generated by their contents. Blow-molded bottles may also contain additives that prevent the product from picking up undesired odors from the surroundings, or that provide for generation of desired odors. Since odors and flavors are highly associated, these same additives can affect taste as well as smell.

5.13   Fillers and Reinforcements

In some applications, it is useful to incorporate non-plastic substances into a plastic object, to reduce its cost or improve its performance in some way. Fillers are typically used to lower the cost of the plastic, and generally consist of minerals of some kind. Reinforcements are often more expensive, per unit mass or volume, than the plastics, but provide improvement in properties such as strength and/or rigidity. They usually consist of either organic or inorganic fibers. Use of fillers and reinforcements is less common in packaging applications than in uses such as automotive components or housewares, but is sometimes significant. In addition, these additives are more commonly used with thermoset polymers than with thermoplastics.

As mentioned, fillers function primarily to reduce cost, although they typically also significantly increase rigidity. They may also be categorized as extenders. Commonly, strength, both tensile and impact, decreases when fillers are present. Filler concentrations are usually in the range of 10 to 50% by weight. Common fillers include wood flour and a variety of minerals such as clay, silica, and talc.

Reinforcements generally are fibers, and properties of reinforced plastics are functions of fiber length as well as the amount of fiber. The primary purpose of reinforcement is to improve the strength or other mechanical properties of the plastic, and adequate adhesion between the fibers and the plastic matrix is essential. Fibrous reinforcements may be grouped into two categories: continuous fibers and chopped fibers. Continuous fibers provide the most improvement in mechanical properties, and have higher cost than chopped fibers. As with fillers, fibrous reinforcements are used more often with thermosets than with thermoplastics. Common reinforcements include glass fibers and carbon fibers. In the last several years, there has been increasing interest in the use of wood fibers. While these are more often used for construction applications such as decks, than for packaging, they have found applications in pallets and other distribution package systems.

5.14   Antimicrobials or Biocides

Antimicrobial agents preserve compounded polymeric materials from attack by microorganisms such as bacteria, fungi, or mildews. Most synthetic polymers in their pure state are not attacked by microorganisms; they are in general non-biodegradable. However, when various low molecular weight additives are compounded within the polymer, conditions for microorganism attack may be created.

Plasticizers, lubricants, or heat stabilizers in the polymeric matrix can be the target of the microbial activity. For example, PVC may contain 40-50 weight % plasticizers and lubricants. Different additives pose different degrees of resistance to biodegradation. If the expected shelf life of a package could allow significant microbial attack, it may be necessary to use an antimicrobial agent, although the use of such agents is rare in packaging applications.

Common preservatives used for polymers include 2-n-octyl-4 isothiazolin-3 and copper-8-quinoleate, in amounts ranging from 0.1-1%. Antimicrobial agents for polymers are considered pesticides by the Environmental Protection Agency under the Federal Insecticide, Fungicide and Rodenticide Act, FIFRA.

5.15   Other Additives

A variety of other additives can be incorporated into polymers, including flame retardants, coupling agents, impact modifiers, and others. Blowing agents, although not truly additives, are used to form plastic foams. Blowing agents can be divided into two categories: chemical blowing agents (CBA) and physical blowing agents (PBA). In the first group, the decomposition of organic compounds generates the blowing gas. In the second group, compressed gases are used and no change in chemical composition of the plastic or its additives is involved. Chlorofluorocarbons (CFCs) were widely used in the recent past as PBAs, but they have been eliminated in most of the world because they cause ozone depletion. PBAs in current use include hydrocarbons such as pentane, carbon dioxide, and hydrofluorocarbons (HFCs). Chemical blowing agents include sodium bicarbonate, azodicarbonamide, and others. Hydrofluorocarbons are potent greenhouse gases, and are not favored for that reason.

It should not be forgotten that, in addition to additives which are deliberately introduced into the resin, plastics may contain substances that are residues of substances produced or introduced at some stage of polymerization or processing, such as residues of catalysts, solvents, or unreacted monomer; or substances which have migrated into the polymer from the contents of the package or from its surroundings.

Study Questions

1. In what types of packaging applications would you expect to find UV stabilizers used in plastic resins?

2. Explain how plasticizers work, and what changes they produce in plastic resins.

3. Why are antioxidants used in nearly all plastic resins?

4. What advantages might an oxygen-absorbing additive in a plastic package have over the use of an antioxidant in the product formulation?

5. Compare the use of antioxidants and UV absorbers.

6. Find the chemical structures of BHT and vitamin E, and find a general structure for HALS.

7. Write a short critical essay analyzing the pros and cons of using additives in packaging.

6   Adhesion, Adhesives, and Heat Sealing

6.1   Adhesion

Adhesion is the process by which two initially separate bodies (called adherends or substrates) are held together by intermolecular forces. Two solid surfaces can be held together by using a substance, generally a liquid, called an adhesive, which is distributed between the adherends. The adhesive may be formed from the solids themselves by heating and firmly pressing them together, as in heat-sealing of thermoplastic materials. Alternatively, the adhesive may be formed by using a solvent to dissolve the interface, permitting intermingling of the substrates at a molecular level. Adhesion is also involved in the coating or priming of a solid flat substrate by a liquid. In all cases, the combination of two substrates with or without adhesive becomes a new composite structure, most often at least as strong as any of its components. Typically, strength develops as the materials at the interface solidify.

For example, paper and aluminum foil can be glued together using a solvent-based liquid adhesive. After evaporation of the solvent, a new flexible structure of paper/adhesive/aluminum is formed. Two sheets of Plexiglas™ (a copolymer of methyl methacrylate) can be bonded to each other by spreading a few drops of liquid dichloromethane between the surfaces. The liquid dichloromethane dissolves a small quantity of the copolymer at both interfaces, promoting superficial molecular entanglement, and the new structure is bonded as the solvent evaporates. Similarly, if one holds a very cold surface (such as a piece of ice or a metal door on a cold winter day) with a bare hand, the hand may stick to the surface when water molecules solidify between the hand and the cold surface. In heat-sealing, thermoplastic materials, such as LDPE, are melted, intermingle with each other, and then adhere to each other when the material is cooled.

The first part of this chapter will cover the adhesion process utilizing an adhesive. Heat sealing of thermoplastic materials will be covered next.

6.2   Adhesives

An adhesive can be defined functionally as any substance capable of holding two materials together. The major mechanism used to join plastics is intermolecular forces, including van der Waals forces and hydrogen bonding. Physical entanglements can also be a significant factor in adhesion, especially for porous materials such as paper. In general, adhesives, typically referred to as glue, are used to join together materials such as paper, glass, metal, plastic, ceramic, or any combination of these materials. Adhesives are vital components of many packages. Applications include the forming and sealing of corrugated cases, folding cartons, and paper bags; winding of paper tubes for cores, cans and drums; affixing plastic and paper labels to bottles, jars, drums, cases and other containers; the lamination of paper, paperboard, foil and plastic films; and coextrusions. Packaging accounted for about 40% of the 5.9 billion lbs of adhesives used in 1999, the largest single market. Pressure-sensitive tapes and labels accounted for an additional 14%. The total value of the adhesives market that year was $9 billion [1].

When an adhesive is used to join two surfaces, such as in the lamination of paper with aluminum foil, or lamination of two plastic substrates, it must: 1) adhere to the surface of each adherend (adhesive bond strength), and 2) provide enough strength within the bulk of the adhesive itself to meet the requirements for the application (cohesive bond strength). When two adherends are bonded together using an adhesive, five different zones can be recognized in the resulting structure (Fig. 6.1):

1.   Adherend body 1
2.   Interface of adherend 1 and adhesive
3.   Adhesive
4.   Interface of adherend 2 and adhesive
5.   Adherend body 2

Adherend 1	zone 1
===========interface============zone 2=========	
Adhesive	zone 3
===========interface============zone 4 =========	
Adherend 2	zone 5

Figure 6.1 Structure of two bodies bonded by an adhesive.

6.3   Adhesive and Cohesive Bond Strength

As mentioned above, the adhesion forces develop at the interface between the adherend and the adhesive, and it is at this interface where interfacial forces play the important role of holding the two surfaces together. These are called the *adhesive forces*. If adhesives are used to join two materials as in Figure 6.1, besides the adhesive forces, the strength and integrity of the bonded structure depends on the strength of each material and of the bulk adhesive. The forces of intermolecular attraction acting within a material are termed *cohesive forces*. The cohesive forces in an adhesive depend on its own molecular and physical structure, and are not influenced by the interfacial forces. Adhesive forces, then, determine the adhesive bond strength at the interfaces, and cohesive forces determine the cohesive strength both within the bulk of the adhesive, and in the substrates being joined. The survival and performance of the composite structure depends on all of these.

Adhesive forces are provided by attractions between neighboring molecules and include the same types of forces discussed in Section 2.2.2. Because these forces require a distance of no more than 3-5 Å to have reasonable strength, the neighboring molecules at the interface must be very close together for adhesion to occur. This has important practical implications for effective adhesion. The adhesive, at the time of application, must be able to completely "wet" the adherend surface, and must have a low enough viscosity to be able to flow into and fill any irregularities in the substrate surface, in order to bring the adhesive and substrate close together on a molecular scale.

To obtain maximum adhesion, the adhesive bond strength between the adhesive and adherend should be greater than the cohesive bond strength of the adhesive, as indicated in Figure 6.2. (Of course, the overall strength is also limited by the cohesive strength of the substrates.)

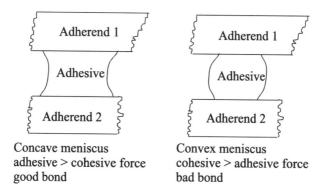

Concave meniscus
adhesive > cohesive force
good bond

Convex meniscus
cohesive > adhesive force
bad bond

Figure 6.2 Cohesive and adhesive forces.

6.3.1   Adhesive Bond Strength

There are several factors that can be used to match an appropriate adhesive to an adherend, including surface tension, solubility parameter, and viscosity.

6.3.1.1   Surface Tension

Solid surfaces have many irregularities, and since adhesion is a surface phenomenon, the adhesive must fill completely all pores and surface irregularities of the adherend at the moment of application. To accomplish this, the adhesive must be applied in a liquid or semi-liquid state. The liquid adhesive must penetrate all the pores and crevices, eliminating any air pockets, to obtain a homogeneous bond between the adherend and adhesive. The adhesive needs to "wet" the adherend surface, and the better the wettability of the adhesive/adherend pair, the better the chance of producing homogeneous spreading of the adhesive.

The wettability characteristics of an adhesive/adherend pair are determined by the relative values of surface tension of the adhesive and adherend. Surface tension of a liquid is a direct measurement of intermolecular forces and is half of the free energy of molecular cohesion. Surface tension is commonly represented by γ (gamma), and is measured in dynes/cm. The value of the surface tension of the solid substrate, or adherend, is called the critical surface tension, γ_c. To ensure that the surface of the adherend will be wetted by an adhesive, an adhesive whose surface tension is less than the critical surface tension should be selected, so that

$$\gamma_{adh} < \gamma_c \tag{6.1}$$

In practice, the surface tension of the adhesive should be at least 10 dynes/cm smaller than γ_c. Selected values of γ are listed in Table 6.1, and published in various handbooks.

The surface tension of plastic surfaces can be measured using a calibrated set of solutions. A more sophisticated, and expensive, method is to measure the contact angle the liquid makes with the surface. This method was first described almost 200 years ago for evaluating the wettability of surfaces. The angle measured is the one formed by the tangent on the surface of a drop of liquid at the point of contact with the solid surface and the surface. If the angle is zero, the liquid is said to completely wet the surface. If the angle is not zero, the liquid is said to be nonspreading, and the surface tension of the surface is related to the surface tension of the liquid and the contact angle.

From the values in Table 6.1, one can conclude the following:

1.   Water does not wet any of these polymers.
2.   Toluene wets PET and nylon 6,6 but not polytetrafluoroethylene (PTFE, Teflon).
3.   The very low γ_c value of PTFE means it will not be wet by most substances, so adhering materials to it is difficult.

Table 6.1 Selected Surface Tension Values

Material	Surface Tension, γ, dynes/cm
Nylon 6,6	42
PET	43
PTFE	18
Water	73
Toluene	27

The critical surface tensions of polymeric materials such as polyolefins can be increased by surface treatment such as corona treatment, chemical etching, flame treatment, and mechanical abrasion, in order to facilitate adhesive bonding.

6.3.1.2 Solubility Parameter

An important criterion for determining the chemical compatibility between an adherend and an adhesive in a solvent is the solubility parameter, δ. The solubility parameter is the square root of the cohesive energy density, CED:

$$\delta = (CED)^{1/2} = (\Delta E / V)^{1/2} \tag{6.2}$$

where ΔE is the energy of vaporization and V is the molar volume. A common unit for δ is $(cal/cm^3)^{1/2}$, which is called a hildebrand.

When the adherend is an organic compound and is not too polar; the solubility parameter is useful in selecting an adhesive, allowing one to pre-screen adhesives for a particular polymer application. According to the laws of thermodynamics, the greater the difference between the solubility parameters of two materials, the less compatible they are. Consequently, good compatibility is favored when the adhesive and adherend have similar solubility parameters.

$$\delta_1 \approx \delta_2 \tag{6.3}$$

where δ_1 and δ_2 are the solubility parameters of the adhesive and adherend. Selected solubility parameters are listed in Table 6.2.

For polar substances, the types of interactions, as well as their strength, becomes significant, and selection of a proper adhesive by solubility parameter alone does not always work well. A more general, simple rule for selection of adhesives is "like sticks to like." In other word, the greater the chemical similarity between two materials, the larger will be the intermolecular forces between them.

Table 6.2. Solubility Parameters

Material	Solubility Parameter, δ (hildebrands)	Critical Surface Tension, γ_c (dyn cm^{-1})
Poly(1H, 1H-pentadeca-fluoroctyl acrylate)	-	10.4
Polytetrafluoroethylene	6.2	18.5
Silicone, polydimethyl	7.6	24
Butyl rubber	7.7	27
Polyethylene	7.9	31
Natural rubber	7.9 - 8.3	-
Natural rubber-rosin adhesive	-	36
Polyisoprene, cis	7.9 - 8.3	31
Polybutadiene, cis	8.1 - 8.6	32
Butadiene-styrene rubbers	8.1 - 8.5	-
Polyisobutylene	8.0	-
Polysulfide rubber	9.0 - 9.4	-
Neoprene (chloroprene)	8.2 - 9.4	38
Butadiene-acrylonitrile rubbers	9.4 - 9.5	-
Poly(vinyl acetate)	9.4	-
Poly(methyl methacrylate)	9.3	39
Poly(vinyl chloride)	9.5 - 9.7	39
Urea-formaldehyde resin	9.5 - 12.7	61
Epoxy	9.7 - 10.9	-
Polyamide-epichlorohydrin resin	-	52
Ethyl cellulose	10.3	-
Poly(vinyl chloride-acetate)	10.4	-
Poly(ethylene terephthalate)	10.7	43
Cellulose acetate	10.9	39
Cellulose nitrate	10.6 - 11.5	-
Phenolic resin	11.5	-
Resorcinol adhesives	-	51
Poly(vinylidene chloride)	12.2	40
Nylon 6,6	13.6	43
Polyacrylonitrile	15.4	44
Cellulose, from wood pulp	-	35.5, 42
Cellulose, from cotton liners	-	41.5
Cellulose, regenerated	-	44
Starch	-	39
Casein	-	43
Wool	-	45

6.3.1.3   *Viscosity*

Once the condition of wettability of the adherend surface is settled, the viscosity of the adhesive has to be considered. Low viscosity of the adhesive facilitates the spread of the adhesive, while high viscosity makes it difficult to apply the adhesive homogeneously over the surface. Viscosity decreases with temperature and increases with increasing values of average molecular weight (MW).

A summary of the main variables affecting adhesion is presented in Figure 6.3.

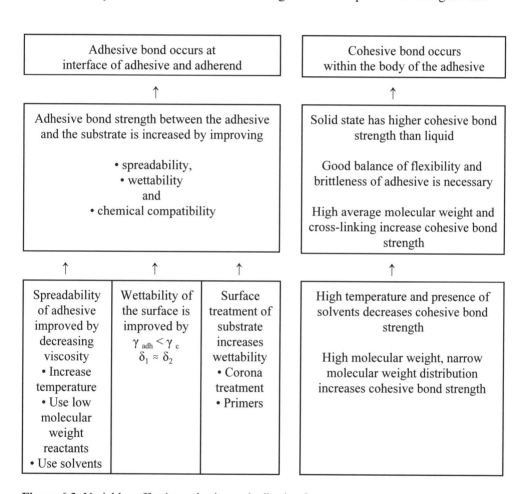

Figure 6.3 Variables affecting cohesive and adhesive forces.

6.3.1.4   Estimation of Adhesive Bond Strength

The adhesive bond strength depends on the ability of the adhesive to wet the adherend surface and is quantitatively determined by the shear strength at the interface. It can be estimated from the following equation:

$$S = \frac{\gamma_1 - \gamma_2 - \gamma_{12}}{d} \qquad (6.4)$$

where γ_1 and γ_2 are the surface tensions of the adhesive and adherend, γ_{12} is the interfacial surface tension, and d is the distance of separation between the molecules at which failure of the adhesive takes place. This corresponds approximately to an intermolecular distance of about 5 Å (5×10^{-8} cm).

Example: Estimate the strength of the adhesive bond produced on bonding PVC with an epoxy adhesive, given the following data:

γ_1 (PVC) = 40 dynes/cm, γ_2 (Epoxy) = 41.7 dynes/cm,
γ_{12} (PVC-Epoxy) = 4.0 dynes/cm; d = 5×10^{-8} cm.

$$S = \frac{\gamma_1 + \gamma_2 - \gamma_{12}}{d} = \frac{40 + 41.7 - 4.0}{5 \times 10^{-8}} = 1.55 \times 10^9 \ \frac{dyn}{cm^2} = 2.24 \times 10^4 \ psi$$

6.3.2   Cohesive Bond Strength

Adhesives are applied in a liquid state to improve the wettability, as mentioned above. In general, the liquid state is obtained by dissolving the adhesive in a solvent (organic solvent or liquid water), by dispersing or emulsifying the adhesive in water to produce a latex, by heating the adhesive, or by applying the adhesive in the form of liquid monomers which later react to form a solid. The adhesive, once applied between the two surfaces to be bonded, solidifies through eliminating the solvent, decreasing the temperature, or allowing time for reaction (curing).

Once it is solidified, the performance of the adhesive depends on its adhesive bond strength, as discussed above, and on its cohesive bond strength. In many applications, the adhesive is selected so that the adhesive bond strength exceeds the cohesive bond strength. In that case, the overall strength of the adhesive joint will be the cohesive bond strength of the adhesive itself, or of the substrates, whichever is less.

Cohesive bond strength depends on both the chemical nature and the physical state of a material. Temperature and the molecular weight of the adhesive are two important factors. Increasing the molecular weight of an adhesive increases its cohesive strength, but also increases its viscosity and decreases wettability.

The cohesive bond strength of an adhesive can be estimated by the following equation:

$$S = \frac{2\gamma}{d} \tag{6.5}$$

where S is the shear stress of the cohesive bond of the adhesive, γ is the surface tension of the adhesive, and d is the distance of separation between the molecules at which failure occurs, approximately 5 Å.

Example: Estimate the cohesive bond strength of a PE based adhesive with a surface tension of 31 dynes/cm and d = 5 x 10^{-8} cm.

$$S = \frac{2 \times 31 \text{ dyn}}{5 \times 10^{-8} \text{ cm}^2} = 1.24 \times 10^9 \frac{\text{dyn}}{\text{cm}^2} = 1.78 \times 10^4 \text{ psi}$$

6.4   Types of Adhesives

Adhesives can be categorized in several ways. One common categorization is reactive, hot melt, solvent-borne, and water-borne. Another categorization breaks adhesives into two groups, natural and synthetic. Additional useful categories of adhesives include pressure-sensitive and cold-seal adhesives. We will use the reactive, hot melt, solvent-borne, and water-borne categories, and then follow those with a discussion of pressure-sensitive and cold seal adhesives. Most natural adhesives are in the water-borne category. Synthetic adhesives appear in all categories.

6.4.1   Reactive Adhesives

Chemically reactive types of adhesives usually involve the polymerization of low molecular weight components that are applied as liquids. These form a polymerized adhesive with good cohesive strength after curing. The adhesive is produced by a condensation reaction, and usually small molecules are eliminated as byproducts. Examples of reactive adhesives include urethanes, cyanoacrylate, polyesters, and epoxies. Uses in packaging include bags, pouches, wraps, and boil-in-bag food pouches

Reactive adhesives typically can be used to bond very dissimilar materials. They are usually flexible, though epoxies are rigid. They provide superior peel strength and ease of application. They are commonly solvent-based, with about 30% solids content, and provide good solvent release. They are also characterized by good stability of color and clarity. In 1999, reactive adhesives accounted for about 10% of adhesive use [1].

6.4.2   Hot Melt Adhesives

Hot melt adhesives are the fastest growing adhesive group for packaging applications. They are 100% solids and do not require any type of solvent.

Hot melts have to be applied hot (above melting temperature) to wet the surface of the substrate. Then on cooling, the molten polymer returns to its solid form, providing good cohesive strength to the bond. They set very quickly after they are applied, do not chemically react with the substrate, and do not generate solvent emissions. The primary disadvantage of hot melt adhesives is poor performance at elevated temperatures.

The most widely used hot melt adhesives are those based on ethylene vinyl acetate. EVA is a very versatile resin and is typically compounded with waxes and tackifying resins, along with stabilizers, antioxidants, and other components if desired. The properties, including melting temperature, of the adhesive depend on the molecular weight of the base polymer and its vinyl acetate content, as well as on the other ingredients in the formulation. Typical VA content ranges from 5% for adhesives intended for non-polar substrates to 30% for those to be used for more polar substrates. EVA adhesives are economical and have very low odor, taste, and toxicity characteristics. In addition to use in hot melts, they are often used in emulsion adhesives.

Other hot melts are based on low molecular weight polyethylene, combined with tackifying agents. These tend to be lower in cost and performance than the EVA-based hot melts, and they are used primarily with paper packaging, such as cartons and multi-wall bags. Atactic PP can also be used as the base for hot melt adhesives. Pressure-sensitive hot melt adhesives for tapes and labels often employ thermoplastic elastomers, consisting of block copolymers of styrene and butadiene or isoprene.

In 1999, hot melt adhesives accounted for about 20% of adhesive use [1].

6.4.3   Solvent-Borne Adhesives

Solvent-borne adhesives consist of a base polymer dissolved in an organic solvent, with other ingredients as needed. The cohesive strength of the adhesive develops as the solvent evaporates. The typical solids content is about 30%. These adhesives have declined greatly in use, due primarily to regulations limiting emissions of volatile organic compounds, as well as to the increasing availability and performance of water-borne and hot melt systems.

Solvent-borne systems still in use include polyurethanes for lamination of plastic films and EVA systems for some heat-seal applications such as lidding stock for plastic food containers.

In 1999, solvent-borne adhesives accounted for about 9% of adhesive use [1].

6.4.4   Water-Borne Adhesives

Water-borne adhesives consist of a base polymer, and other ingredients, suspended in water in the form of an emulsion. Since water is the only volatile ingredient, these adhesives do not create the solvent emission problems characteristic of solvent-borne adhesives, and consequently have grown considerably in use during the past two decades or so. The cohesive strength of the adhesive develops as the water evaporates.

Water-borne adhesives can be grouped in two categories: natural and synthetic. Natural waterborne adhesives include starch, animal glue, casein, and natural rubber. Nearly 1.8 million tonnes (4 billion lb), about one third of all adhesives produced in 1996, were natural adhesives, primarily starch and dextrin based. The remaining two-thirds were based on synthetic polymers [2]. Within the packaging industry, 58% of the adhesives used are starch and dextrin-based, although the trend is towards increasing reliance on synthetic polymers, fueled by the growth of plastics in packaging. Natural adhesives do not generally adhere as well to plastics as to paper.

Starch, based on polysaccharides from plants, is the most widely used base for natural adhesives. The starch, most often corn starch, can be used in its original form, but is more often hydrolyzed to smaller chain segments, yielding modified starches or dextrins. Additives are used to modify properties. The largest use of starch-based adhesives is in the manufacture of corrugated board for shipping cases. These adhesives have good adhesion to paper, are low cost, easy to handle, and easy to clean up. Disadvantages include slow rate of bond formation, limited applicability to plastics, and poor water resistance.

Animal glues are made from animal skins and bones. They have largely been replaced by synthetic adhesives, but they still find limited application in packaging as the re-moistening adhesive on gummed tape used for box sealing, and in forming rigid set-up boxes.

Casein adhesives are based on milk protein. They are used for labeling refillable beer bottles because they resist cold water immersion but are readily removed in an alkaline wash when the bottles are returned. Another use is as an ingredient in adhesives for laminating aluminum foil to paper.

Natural rubber adhesive, or latex, is used in self-seal applications, where it makes use of its unique ability to bond only to itself with pressure. Applications include self-seal envelopes, press-to-seal cases, and cold-seal candy wraps. Natural rubber latex is also used in adhesives for lamination of polyethylene film to paper, such as in multiwall bags.

Synthetic waterborne adhesives are the most widely used types of adhesive in packaging. Almost all consist of an emulsion of a base polymer in water, with additional ingredients that may include protective colloids, plasticizers, fillers, solvents, defoamers, and preservatives.

The most common type of synthetic water-borne adhesives are based on vinyl acetate, and include the homopolymer, *polyvinyl acetate* and copolymers with ethylene, *ethylene vinyl acetate*, the popular white glue. Copolymers with other monomers, such as acrylics, are increasing in use, particularly for adhesion to plastics.

Other synthetic emulsions include *acrylics,* which are widely used in pressure-sensitive adhesives for labels, *polyurethanes*, and *synthetic rubbers*. *Sodium silicate* was once widely used in paper packaging, but its only remaining major use is in tube winding, especially for large drums or cores.

Water-borne adhesives are by far the largest category of adhesives, accounting for about 59% of adhesives used in 1999 [1].

6.4.5   Pressure Sensitive and Remoistenable Adhesives

For labels and tapes, where the adhesive is commonly applied to one of the adherends and then much later used to affix that substrate to another, such as a container, it is common to classify labels as either remoistenable or pressure-sensitive.

Remoistenable adhesives are applied as water-borne adhesives and then dried. When dry, they have little or no tack. To apply the tape or label to the box, bottle, or other package structure, water must be applied to reactivate the adhesive. Then the full strength of the bond will develop as the water is removed. Such adhesives are commonly used for adhering paper to paper or to other materials.

Pressure-sensitive adhesives have been available for many years. Scotch tape was one of the early uses, as was "adhesive tape" for medical uses. Use of pressure-sensitive labels is considerably newer, and has increased dramatically in the past decade or so. While growth in labels remains the most rapid, pressure-sensitive adhesives are also appearing in other packaging applications, such as providing a reclosable opening in flexible packages. Growth in non-packaging applications has been rapid, as well. For example, full-scale production of pressure-sensitive postage stamps began only in 1992, and by 1997, over 80% of the 40 billion postage stamps sold in the U.S. used pressure-sensitive adhesives [2]. Much of this growth is fueled by declining costs of pressure-sensitive adhesives, as formulating technologies and manufacturing processes are improved.

Pressure sensitive adhesives are applied from a liquid base, usually as hot-melts, and retain tack when they are cool and dry. To keep pressure-sensitive labels and other components from sticking to something before they are supposed to, they generally remain on a carrier web until they are ready to be applied. The carrier web is coated with a material that provides easy release, usually a silicone.

A variety of different pressure-sensitive formulations are available, many of which have already been mentioned. Some formulations provide permanent adhesion, while others may be repositioned. The Post-it Notes produced by 3M use a unique acrylic emulsion, and earned its inventor the American Chemical Society Award for Creative Invention in 1998 [2]. Common formulations include rubber-based systems (both natural and synthetic rubber), and acrylic systems. Substrates used in labels (or other components) and the packages to which they are applied include paper, cloth, rubber, metal foil, and, most often, plastics.

6.4.6   Cold-Seal Adhesives

Cold-seal adhesives are members of a general category sometimes termed co-adhesives - substances that have a great tendency to stick to themselves, but often not to much else. Cold-seal adhesives are typically based on natural rubber, which has been applied in a latex (suspension in water) form. A major application is in plastic packaging for chocolate candy. The seal between the two parts of the wrap can then be activated with pressure, as an alternative to heat-sealing, which is problematic with the low-melting temperature candy.

6.4.7   UV and E-Beam Curing

One of the newer developments in adhesives is the growing use of ultraviolet light or electron beam radiation to cure adhesives. Adhesives designed for UV or E-beam curing are usually pressure sensitive or hot-melt systems based on acrylates, functional rubbers, or epoxidized rubbers, and use special UV or EB lamps to provide the cure. Use currently is quite small, but may increase substantially in the future. These systems can provide greatly improved heat resistance compared to hot melts, and avoid the solvent emission problems of some of the solvent-based systems with which they compete.

6.5   Application of Adhesives

Adhesives can be applied to adherends in a variety of ways. In some cases, a system of rollers is used to transfer metered amounts of the adhesive to the substrate. In these wheel and roll dispensers, wheels or rollers containing a desired pattern rotate in a reservoir of adhesive. The adhesive is picked up in amounts and configurations determined by the patterns, and then transferred to the substrate directly, or with the intervention of an additional roller. More often, applicator heads deliver the adhesive directly to the substrate in the form of beads, sprays, or droplets.

Systems for applying water-borne adhesives, also termed cold-glue systems, generally consist of a tank of adhesive, a system of hoses to deliver the adhesive to the applicator head, and the head, or gun, itself. Multiple-gun configurations are used as well as single guns. The tips may make close contact with the surface, for bead or ribbon dispensing patterns, or may be farther away, in order to deliver a mistlike pattern.

For applying hot-melt adhesives, it is necessary, of course, to provide for heating of the adhesive to melt it. The most common method is to use a tank melter, which consists, in essence, of an open heating pot with a lid that is used for loading in the adhesive. The tank is generally electrically heated. Adhesive first melts along the wall of the tank, then currents from the pumping action help in transferring the heat to the rest of the adhesive in the tank. An alternative to tank melters is grid melters, in which the solid adhesive is above the grid, and melted adhesive flows through the grid to a reservoir. Grid melters, because they have much larger heating surfaces, and because greater temperature differences can be accommodated, provide a faster melting rate than tank melters. They also produce more uniform temperatures within the adhesive melt, which can help to minimize thermal degradation. In these systems, heat is also provided to the reservoir, to maintain the adhesive at the desired temperature.

Once the adhesive is melted, in either the tank or the grid melter, it must be conveyed to the dispensing apparatus. A pumping unit is necessary to provide the required pressure. In many systems, the adhesive travels from the pump through hoses to the applicator. The hoses are generally insulated and heated for temperature control, as well, and must be able to withstand high pressure. The most common applicator in high-speed hot melt processes is a pressure-activated automatic gun; manual, or handguns, are also available. The gun contains heating elements for temperature control. The pattern and amount of adhesive that is deposited is controlled by the extrusion nozzle attached to the gun. Some systems provide for filtering of the melt either between the hose and the gun, or within the gun itself. There are also zero-cavity guns that do not need a separate nozzle, controlling the flow using a tapered needle and a matched nozzle seat. Movement of the needle, which can be adjusted as needed, controls the flow. These guns are very good at preventing nozzle clogging and drool (adhesive coming out of the nozzle after flow is stopped). In addition to beads or a spray of adhesive, nozzles are also available to deliver a continuous film of adhesive. These systems are called slot nozzles, coating heads, or web-extrusion guns.

One melting device and pump may serve several dispensing devices. For automatic applications, controls must be available which detect the presence of a substrate, and insure that the adhesive is deposited in the desired location. Both mechanical and optical sensors can be used for these purposes.

6.6   Adhesive Terminology

Several terms are used to describe important aspects of the formulation or functioning of adhesives, and of equipment for doing adhesive bonding. Some of these terms, and their definitions, are the following:

Solids content refers to the amount of the adhesive that remains after solvents or carrier liquids are removed. It is usually reported as percent solids, defined as solids/(initial weight) x 100%.

An *emulsion* is a stable suspension of adhesive solids in a fluid; the fluid is usually water.

A *solution* is an adhesive in which the solids have been dissolved in the fluid; the fluid is usually an organic liquid, not water.

The *setting time* is the time required for an adhesive to form a bond sufficient to allow the bonded substrate to undergo further handling. This initial bond is generally weaker than the final bond that will develop over time.

The *open time* refers to the elapsed time between application of the adhesive to one or both substrates, and the time the substrates are brought together to form the adhesive joint. Thus it is a function of the machinery and operating conditions, not of the adhesive or substrate.

The *range* is the time over which an adhesive, once applied to a substrate, retains a useful degree of tack. It is a function of the adhesive formulation. For an adhesive to form a useful bond between two substrates, the range of the adhesive must be at least as long as the open time.

Tack is the stickiness of the adhesive, its ability to form an initial bond of measurable strength immediately after the adhesive and adherend are brought into intimate contact.

Blocking, as defined in Chapter 6, is undesirable adhesion between adjacent layers of a material, such as that which can occur under moderate pressure during storage, causing them to stick together.

Shortness of an adhesive is defined as the lack of stringing, cobwebbing, and formation of threads during the application of adhesives to a substrate.

The *pot life* refers to the stability of the adhesive, and is related to its acceptable residence time in the application machinery.

6.7   Adhesive Additives

Adhesives often have very complex formulations, which are regarded as highly proprietary. Additives used in adhesion formulations include:

Tackifiers can be used to improve the tack properties of thermoplastic polymers, the compatibility between the polymer and substrate, wetting power, or the strength of the permanent bond. Tack is the bond strength that is formed immediately when a given material comes in contact with another surface, or "instantaneous adhesion." Types of tackifiers include natural resins, phenolic resins, and polyterpenes.

Solvents are used to modified the viscosity of an adhesive. Common solvents include methyl-ethyl ketone (MEK), methyl isobutyl ketone (MIBK), and ethylene chloride.

Fillers are solid inclusions, often mineral-based. They can be used to reduce cost, improve cohesive strength, reduce contraction at the adhesive line as the adhesive solidifies, and to help in controlling viscosity. Commonly used fillers, with their major contribution to adhesive properties, are listed in Table 6.3.

Primers, coatings applied to a surface prior to the application of an adhesive, are not part of the adhesive formulation, but can be used to improve bonding of the adhesive.

Table 6.3 Commonly Used Fillers in Adhesives

Type	Contribution to adhesive properties
Carbon black	Excellent reinforcement
Kaolin clay, sand	Inexpensive
Titanium dioxide T_iO_2	White color
Zinc oxide	Good thermal conductivity
Chalk, $CaCO_3$, talc	Inexpensive, white color
$BaSO_4$	Acid resistance
Glass fiber	Strength
Hollow microspheres	Lower density
Powdered copper	Electrical conductivity

6.8   Heat Sealing

Heat sealing is the process by which two structures containing at least one thermoplastic layer are sealed by the action of heat and pressure. This process can be applied to flexible, semi-rigid, and in some cases rigid packaging structures. The following discussion considers flexible structures, but the principles of heat sealing can be extended to other cases. Flexible structures can be classified in two groups, according to the type of material employed in their construction: supported and unsupported structures. Supported structures consist of laminations containing one or more non-thermoplastic layers (such as paper or foil), bonded to thermoplastic layers, at least one of which is used for sealing. Unsupported structures consist of one or more thermoplastic layers and do not contain a non-thermoplastic layer.

When sealing a flexible structure to make a package, the heat sealing layer is located in the interface, typically contacting another heat sealing layer. When heat and pressure are applied to the external surface to make the seal, the heat is transmitted by conduction or radiation to the packaging material, and then is transmitted through the material by conduction to the sealing layers (Fig. 6.4). Conduction is used more frequently than radiation as the heat input. The heat at the interface must be sufficient to melt the interface materials in order to produce a seal. The external pressure is needed to bring the thermoplastic sealing layers very close to each other, around a distance of 5 Å. A good seal is obtained when enough molecular entanglement has taken place within the polymer chains from the two thermoplastic heat sealing layers to destroy the interface and produce a homogenous layer that remains homogeneous after cooling. *Dwell time* is the time during which the external pressure holds the two structures together to allow molecular entanglement to take place. The pressure is released at the end of the dwell time. Often, the heat seal materials are still molten at this point, and the molecular interactions in the heat seal polymer(s) must be able to keep the sealing surface together against the forces that may act to pull them apart. This strength during the cooling phase is called hot tack.

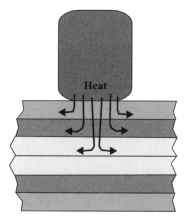

Figure 6.4 Heat conduction in heat sealing.

6.8.1 Sealing Methods

The method for heat sealing a particular structure depends on the type and form of the structures being sealed, as well as the type of package and product. The following are the most important sealing methods used in packaging:

6.8.1.1   Bar or Thermal Sealing

Thermal sealing uses heated bars to press together the materials to be sealed, with heat from the bars conducted through the materials to the interface, melting the heat seal layers and fusing them together (Fig. 6.5). When sufficient time has elapsed, the bars release and the material is moved out of the seal area. At this point, the materials are still hot, and the seal does not have its full strength, but the materials must be able to adhere to each other well enough to insure the integrity of the seal. This ability of the materials to remain together while they are still hot is known as hot tack. The full strength of the seal develops as it cools to ambient temperatures. Proper seal formation requires the correct combination of heat, dwell time (the time the material is held between the sealing bars), and pressure. Too little of any of these will prevent an adequate seal from forming. On the other hand, excessive heat, time, or pressure will result in too much flow in the heat seal layers, weakening the material.

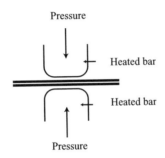

Figure 6.5 Bar sealing, or thermal sealing (reprinted with permission from [3]).

The edges of the heat-seal bars are often rounded so that they do not puncture the packaging material. Often the contact surface of one of the bars contains a resilient material to aid in achieving uniform pressure in the seal area,. Bar sealing is the most commonly used method of heat-sealing packaging materials, and is often used in form-fill-seal operations.

A variation on bar sealing uses only one heated bar, with the other bar not heated, resulting in heat conduction occurring only in one direction. Another variation uses heated rollers instead of bars, with the materials sealed as they pass between the rollers. In this type of system, preheating, slow travel through the rollers, or both, are generally required due to the very short contact time between the rollers. A third variation uses shaped upper bars for sealing lids on cups and trays.

6.8.1.2   Impulse Sealing

Impulse sealing (Fig. 6.6) is another common heat-seal method. Impulse sealing uses two jaws, like bar sealing, but instead of remaining hot, the bars are heated intermittently by an impulse (less than one second) of electric current passed through a nichrome wire ribbon contained in one or both jaws. The jaws apply pressure to the materials both before and after the current flow. The current causes the ribbon to heat, and this heat is conducted to the materials being sealed. After the pulse of current is passed through the wire ribbon, the materials remain between the jaws for a set length of time, and begin to cool. Thus, impulse sealing provides for cooling while the materials are held together under pressure. This method allows materials with a low degree of hot tack to be successfully sealed, as well as permitting sealing of materials that are too weak at the sealing temperature to be moved without support. The sealing jaws can be water-cooled for faster cooling of the materials being sealed. Shaped impulse seals are used for sealing lids on cups and trays.

Impulse sealing produces a narrower seal than bar sealers, resulting in a better looking but weaker seal. Maintenance requirements tend to be heavy, since the nichrome wires often burn out and require replacement. A fluoropolymer tape on the jaws, covering the nichrome wire, is often used to keep the plastic from sticking to the jaws, and may also require frequent replacement.

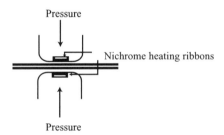

Figure 6.6 Impulse sealing (reprinted with permission from [3]).

6.8.1.3 Band Sealing

Band sealing, illustrated in Figure 6.7, like impulse sealing, provides a cooling phase under pressure. This high speed sealing system uses two moving bands to provide pressure and convey the materials past first a heating station and then a cooling station. The primary disadvantage of this method is the tendency for wrinkles in the finished seals. Preformed pouches that are filled with product are often sealed using this method.

6.8.1.4 Hot Wire or Hot Knife Sealing

This method, as its name describes, uses a hot wire or knife to simultaneously seal and cut apart plastic films. The wire or knife causes the substrates to fuse as it is pushed through, cutting them off from the webstock. The seal produced is very narrow and often nearly invisible. It is also relatively weak, and does not provide a sufficient barrier to microorganisms to be used when a hermetic seal is required. However, it is very economical due to its high speed, and is an excellent choice for relatively undemanding packaging applications with materials that seal readily, such as LDPE bags used in supermarket produce sections.

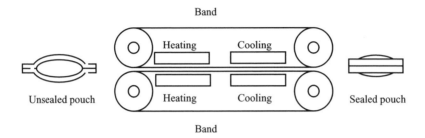

Figure 6.7 Band sealing (reprinted with permission from [3]).

6.8.1.5   Ultrasonic Sealing

In ultrasonic sealing, two surfaces are rubbed together rapidly. The resulting friction generates heat at the interface, melting the surfaces of the substrates and producing a seal. Since the heat is generated only in the seal area, ultrasonic sealing is particularly useful for thick materials where conduction is inefficient. It is also useful when exposure to heat for a sufficient time to conduct heat to the seal can damage the substrates, such as in sealing highly oriented materials, which can lose their orientation and shrink when heated.

6.8.1.6   Friction Sealing

Friction sealing, often called spin welding, like ultrasonic sealing uses friction to produce heat. It is most often used for assembling two halves of a rigid or semirigid plastic object, such as a deodorant roller or a container, or sometimes for sealing caps to bottles. The two halves are most often circular in cross section, and one is rotated rapidly while the other is held in place. The halves are designed to fit together only with some interference, so there is considerable friction, generating heat that welds them together. The sealing mechanism usually has a sensor that measures the amount of resistance to rotation, and the object is released when the resistance reaches the level determined. A variation of this method allows sealing of non-circular materials, using an oscillating motion rather than rotation.

6.8.1.7 Hot Gas Sealing and Contact Sealing

In hot gas sealing, the sealing surfaces are exposed to a gas flame or to hot air, also to avoid the need to conduct heat through the materials being sealed. The heat melts the sealing surfaces, and then the two materials are pressed together between cooled jaws. Contact sealing is similar to hot gas sealing, except that the sealing surfaces are touched to a heated plate.

6.8.1.8 Radiant Sealing

In radiant sealing, heat is transferred to the sealing surfaces primarily by radiation rather than by conduction or convection . It is most commonly used for materials which distort excessively under pressure. The most common application is for spun-bonded high density polyethylene (e.g. Tyvek™ from DuPont) used for medical device packaging. Other applications include sealing highly oriented materials, sealing uncoated PET, or for producing shaped seals. Radiant heating is often used in thermoforming, as well.

6.8.1.9 Dielectric Sealing

In dielectric sealing, an oscillating high frequency electrical field is used to seal polar materials. Polar molecules exposed to the field attempt to line up with the electrical charge, so they oscillate as they are attracted first in one direction and then in the other. This generates heat, melting the materials and producing a seal. Dielectric sealing is ideal for heavy PVC materials, especially for textured PVC, as it experiences considerable distortion with methods that rely on conduction. Non-polar materials, such as polyolefins, are not affected by the field, so cannot be sealed in this manner.

6.8.1.10 Magnetic Sealing

Magnetic sealing relies on a similar idea, with an oscillating magnetic field that causes magnetic iron compounds to attempt to line up with it, producing heat as the field, and consequently the iron particles, oscillate. Non-magnetic materials are not affected. Since magnetic iron compounds are not normally found in plastic packaging materials, this method relies on the use of special gaskets or coatings containing magnetic iron. Magnetic sealing is rarely used, but does have some applications with cap liners or lids.

6.8.1.11 Induction Sealing

Induction sealing, as shown in Figure 6.8, is also a method to generate heat near the sealing surface. It is often used to apply tamper-indicating inner seals on plastic bottles and jars. It uses an alternating magnetic field to induce an electric current in any metal within the field, most often a layer of aluminum foil. The electrical current heats the foil and is conducted to a neighboring heat-seal layer, resulting in the sealing of the liner to the container. The original inner seal structure for induction sealing, which is still sometimes used today, consisted of a heat seal coated foil attached with wax to a paperboard backing glued into the closure. When the assembly is placed in an alternating magnetic field and current induced in the foil, in addition to sealing the liner to the bottle rim, the heat melts the wax layer, releasing the foil liner from the cap, and thus allowing the consumer to remove the cap to open the container. This structure has now generally been replaced by other designs. These inner seals provide an excellent barrier against gain or loss of moisture, oxygen, and other components, as well as providing a useful tamper-indicating feature.

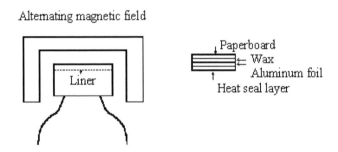

Figure 6.8 Induction sealing (reprinted with permission from [3]).

6.8.1.12 Solvent Sealing

Solvent sealing dispenses with the use of any heat in forming the seal. If a plastic material is soluble, addition of solvent at the interface produces a solution that permits intermingling of the materials across the interface. Removal of the solvent, typically by migration through the bulk material followed by evaporation, results in bonding of the substrates. Most plastics are not at all soluble in water, and only sparingly soluble in other solvents, so appropriate selection of an organic solvent is required, along with sufficient residence time to form an initial solution at the interfaces. The surfaces are then pressed together to achieve blending at the interface. At this point the bond has very little strength, so the materials must be held together under pressure long enough for much of the solvent to diffuse away from the seal area into the bulk of the material. The interface will then be more solid, and when the seal has enough tack to hold together, the pressure is released and the material moved out of the jaws. The seal strength increases as the residual solvent continues to migrate away from the interface. Eventually most of the solvent diffuses out of the plastic and escapes into the air. Residual solvent in the material can alter its properties, and solvent emissions are a concern, so this sealing method is generally avoided if other sealing methods can be used effectively. For materials such as polyvinyl alcohol which are water-soluble, solvent sealing with water can be quite attractive.

6.8.2 Heat Conduction in Multilayer Flexible Materials

Frequently in heat sealing, the flexible packages being sealed are formed from more than one material. In this case, the heat conduction across the material is not uniform. Each material has its own heat capacity and thermal conductivity, as well as its individual thickness. As shown in Fig. 6.4, heat flows from the heater bar (or other heat source) through the various layers of the structure. Paper and polymers are excellent insulators, so the heat transfer through such materials is slow. Dwell times in heat sealing machinery are typically very short, perhaps one-quarter second. To transmit enough heat through several layers of insulators, a high temperature on the heat seal bar is necessary. On the other hand, if the middle layer of a structure were aluminum foil, it would transmit heat easily, both through itself and in the lateral direction. When aluminum layers are too thick, they can cause seals to weaken during the cooling phase because of the heat that they transmit through the insulating layers.

The critical temperatures for sealing are those at the sealing interface, since this is where the heat seal is formed. Temperatures measured on the outside of the material may not give useful information about the conditions at the interface. Measurements of interface temperatures can be made, however, by sealing a very fine thermocouple into the seal interface and monitoring the response. This is not often done on a packaging line, except when trouble-shooting very difficult problems. Normally, a heat seal curve for the material is determined in the laboratory by sealing at various temperatures and measuring the strength of the seal on a tensile tester. The type of data obtained is shown in Figure 6.9. As can be observed, the seal strength first increases as the heater temperature is raised, and then decreases. At the lower temperatures, insufficient melting and intermingling of the materials is occurring. At the higher temperatures, the heat seal layer is flowing out of the heat seal area, thus decreasing the strength of the seal.

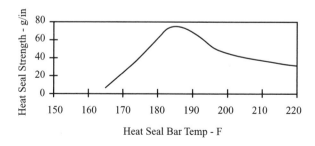

Figure 6.9 Heat seal strength as a function of temperature.

6.8.3   Hot Tack

Hot tack, as mentioned earlier, is the ability of the seal to withstand strains while it is molten and soft, and is an important aspect of the performance of a heat seal layer. Until fairly recently, there was no standard ASTM or TAPPI procedure for measuring hot tack of heat seals. In 1998, ASTM published standard F1921, Standard Test Methods for Hot Seal Strength (Hot Tack) of Thermoplastic Polymers and Blends Comprising the Sealing Surfaces of Flexible Webs, which was revised in 2004. A related standard is D3706, Standard Test Method for Hot Tack of Wax Polymer Blends by the Flat Spring Test. Many companies now manufacture equipment designed to measure hot tack of heat seals in accordance with the ASTM F1921 procedure.

An example of such data is plotted in Figure 6.10, along with the heat seal data for the same polymer. The hot tack strength is very low compared to the heat seal strength, since the full seal strength develops only after the material is completely solidified and cooled to ambient conditions.

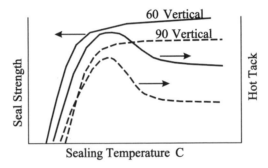

Figure 6.10 Sample heat seal and hot tack data.

6.8.4   Heat Seal Jaws

On most modern horizontal and vertical form-fill-seal equipment (see Sec. 9.3), seal jaws are not flat bars because these give weaker seal strengths than jaws designed with serrations. If one looks at flexible packages in the marketplace, one will find vertical and horizontal serrations as shown in Figure 6.11.

Vertical serration gives the strongest heat seals, but can also suffer from small channel leaks caused by folds that occur in the material in the heat seal area. These are often referred to as serum leakers. If horizontal serrations are used, these cut off the serum leakers, but because the serrations run parallel to the peeling forces, the seals are weaker.

The serrations can be cut at various included angles, and have various numbers of teeth per inch, and are usually truncated as shown in Figure 6.11. This approach minimizes the cutting of the structure during heat sealing. There are many different serrated seal jaw designs available from equipment manufacturers and film suppliers. A thorough discussion is beyond the scope of this book.

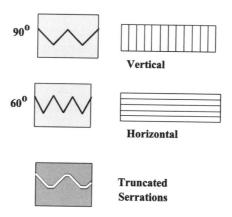

Figure 6.11 Serrated seal jaws.

6.8.5 Heat Seal Failure Modes

Once the seal is made and the package transported, the end-user needs to be able to open the package. Some heat seals are made to be *lock-up* seals, where the package must be destroyed or cut with scissors to open. Many consumer food packages are supposed to have easy-open, or peel seals. There are two common ways to achieve peel seals.

One method is to add a second polymer to the heat seal layer that melts at a higher temperature than the sealant; it should also melt at a higher temperature than is expected to be used on the packaging equipment for sealing. If this polymer is dispersed in the heat seal polymer in tiny domains, it acts to disrupt the continuity of the heat seal and weaken the seal strength to the point where the seals can be peeled apart easily. A second method of achieving a peel seal is to have the sealant break away from the next layer in the package and fail under tension.

Drawings of examples of these types of packages are shown in Figure 6.12. In this drawing, the EVA blend fails due to cohesive failure within the sealing layer, the first type of mechanism for achieving peelable seals. In the second sketch, the Surlyn™ heat seal layer breaks away from the HDPE layer and then fails across the Surlyn layer. This type of peel seal is used, but it has its problems. If the inter-layer adhesion is too much weaker than the strength of the Surlyn film layer, one ends up with a bag in a bag, which does not make the consumer happy.

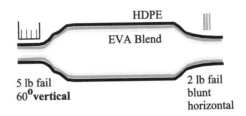

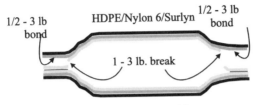

Figure 6.12 Seal failure mechanism examples in cereal bags.

6.8.6   Evaluation of Seals in Flexible Packaging Materials

Tensile testing is the most common method for evaluating seal strength. Both the strength value and the location and type of failure are important. Usually the goal is to produce a seal that is stronger than the packaging materials being used. In tensile testing, this results in rupture of the substrate rather than peeling of the seal. Because the sealing process usually produces some flow and weakening of the substrate, failure most often occurs immediately adjacent to the seal. If a peelable seal is desired, either the adhesive forces within the seal or the cohesive strength of the seal layer should be weaker than the cohesive strength of the substrates, resulting in peeling rather than substrate rupture.

ASTM publishes a variety of standard adhesion tests, including lap shear, peel, cleavage, creep, fatigue, and impact, in addition to tensile tests. Examining the birefringence patterns produced by polarized light as it passes through a seal can give a visual guide to seal consistency, revealing gaps and stress areas.

Wrinkles in seals are a significant problem when hermetic seals are needed. Small wrinkles often do not significantly affect the strength of seals, but can affect their ability to keep out microorganisms. The most effective way to prevent wrinkles is to exert tension in two perpendicular directions on the material during the sealing process. Contaminants in the seal area, such as grease or particulates, can significantly reduce seal strength and integrity. Ionomers are often used for packaging products that are prone to these problems, since they are excellent at producing good seals through contaminants.

References

1. McCoy, Michael, "Adhesives: Sticking to Growth," *C&EN*, Vol. 78, May 29, 2000, pp. 21-32.
2. Morse, Paige M., "Adhesives," *C&EN*, Vol. 76, April 20, 1998, pp. 21-32.
3. Selke, Susan E., *Understanding Plastics Packaging Technology*, Carl Hanser Verlag: Munich, 1997.

Study Questions

1. Why do adhesives generally contain polar structures?

2. Give an example of a situation in which cohesive failure of a structure is desired. Give an example where adhesive failure is desired. Explain why different types of failure are wanted in these situations.

3. Using Table 5.1 and 5.2, determine whether water and toluene will wet polyethylene and polyacrylonitrile. Explain why wetting the surface is important for proper functioning of an adhesive.

4. Use of organic solvent borne adhesives has declined considerably in recent years, while use of hot melt and water borne adhesives has increased. Why?

5. Compare and contrast bar sealing and impulse sealing.

6. What type of sealing can be used for PVC but not for HDPE?

7. Why is radiant sealing, rather than bar or impulse sealing, used for spunbonded polyolefin?

8. What is a hot melt adhesive?

9. What are adhesive strength and adhesive bonds? What are cohesive strength and cohesive bonds?

10. What methods can be used to increase the wettability of an adhesive on a particular surface?

11. How can you use temperature to increase or decrease cohesive bond strength?

12. How do you think dirt or dust will affect a plastic surface ready to be glued? Explain.

13.   Discuss the use of solvents versus temperature in improving the wettability of a substrate.

7   Extrusion, Film and Sheet

7.1   Extrusion and Extruders

In nearly all applications of plastics in packaging, the first step is to convert the solid plastic, usually in pellet form, into a melt. This melt can then be shaped using heat and pressure into a useful form. The equipment used to do this is an extruder. It is used for film and sheet, and it is part of a blow molder for bottles, and of an injection molding machine for injection molded or injection blow molded packaging. The extruders used in all of these applications work in a similar manner, but they deliver the melt to the shaping operation differently. We will explore these differences as we proceed.

The purpose of an extruder is to use heat, pressure, and shear to transform the solid plastic into a uniform melt, for delivery to the next stage of processing. This frequently involves mixing in additives such as color concentrates, blending resins together, and incorporating regrind. Regrind is the granulated scrap from the conversion process. The final melt must be uniform in temperature and in composition. Because single screw extruders are often not very good mixers, an additional mixing device may be needed. The pressure of the melted viscous polymer as it exits the extruder must be high enough to force it through a die to produce a desired shape, or to force it into a mold chamber.

The extruder accomplishes all this by using a *barrel*, a hollow tube, containing a *screw* with helical channels. A simplified extruder diagram is shown in Figure 7.1, and details of a screw are shown in Figure 7.2. The screw is generally divided into three sections: (1) the solids conveying section, (2) the compression or melting section, and (3) the metering or pumping section. The standard single screw extruder has a right-hand helix on the screw, and the screw rotates in the counterclockwise direction. If one thinks of a standard wood screw, which is also right-handed, being turned counterclockwise while it is held stationary in space, the wood moves toward the screw tip. The same action occurs in the extruder. The basic screw design has only a single flight, but other designs have double flights along part or all of the screw length. Other important components are the hopper, which feeds the plastic or other components into the extruder through the feed port, and the die or nozzle, through which the melted plastic exits the extruder.

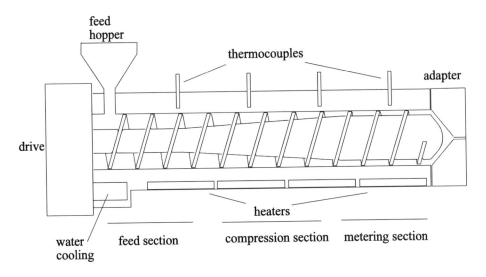

Figure 7.1 Extruder, change in screw diameter exaggerated.

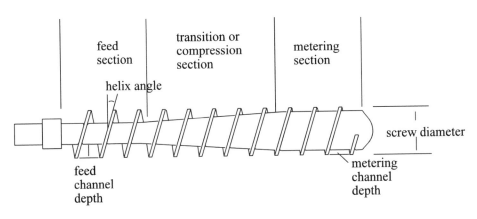

Figure 7.2 Gradual transition extruder screw.

7.1.1   Hopper and Feed Port

The pellets of plastic are first transferred from storage to the hopper on the extruder, which serves as a reservoir to feed the solid particles into the barrel. The hopper can be a simple one like the one shown in Figure 7.1. Most extrusion operations are automated at least to the point of having a vacuum hopper loader to bring material to the hopper. Specialized equipment is available to mix in color and other additives, such as antioxidants and slip agents. Hoppers are usually circular in cross section, to avoid stagnant areas in the flow path. The diameter gradually decreases as it nears the feed port. A too-rapid decrease in diameter can cause bridging, where a compressed plug of material forms that is strong enough to support the material above it, cutting off flow into the extruder.

Gravity is usually the driving force for the flow of the plastic into the extruder. However, hoppers can also be equipped with augers to help in forcing plastic material into the throat of the extruder. This is particularly important for systems running regrind, because the regrind can cause the bulk density of the feed to vary. Hoppers can also be equipped with stirrers, which may even include blades to wipe the material off the hopper walls. The resin passes from the hopper through the feed port into the extruder barrel. The feed section of the screw must have a constant rate of incoming material if it is to perform uniformly.

Some extruders are designed to permit feeding at more than one location of the extruder, and therefore contain multiple feed ports, each with its own hopper. This permits, for example, fillers to be introduced into the extruder after the plastic resin has already been melted.

Cooling is often provided at the hopper and feed port, to prevent premature melting of the plastic, which can create blockages which interfere with a consistent rate of flow of the plastic into the extruder.

7.1.2   Feed Section

The major purpose of the feed section is to convey the plastic pellets forward through the extruder, once they have been fed into the barrel at one end through the hopper and feed port. During the conveying of the pellets through the feed section of the extruder, unevenness in the feed rate is smoothed out, and the solid bed is compacted and its density increased. This solids feeding action is a very important step in the extrusion process. If the feed rate is too low, the screw is "starved", and the extruder output is lower than expected. If the feed rate is too high, melting farther down the extruder may not be complete, and the output may contain unmelted particles.

While the feed and other sections of the screw are defined by the screw geometry, the actual locations within the extruder where melting begins and ends are also affected by the resin and processing conditions being used.

Within the feed section, as well as elsewhere in the extruder, friction between the plastic and the barrel is responsible for conveying the plastic forward. Without this friction, the plastic would simply rotate with the screw, staying in place rather than moving forward. Grooving of the feed section of the barrel is sometimes used to increase the frictional force, thus increasing the conveying force. The grooves typically run in the axial direction for a distance of several screw diameters. Spiral grooves can also be used. It is important to cool the grooved section to prevent melt from collecting in the grooves and undergoing thermal degradation. The major advantages of grooved feed extruders are increased output, reduction in the dependence of the output on pressure, resulting in improved stability, and improved ability to extrude some very high molecular weight plastics, such as high molecular weight polyethylene. Disadvantages include higher pressures, higher energy requirements, and higher wear rates. Special screw designs are also required. In these extruders, the groove depth typically decreases with distance from the feed port, and most of the compression is provided by this decreasing depth, rather than by change in the depth of the screw flights, which therefore remains nearly constant. The screws often contain a decompression section after the melting, compression, and metering zones, to provide lower melt temperatures. Mixing sections are needed to provide sufficient uniformity of the melt. In some cases, grooving of the feed throat can provide improved stability and other benefits without the complexities of grooving the feed zone itself.

The conveying action of the screw can also be improved by reducing the friction of the screw itself, through modifying the screw design, temperature, or material. Single flighted rather than multiple flighted screws and a large helix angle both act to reduce screw friction. In some cases, internal screw heating, by coring the screw and circulating heating oil, or by using an internal cartridge heater, is employed. This, of course, complicates the overall extruder design. The surface of the screw may be modified using coatings or surface treatment. Low friction coatings reduce the tendency of the plastic to build up on the screw, and thus simplify cleaning as well as improving conveying.

7.1.3   Compression Section

In the compression section of the screw, the main function is melting of the plastic. The depth of the screw channels decreases, building up more pressure on the plastic and forcing the solid against the barrel wall. External heating, as well as internal heating from friction between the plastic and the barrel and especially from the shearing of the plastic itself, causes the plastic to soften and melt. For small extruders, with barrels less than 5 cm (2 in) in diameter, most of the heat usually comes from the external heaters. For larger extruders, most of the heat is provided by shear. In these large extruders, the downstream sections of the barrel may actually have to remove heat from the melt with water or air cooling jackets, especially if they are processing heat-sensitive materials such as PVC.

In most cases, single screw extruders are designed so that the solid particles are compacted together and form a solid plug, with most of the melting occurring at the interface between the solid bed and the melted material, which accumulates in a melt pool. The solid bed ideally should stay intact until it is completely converted into melt, as shown in Figure 7.3, or else unmelted particles may end up mixed with the melt, causing problems during the shaping operation. Maintaining consolidation of the unmelted particles can be improved by using a barrier screw design in this portion of the extruder, where the solid particles stay in a decreasing size solids section while melted materials pass over the barrier to an increasing size melting section. High speed twin screw extruders (see Section 7.1.10) and single screw extruders designed for compounding are often designed to keep the solid material dispersed in the melt. In general, double flighted screws produce faster melting than single flighted screws.

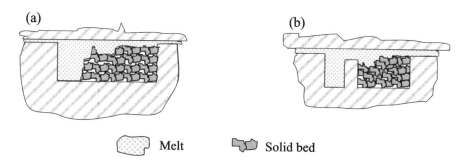

Figure 7.3 Melting (a) in a conventional screw and (b) in a barrier screw.

7.1.4   Metering Section

In the metering section of the screw, the channel is again a constant depth. This part of the screw is responsible for delivering molten polymer at a desired and uniform rate to the shaping device. Thus its function is primarily to convey the melted polymer.

7.1.5   Mixing Devices

Because single screw extruders are essentially plug flow devices, they cannot mix in additives or color very well, especially when the polymer viscosity is high. They therefore depend on the mixing done in the solid state prior to entry of the materials into the hopper. To achieve better mixing capability, special mixing devices are often incorporated into the screw design. These are important to correct for any nonuniformity in either temperature or composition within the melt. The mixing devices function to create back-flow and increased shearing to aid in the dispersion of the nonuniformity, and are typically located within the metering section or just after it.

A variety of geometries of mixing devices are available. Pin mixing sections are common. In these, a series of pins, arranged in an axial or a circumferential orientation, is placed on a section of the screw to break up the flow in the channel. Designs are also available which place the pins on the inside of the barrel rather than on the screw, but these are less common. Pineapple mixing sections have a resemblance to the surface design of the pineapples for which they are named. Other common mixers include the Maddock and Dulmage designs. A wide variety of other designs are also in use. It is possible to have more than one mixing section, of similar or different designs, on a single screw. In selection of the optimal mixing design, the goal is to provide sufficient mixing by splitting and redirection of the flow and high stress, especially elongational stress, while minimizing pressure drop and dead spots, as well as cost.

7.1.6   Extruder and Screw Design and Size

The designs of extruders and extruder screws are frequently tailored to the needs of particular polymers and applications. The initial designs are done using computer models, and then checked in laboratory systems to verify the predicted performance. The parameters that enter into the screw design are the desired output, the properties of the polymer (thermal conductivity, melting point, viscosity as a function of temperature and shear, coefficient of friction as a function of temperature and pressure, etc.), and the screw dimensions (diameter, helix angle, length). The best way for the non-specialist to choose a screw design is to discuss the application with both the extruder manufacturer and the polymer resin supplier.

The size of extruder needed is determined by the output needed for the converting process. Extruder output is proportional to the cube of the inside diameter of the barrel. This diameter is often used as a designation of the extruder size. Another important parameter is the ratio of the length of the screw to the diameter of the barrel, or *L/D ratio*. This ratio determines the residence time of the extruder, and affects the range of outputs that can be obtained. The diameter of the screw at the top of the flights is only slightly less than the diameter of the inside of the barrel, to limit backflow of the plastic. The *root diameter* of the screw refers to the diameter of the solid core, between the bottoms of the channels, which decreases in the compression section, either gradually or abruptly, while remaining constant in the feed and metering sections. The ratio of the depth of the channels in the feed section to the depth of the channels in the metering section, called the channel depth ratio, typically is 2-4, with 3 the most common value.

A typical extruder for packaging applications would have a barrel diameter of 5 to 20 cm (2 to 8 inches) and a length to diameter (L/D) ratio of between 16:1 and 32:1, with most falling in the 20:1 to 30:1 range.

Extruder screws are typically built of steel alloy, and the channels may be chrome-plated to minimize scratching and other damage. The tops of the flights are often hardened with special alloys to reduce wear.

7.1.7   Dies

At the end of the barrel, the melted plastic leaves the extruder through a die, which has been designed to impart the desired shape to the stream of melted plastic, or through a nozzle into an injection mold. Dies generally have at least one temperature control zone, independent of the extruder temperature settings, and may have several. Sometimes dies are treated with low friction coatings to reduce pressure drop across the die, and limit the tendency of the plastic to build up on the die surfaces. Since the design of the die depends on how the plastic is to be shaped, die designs will be discussed further in subsequent sections.

7.1.8   Melt Filters

Extruders generally contain screens for filtering impurities out of the melt. A screen pack typically contains a series of screens of decreasing coarseness, supported by a breaker plate, which is a thick metal disk containing closely spaced holes. Screen packs need to be changed periodically as impurities build up. Usually the pressure drop through the screen is monitored to provide an indication of the need for a change. Provisions may exist for automatic screen changes, to avoid downtime. This is particularly important when contaminated streams are being processed, such as recycled material. Some screens are designed to be semi-continuous, or even have a continuous moving screen, to minimize changes in pressure associated with screen changes.

7.1.9   Drive Mechanisms and Screw Speeds

The drive mechanism is an important part of the extruder, as it provides the energy for and controls the speed of rotation of the screw, generally through a variable speed motor. The motor typically runs at about 1800 rpm, so a speed reducer is needed between the motor and the screw, which usually runs at about 100 rpm. The screw speed is the primary determinant of the output of the extruder. The motor may provide much of the energy to melt the plastic, as well as the energy required to convey it though the barrel. The amount of power required is affected by the viscosity (resistance to flow) of the polymer and the back pressure generated by the die. Polymer viscosity is dependent on temperature, flow rate, polymer molecular weight, and molecular weight distribution, as well as polymer type, as discussed in Chapter 3. If the extruder is under-powered, output may be limited by the lack of ability to melt the polymer.

7.1.10   Special Designs

Not all polymer melting operations can be carried out effectively with a simple single screw extruder. For example, polystyrene may contain sufficient residual volatiles from the polymerization that a *vented extruder* will deliver a higher quality melt. In this type of extruder, a vacuum is applied to the melt through a specially designed vent port on the side of the barrel. Of course, the extruder must be designed to prevent flow of plastic through the vent. One common way to do this is to use a two stage screw, where the root diameter of the screw decreases adjacent to the vent, so that pressure on the melt is reduced.

Another example of a special extrusion operation is a *twin screw extruder*. Twin screw extruders contain two screws. Because they can mix components much more effectively than single screw extruders, they are often used for mixing operations such as compounding in additives to make concentrates of color, antioxidant, etc. If the output of a twin screw extruder is to be fed directly into a shaping operation that requires high pressure, the melt is often discharged into a secondary single screw extruder to generate the needed pressures. The pressure that can be generated by a twin screw extruder is limited by the thrust bearing design.

Twin-screw extruders can be corotating, with both screws turning in the same direction, or counterrotating, with the screws turning in opposite directions. The screws can be either intermeshing or non-intermeshing. Intermeshing screws provide better mixing. While the screws are usually parallel to each other, some designs use conical screws that are not parallel.

7.1.11   Extrusion Temperatures

As mentioned, proper control over temperature is important in processing plastics. Extruders typically have at least three temperature zones along the barrel, often with provision for both heating and cooling in each zone. The heating is most often done with electrical band heaters, and the cooling with forced air or with water-filled tubing. Large extruders may have eight or more different temperature zones that can be individually controlled. Temperature in each zone is often measured in the barrel, and the actual temperature of the plastic can differ significantly from that reading. Immersion probes for measuring the temperature of the melt itself are also available. Ideally, the screw provides most of the energy needed in the extruder, so little additional heating or cooling is necessary. Dies have an additional one or more temperature zones, but usually contain provision only for heating, not cooling. Temperature measurements commonly utilize thermocouple devices, though resistance and infrared detectors are also available. Proportional control of heaters is preferred to on-off control devices, since they result in much less variation in temperature of the melt.

The proper temperature for plastics extrusion is dependent, of course, on the type of plastic and its molecular weight, as well as on the requirements of downstream processing. In general, two competing factors must be balanced. As temperature increases, the viscosity of the plastic decreases, so throughput increases and power requirements decrease. On the other hand, the higher the temperature, the greater is the requirement for downstream cooling, which can lengthen cycle times; also, the potential for thermal degradation increases. A rough guideline for crystalline plastics is to process them at about 50°C above their melting point. Plastics which are heat-sensitive should be processed at lower temperatures, while for highly viscous plastics, higher temperatures are preferred. For amorphous plastics, processing at 100 °C above the glass transition temperature provides a rough guideline.

Recommended extrusion temperatures for some common packaging plastics are shown in Table 7.1.

Table 7.1 Processing Temperatures for Some Common Plastics

Polymer	Processing Temperature (°C)
EVA	150-205
EVOH	200-220
HDPE	200-280
Ionomer	180-230
LDPE	150-315
LLDPE	190-250
Nylon	240-290
PC	245-310
PET	260-280
PP	205-300
PS	180-260
PVC	160-210

7.1.12 Extrusion Pressures

Proper control over pressure is also essential. The amount of pressure required depends on the polymer viscosity and the requirements of downstream processing. For example, pushing a plastic into a mold generally requires considerably more pressure than forcing it through a die. Fluctuations in pressure lead to corresponding fluctuations in output. Further, rapid buildups in pressure can present serious safety concerns, including the potential for explosion. All extruders should be equipped with an over-pressure safety device, which usually consists of a rupture disk or a shear pin in the clamp which holds the die against the barrel of the extruder, as well as with an automatic shutoff if the pressure reaches a critical value. There are a number of devices that can be used to measure pressure, including pneumatic pressure transducers, capillary, and pushrod strain gauges, and piezo-resistive transducers.

7.2   Cast Film and Sheet

For the production of cast film or sheet, the melted plastic leaves the extruder through a slit-shaped die, producing a rectangular profile with the width much greater than the thickness. The processes for manufacturing of cast film and cast sheet are essentially identical. Film is differentiated from sheet by its relative stiffness or flexibility, with no clear line between them. Materials with thickness of 0.003 in or less are considered film, and materials with thickness of 0.010 in or greater are considered sheet. Between these values, materials are considered film if they are relatively flexible, and sheet if they are relatively rigid. The usual method of producing thin cast film is the cold cast or chill roll process.

7.2.1   Cold Cast or Chill Roll Cast Process

Cast film is generally produced by downward extrusion of the melt onto chilled chrome rollers, which are highly polished to impart good surface characteristics to the film, as shown in Figure 7.4. The extrudate contacts the first chill roll tangentially, and then typically travels in an S-pattern around two or more chill rolls. The first chill roll typically operates at a temperature of at least 40°C (104°F), with subsequent rolls operating at successively lower temperatures to cool the film enough that it can be trimmed and wound. An air knife is typically used to pin the plastic against the first chill roll. The film dimensions are controlled primarily by the die dimensions, extrusion rate, and take-off speed. Film produced in this manner is characterized by good transparency and stiffness, and output rates are typically higher than for quench tank cooled film.

7.2.2   Roll Stack and Calendering Processes

Sometimes, especially in sheet applications, the plastic is extruded, either horizontally or vertically, onto one roll in a three-roll stack of polishing rolls, rather than onto a chill roll. The roll stack is designed to exert pressure, and to impart desired surface characteristics to the sheet - either smooth or textured. The gap between the rolls helps in controlling the sheet thickness. The rolls are temperature-controlled, often with circulating hot oil, and the plastic next travels through a cooling section consisting of a number of smaller rolls in a frame.

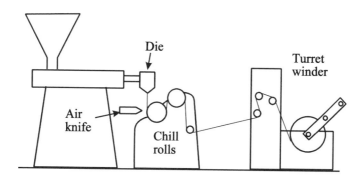

Figure 7.4 Cast film line.

Calendering can also be used to produce film and sheet with desired thicknesses and surface textures. In this process, the polymer melt is squeezed between pairs of co-rotating high precision rollers, followed by a series of chilling rolls. Very tight tolerances in film or sheet thickness can be attained. PVC sheet is usually made in this manner.

7.2.3   Quench Tank or Water Bath Process

For some operations, the chill roll method does not provide rapid enough cooling. In that case, a water-filled quench tank may be used for cooling and solidifying the plastic, as shown in Figure 7.5. After solidification, the film is dried, trimmed, and rolled up. Drying may be accomplished by evaporation alone, or air jets, heated rolls, or radiant heat may be used. The film characteristics are controlled by the die dimensions, extrusion rate, melt temperature, drawdown, and water temperature. This method used to be widely used for polyethylene and polypropylene, but is now much less common, since chill roll casting can provide better control over optical properties and thickness.

7.2.4   Nip Rolls and Winding

After the plastic is cooled, a set of nip rolls pull the film or sheet along, exerting an even tension and feeding the winder, which rolls the sheet on a core. A variety of winder designs are available, and some provide an automatic change to a new core when one fills. Winders may be driven at the core, on the surface, or by a combination. Winders are discussed further in Section 7.3.6.

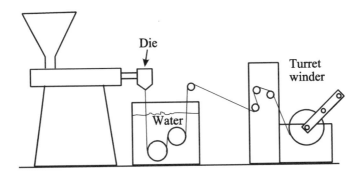

Figure 7.5 Cast quench line.

7.2.5   Gauge Control

Due to thermal contraction and elastic relaxation, the film or sheet produced by the cold cast process is narrower than the die dimensions, and tends to thicken at the edges, requiring them to be trimmed off. Any irregularities in the gauge of the film tend to be magnified when it is rolled up. Gauge variations of ±3% are common in cast film, and this can produce gauge bands in the roll that can cause difficulties in later converting operations. There are two major approaches to minimizing the problem of gauge bands. The oldest method is oscillating the film as it is wound, to produce some randomization of the thickness variations. However this method significantly reduces the width of the web and increases scrap generation. Although the scrap is generally fed back into the extruder in-line, this is undesirable. The more modern approach uses sensors to monitor the thickness of the web on-line, using a scanning measuring head that travels across the film so that thickness can be measured along both the length and the width dimensions. Sophisticated cast film lines can feed results of the thickness measurements back to the die, permitting automatic computer-controlled adjustments to be made in the die dimensions to minimize unevenness in the thickness profile.

7.2.6   Orientation

Orientation, stretching the film to produce some molecular realignment in the direction of the stretch, is often used to modify cast film properties. Orientation tends to increase crystallinity, barrier properties, and strength in the stretched direction(s), while decreasing strength in the direction(s) perpendicular to the orientation.

If cast film is not stretched significantly in the machine direction (the direction of travel through the equipment) during production, it is relatively unoriented and has mechanical properties that are similar when measured in the machine and cross-machine (perpendicular to travel) directions. If the takeoff speed, the speed at which the plastic is wound up, is significantly higher than the rate of extrusion, the plastic is stretched and uniaxially oriented. The stretching can occur at the initial contact with the chill roll, but is more commonly done after the first chill roll, and usually requires some reheating of the film before it is stretched. Biaxial orientation involves stretching the film in both the machine and cross-machine directions. This can be accomplished in a single step or, more commonly, in two consecutive steps. The film is referred to as "balanced" if the orientation is equal in the two directions, or "unbalanced" if it is more highly oriented in one direction than in the other.

Two-step orientation is widely used to make oriented PET and PP film. First the film is drawn in length, and then it is drawn in width. In one-step orientation, the film is stretched simultaneously in the length and width directions. Both processes are known as tentering (Fig. 7.6).

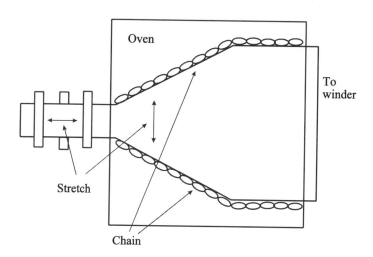

Figure 7.6 Tentering process for biaxial orientation

A crystalline polymer must be oriented at a temperature below its melt temperature, but high enough to provide some mobility to the molecules. The higher the orientation temperature, the more the material tends to flow, and the less orientation is produced. On the other hand, if the material is too cold, uneven stretching, with thin spots, and rupture will result. One design guideline is to orient at 60 to 75% of the range between the glass transition temperature and the melting temperature. After the stretching, or drawing, the film is annealed to improve its thermal stability, if this is desired, and then cooled to "freeze" the orientation before the tension is released. For successful orientation, the crystallization within the polymer needs to be homogeneous and not excessive. The temperature of the polymer can be controlled by contact with oil-heated rolls, or using other heating mechanisms.

7.2.7   Cast Film Dies

The main components of a die include the inlet channel, which delivers the plastic from the extruder to the die; the manifold, which is designed to evenly distribute the melt within the die; the approach or land, which carries the melt from the manifold to the die opening; and the die lips, which perform the final shaping of the melt as it exits the die. For cast film or sheet, the die opening is slit-shaped, producing a thin, wide exiting stream of plastic.

The most common die configuration for production of cast film and sheet is a coat-hanger die (Fig. 7.7), named because of its resemblance to a common coat hanger. T-shaped dies are also used. The die opening is wider and thicker than the finished film. As the film is drawn down between the die and the chill roll, it contracts in width and thins, due to tension on the film being produced and polymer relaxation effects.

Flow

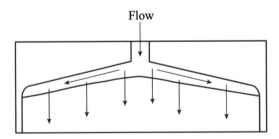

Figure 7.7 Internals of coat hanger die for cast film and sheet.

The dies are generally equipped with flow modifiers such as restrictor bars and adjustable lips, to adjust for processing variations such as changes in resin, extrusion temperature, and flow rates. As mentioned, sometimes the lip adjustments are computer-controlled and linked to in-line thickness measurements.

The dies themselves are generally constructed of medium carbon alloy steels, with flame-hardened lips and chrome or nickel plating on the flow surfaces. Insulation of the die body is common to prevent temperature variations that could result from air drafts. Dies can be electrically heated, or steam or oil heat can be used. Cooling is usually provided by natural convection, though forced air can also be used.

7.3 Blown Film

Blown film extrusion is a continuous process in which the polymer is melted, the melt is forced through an annular die, and the resulting tube is inflated with air into a "bubble" and cooled (Fig. 7.8). Air is always blown on the outside of the bubble to cool the film; to increase production rates, internal bubble cooling can also be used. The film is stretched in the longitudinal and circumferential directions during production, resulting in biaxial orientation of the film. The amount and relative degree of stretching determine the degree of orientation. The circumferential stretching is inherent in the blowing process. Longitudinal stretching is imparted by drawing of the film between the extruder and the nip rolls.

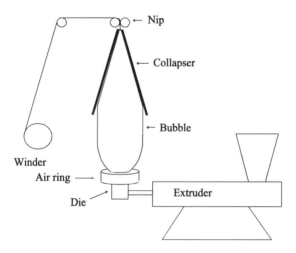

Figure 7.8 Production of blown film.

The principal polymers used in blown film production are polyolefins, although other polymers can also be used. The major applications are those that require biaxial strength, and include bags of all kinds, as well as agricultural and construction film. In food packaging, coextruded structures with three to five layers, or even more, are common, with major markets including packaging for cereal, meat, snacks, and frozen foods.

The properties of the film are determined by the *blow-up ratio* and the linear line speed. The blow-up ratio is the ratio between the diameter of the final tube of film and that of the die. The internal air pressure that expands the tube into the bubble is typically supplied through a port into the mandrel, the interior part of the die. Once the process is running steadily, little air is usually lost, so make-up requirements are small. When internal bubble cooling is used, air is constantly being exchanged inside the bubble.

The travel of the film through the blown film tower is aided by various guiding and sizing devices. The film turns from molten to semi-solid at the "frostline" but is still easily deformed as it moves up the tower. However, the orientation of the film is generally complete at this point. When the film is cool enough, the bubble is collapsed by plates and rollers (pinch rollers), and wound up, with or without slitting, gusseting, or other treatment. Thus, the blown film process can produce tubular as well as flat film.

Blown film tends to be of somewhat lower quality than cast film in terms of transparency and uniformity of gauge. The slower cooling of blown film is largely responsible. Slower cooling allows time for a greater degree of crystallinity and larger crystallites to develop, making the film hazier, though producing better barrier properties. The cooling can be more uneven than in the cast process, contributing to greater variation in thickness. Since even a slight lack of perfect concentricity and freedom from defects in the die opening can result in significant unevenness in the profile of the film roll, it is common to rotate or oscillate the die during extrusion, in order to randomize the thickness variation and produce a uniform roll of film. Uneven air flows caused by air ring imperfections or drafts in the room also contribute to gauge variations. It is also possible to rotate or oscillate the nips to randomize the variation. Variation in blown film gauge of ±7% is common, and even higher variation is sometimes encountered.

Once the blown film process is up and running, little scrap is produced, since there is no need to trim edges. However, getting the line running properly is more complex than with cast film, so more scrap is produced during start-up. Thus, blown film lines are more appropriate for high volume production than for situations where starting and stopping the line is frequent.

Most blown film operations extrude the resin in an upward direction. However, blown polypropylene film is generally extruded downwards and water or mandrel quenched. The extruded tube is then reheated, to a point still below its melt temperature, before it is blown. The collapsed bubble can be fed over a series of heated rollers to reheat it and relieve thermal stresses if a heat-stabilized film is wanted; or it can be heated and reinflated in what is known as the "double bubble" process, which will be discussed in Section 7.3.7. In either case, the film is restrained until cooling is complete, to keep it from shrinking.

7.3.1   Blown Film Extrusion

The extruders used in blown film are like those described in Section 7.1. Typical extruder sizes for most blown film lines are 5-11.4 cm (2.5-4.5 in). After the polymer leaves the extruder, it enters a transfer pipe and the adapter section of the die. The transfer pipe is not usually technologically advanced. The only concern is to keep it as short as possible, and to avoid dead spots (areas where flow is interrupted). The adapter section of the die changes the direction of flow of the polymer from horizontal to vertical, and in the case of coextrusion, keeps the polymer streams separate until they are delivered to the die. If the die rotates or oscillates, the adapter section must also provide the necessary bearing surfaces for this motion and seals for the moving polymer channels. Seal design technology is a challenge, and an area where many developments are proprietary. When coextruding polymers with wide variation in melt temperatures and heat stability characteristics, it is necessary to provide some insulation between the melt channels through dead air spaces, etc. These approaches are only partially successful, and more system cleaning is necessary when running such structures. If stationary dies are used, a subsequent step is required to spread any gauge non-uniformities across the rolls. Such steps, i.e. oscillating nips and rotating winders, will be discussed in Sections 7.3.5 and 7.3.6.

The difficulty of designing, manufacturing and maintaining the seals for a rotating or oscillating die is one of the reasons for manufacturers going back to stationary dies, particularly for coextrusion systems. The object of rotation is to spread any temperature or thickness variation of the polymer stream across the film rolls. The mixing head of the extruder or the die design may not completely erase temperature streaks in the melt, and these will result in thickness variation in the final film, just as will an improperly adjusted die gap. Steady rotation of the die may not spread the temperature streak across the bubble circumference, but may only cause it to be dragged to some equilibrium position. Oscillating dies and static mixers in stationary adapters can correct this problem more effectively.

7.3.2   Blown Film Dies

Most dies produced today are spiral channel dies, though some antique side entry and spider dies are still in use. In spiral channel dies, the polymer melt is introduced through several ports into spiral channels, as shown in Figure 7.9. It then flows partially up the channel and between the channel and the die wall. This type of smearing flow reduces any existing temperature differences in the melt. As the flow moves up the die channel, the spiral disappears and the melt flows into a relaxation chamber designed to allow the melt to lose the memory of its previous shear history. Beyond the relaxation chamber is the die land itself, which shapes the melt to its final dimensions as it leaves the die.

In a coextrusion process, the spiral section and relaxation chamber of the die are replicated for each polymer layer in the final structure. For this reason, coextrusion dies can be very large. Above the relaxation chamber, the melt streams must be combined. Combining sections vary in design depending on the number of layers and the materials expected to be run. In the simplest form, all the streams are brought together in a short section just before the die land. If the die is large, there are many layers, or polymers of widely varying viscosities are involved, several combining sections may be used for improved control. Polymers of widely varying viscosities do not flow together in a stable manner, so the channel length must be kept as short as possible in such situations. The polymer with the lowest viscosity tends to move towards the outer perimeter.

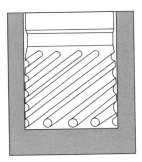

Figure 7.9 Spiral die.

Die channel design, like extruder design, is often a compromise because of the wide array of materials that must be run through the system. Channels designed for tie layers and barrier polymers are generally small because of the small amount of these materials that is needed. Problems may be encountered if a polymer used in small flows is switched into a large extruder or die channel, since the residence time may exceed the polymer's heat tolerance.

The size of the die is specified by the diameter at the die land opening, known as the *die gap*. Selection of the appropriate gap size depends on the types of polymers to be run in the system and the final product dimensions. One important parameter is the *drawdown*, which refers to the ratio between the speed at which the extrudate is pulled away from the die, and the speed at which the melt emerges from the die. For monolayer films, a rule of thumb is that branched polyolefins can only withstand a drawdown of 16:1 because of the strain hardening of the melt due to chain entanglements; while a linear polymer such as HDPE or LLDPE can be run at a 160:1 drawdown. If the drawdown is too high, the bubble tears off. In a coextrusion system, the choice of die gap is more complicated because the various polymers are not behaving individually. However, a weighted average of the tensile strengths generally gives a good estimate, and the melt properties are dominated by the characteristics of the heaviest component.

The die land refers to the section of the die where the die gap is constant, just before the polymer exits the die. The selection of the length of the die land is determined by the amount of shear orientation that is required. Typical values are a land length to die gap ratio of between 5:1 and 30:1, with a 10:1 ratio the most common.

Small die gaps impart greater shear stress on the flowing polymer, which improves mechanical properties by imparting orientation. However, if the shear is too high, melt instabilities can arise, which detract from the film's appearance or may even lead to melt fracture. Maintaining the die lips at a significantly higher temperature than the remainder of the die can help minimize these problems.

In addition to the die gap itself, the centering of the die gap is important. Since the polymer is being extruded in an annulus, the centering of the external wall with respect to the internal wall is essential for production of a uniform material. One method is to adjust the position of the outer ring with bolts. This makes machining of the die simpler, but may result in too many adjustments available to the operators for quality production. The alternative is to machine the die parts so that precise alignment can be made during assembly, and provide sufficient sturdiness to keep the parts from moving once assembled.

Because all the temperature variations that occur in the extruder and the rest of the piping and die cannot be completely removed from the melt by the die channels, as mentioned, the die is often rotated or oscillated to spread the non-uniformity over the film rolls being made. While this creates even rolls of film, the point to point gauge variation can still be ±10%. Therefore, the most modern die designs include automatically adjustable die lips. In these designs, the flexible lip technology used in cast film dies has been translated into blown film dies. A downstream gauging system senses spots in the film that vary from the set-point gauge, and feeds a signal back to an adjustment device on the die lip. These types of dies, while allowing better gauge control, are very complex, and introduce additional maintenance problems.

Blown film dies, like those for cast film, are generally manufactured from steel, and are often plated with chrome or nickel to harden the surface and increase its durability, as well as making it easier to clean. In some cases, the die lips are coated with PTFE-based coatings or special alloys. For processing of highly corrosive plastics such as PVDC, special high-nickel alloys can be used.

7.3.3   Air Rings and Internal Bubble Cooling (IBC)

The cooling of the bubble is the next important step in determining the final properties of the film. The original air ring was a pipe with holes drilled in it that wrapped around the bubble. Modern technology has used an understanding of aerodynamics to develop dual lip air rings and internal bubble cooling packages (IBC) that greatly increase production rates and uniformity of the film.

The dual lip concept of an air ring is shown in Figure 7.10. The first air channel is designed to lock in the correct bubble shape and to give the bubble stability in the first few diameters above the die. The second air channel is the one that produces the main cooling of the outer surface of the bubble. With older air rings where both functions were done by one airstream, it was difficult to accomplish both tasks correctly. If the air volume was high enough to give adequate cooling at high production speeds, the velocity was generally too high near the neck of the bubble for stable flow. This resulted in unstable bubbles and non-uniform gauge.

The application of IBC further improves gauge uniformity by providing heat transfer from both sides of the bubble. For bubbles with very large widths, IBC is the only way to obtain high production rates. For any size bubble, the use of IBC can increase maximum line output. The latest IBC packages also utilize aerodynamic design for the application and extraction of air within the bubble. Automated controls for bubble cooling are now available, and assist in obtaining accurate bubble dimensions. The use of a sizing cage or an ultrasonic gauging ring to control the exhaust air volume flow and maintain the bubble size is now common.

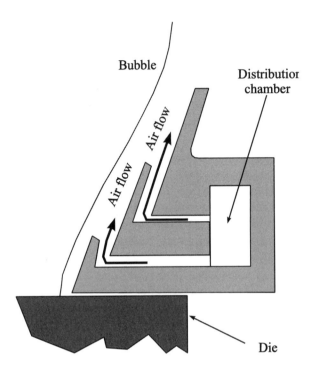

Figure 7.10 Dual lip air ring.

The extrusion of HMW-HDPE has resulted in the development of new specialized cooling technology to produce oriented films. Because of the linearity of the polymer, it has a pronounced tendency to relax and destroy orientation if temperatures are too high. Therefore, HMW-HDPE bubbles are run with a long stalk (neck) before the bubble is forced to stretch in the circumferential direction. This allows the melt temperature to drop sufficiently to maintain the cross-direction orientation, which is introduced just before the frost line. The air ring design had to be modified to allow this type of processing, and an additional bubble stabilizing iris is used just above the frost line. An internal bubble stabilizer is often included even if IBC is not used. Newer developments include special air flow guides to produce a more stable bubble and minimize the effect of drafts on the neck region. These efforts are especially important because these films are often used with very thin gauges, 12.5 μm (0.5 mil) or less. A variation of ±10% is very significant at these low thicknesses.

7.3.4   Collapsing Frames

As the bubble approaches the nip rolls, the bubble shape must be transformed from a round tube to a flat material. While the bubble looks symmetrical, a thorough examination of bubble geometry reveals that the path traveled by a particle in the bubble is not the same at all points in the circumference. Knittel [1] showed that the shortest path up the bubble is at a location 15.5° from the crease (Fig. 7.11). The path differences are not much of a problem for polymer films that are flexible, such as LDPE or LLDPE. For HDPE and some stiff coextrusions, however, this path difference is noticeable and results in wrinkling of the film.

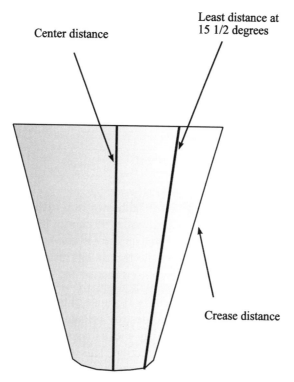

Figure 7.11 Bubble collapsing geometry.

The oldest type of collapsing frame, which is still in common use, is made from maple slats. Since the film drags across the maple surface, wrinkles from stretching may be induced, as well as scratches in high gloss films. A better approach is to use rollers in the collapsing frame. Better yet is an air collapsing frame, which uses an angled bearing surface with a layer of air, so the surface never actually touches the film. Some additional cooling is also accomplished by the air flow. These air collapser frames reduce the incidence of scratches and result in less wrinkling and stretching.

Other approaches have been employed to overcome the wrinkling of HDPE as well as gauge variations in some coextrusions that are caused by differences in path length. For HDPE, one approach is to lower the nips (the collapsing rolls) so that the film is collapsed before it has been cooled completely, when it is more flexible. This does not work well for heavier gauge films or for films containing lower melt temperature sealing layers. Knittel suggested installing a bubble shape modifier before the collapser to change the bubble from round to square. This reduces the path differences considerably, but it introduces other problems. The best solution to wrinkling and gauge variations due to path differences is to use very tall towers, since the longer the path to the nips, the less the path difference between bubble sections.

It is also possible to collapse the bubble and heat it in order to seal the two halves of the tube together, producing a multilayer film of double the thickness.

One of the latest innovations in collapsing frames is using carbon-fiber rollers rather than aluminum ones. Because of the much lower heat transfer rate of the carbon-fiber, the rollers do not act as a significant heat sink, which reduces wrinkling problems. In addition, the carbon-fiber rollers produce less inertia so there is less drag on the film.

7.3.5   Nips

The nip section of the tower on a conventional blown film tower consists of a pair of rollers, one metal and one rubber, which provide the energy to stretch the melt in the machine direction and to transport the film up the tower. Accurate speed control is necessary, but that is about the extent of the technology usually required. However, oscillating and rotating nips are now being used increasingly. In some operations, the rotating or oscillating die cannot successfully spread all of the gauge variations over the rolls. There may be drafts in the room causing gauge variations, or there may be temperature histories in the melt that are not successfully randomized. The use of an oscillating nip will spread these variations over the rolls. Another pressing economic reason for their use is that stationary dies used with oscillating nips allow for less expensive die construction, particularly in multilayer applications. Stationary adapters do not need expensive and problematic bearings and seals. Especially when the number of layers reaches five or more, the use of a stationary die produces a much more reliable design.

Fully rotating nips with the winder attached are used for HMW-HDPE film because they result in better roll formation for the very low gauge film. If these are used with upward extrusion, it requires dealing with roll handling on the second story. If extrusion is downward, roll handling is easier, but extrusion is complicated.

7.3.6 Slitting and Winding

As discussed, blown film, once collapsed, can simply be wound up, resulting in a tube of film which can be used for bags or pouches. An alternative is to slit the film so that a flat film can be produced. Typically slitting is done by trimming a small amount of material at the edges of the collapsed tube. Then two equal sized rolls of flat film can be produced. Another option is to trim the film at only one edge, producing a roll of folded over flat film. This is particularly suited for making bags or pouches where the fold will serve as the bottom of the bag.

The winder produces the rolls of the final product and can affect quality. The ease of use of the winder will dictate how much attention is given to it by the production crew. Both are important factors in producing a quality product. Winder designs vary depending on the features desired. Just as for cast film, there are (1) center drive winders, where the core is driven; (2) surface drive winders, where the roll is driven from the outer surface; and (3) surface/center drive winders, which provide control on both surfaces. Most blown film applications use the center drive design.

The main function of a winder is to guide the film evenly into the roll form. To accomplish this, it must provide tension control on the film to produce evenly wound rolls. A secondary nip, either as part of the winder or just prior to the film entering the winder, is used to provide this tension control. Tension must be sufficient that the roll will not telescope when handled, but must not be so much that it causes crushed cores or center-roll bagginess. Tensions on center drive winders are generally tapered from the core to the outer wraps to compensate for the change in diameter of the roll. On a surface winder, the compensation is automatic because of the drive mechanism. When tension is being adjusted on a winder, it must be kept at or below 0.5% of the secant modulus of the film to avoid excessive stress in the film. One reason for this low value of recommended tension is that the film is not totally at equilibrium at the time it is wound into the roll. Dimensional changes occur as the film crystallizes or cools, and induce stress.

Winders can use different methods of roll changing. Most winders used in the film industry are turret type, where the empty core is rotated into position by a turret mechanism and an automatic cutoff is made.

Many variations in winder design are available from manufacturers, and one needs to determine the requirements of the products to be produced before winder selection. Loonsbury [2] and Knittel [1] have published discussions about winders that offer insight into the design parameters that are important. Winding technology is an area of study that has received more attention in the paper industry than in the film industry, but one can learn the principles from paper and apply them to film. One major difference between the two materials is the much greater elasticity of films, which must be considered in analysis.

7.3.7   Double-Bubble Process

The double-bubble process involves the extrusion and blowing of a tube of molten plastic in a downward direction. The tube is then cooled, most often using a water bath, reheated to just below the melt temperature, and reinflated. The reinflation along with the increase in haul-off speed provides biaxial orientation. Typically the next step is annealing to relieve thermal stresses and stabilize the film. The double-bubble process is most often applied to PP film, but is also used with multilayer PP or PE-based films. One of the major advantages is that this process can deliver a high-clarity film with precise shrink characteristics and very uniform flatness. The flatness, in particular, is critical for high quality printing systems. The ability to produce uniform shrinkage in both directions is an important property for shrink wrap applications, particularly for printed shrink materials, including labels.

A German company, Kuhne GmbH, has introduced a triple-bubble process, which uses infrared heat, rod heaters, and other annealing techniques in a "thermofixation" station to impart precise shrink rates to coextruded film.

7.4   Stretch and Shrink Wrap

A wrap is the simplest form of plastic package, using a flat piece of plastic film that is folded or wound around the packaged item. Wraps are used to an extent for retail packages, but the largest market is stretch wrap for pallet loads of products, serving to unitize and stabilize them during distribution.

7.4.1   Stretch Wrap

Stretch wrap exerts holding power on its contents because the molecules in the film attempt to return to their original conformation after they have been stretched. As long as the elastic limit of the material has not been exceeded, the holding force increases with increasing amounts of stretch, and thus with increasing tension on the film. Because the extent of stretching done during wrapping of products is limited by the potential for product damage and other factors, stretch film is commonly pre-stretched before it is wound on the load. This permits elongations of up to 250% to be achieved, and minimizes necking-in of the film. Necking-in refers to the tendency of a material's width to decrease when it is stretched.

The end of the stretch film is usually secured to the load simply by self-cling. Tackifiers such as polyisobutylene are usually added to the base resin to improve cling. A very smooth outer surface of the film also improves cling. Stretch film should have very high elongation, low neck-in, high tensile strength in the machine direction, high elasticity, good puncture resistance, resistance to tear propagation, low creep, and good fatigue resistance. The last two properties determine the ability of the film to sustain its holding power over time. The holding force exerted by stretch wrap decreases over time, especially at elevated temperatures, due to stress relaxation in the plastic.

Most stretch films are LLDPE, modified with various additive packages to provide the desired properties. EVA copolymers and PVC are also used. Coextrusions are often used to provide features such as stronger self-cling on one side than on the other, so that the film sticks to itself as it is wound on the pallet, but does not adhere to adjacent pallet loads.

7.4.2   Shrink Wrap

Shrink films, like stretch films, use the tendency of a film to try to return to a smaller dimension after deformation to provide a tight wrap around a packaged object. In shrink film the product is loosely packaged, and when it is exposed to heat, the film shrinks. The resistance of the product on the film provides the holding force.

To make shrink film, a polymer film can be oriented at an elevated temperature and the orientation "frozen" by rapid cooling. When the film is subsequently heated, the molecular "memory" of the polymer causes it to attempt to return to its original dimensions. Lightly cross-linked materials are often used to increase the tendency to shrink. In that case, electron beam irradiation of the plastic film produces free radicals, which then react to produce cross-links between adjacent molecules. The presence of these cross-links means the material will no longer become liquid and flow at its normal melting temperature. That, in turn, allows the shrink film to be exposed to high temperatures, at or above its former melt temperature, without flow, so these elevated temperatures can be used to promote shrinking.

Shrink wrapping starts with loosely sealing the plastic around the product, followed by passing the package through a shrink tunnel, where it is exposed to heat. If the temperature, residence time, and size of the package and product are chosen properly, then a tightly wrapped package comes out of the shrink tunnel. Because exposure to the heat is very brief, even products with some temperature sensitivity can be packaged in this way.

Over time, shrink wrap, like stretch wrap, tends to loosen somewhat, due to creep and stress relaxation, with the loss of holding power increasing at higher temperatures.

Shrink wrap materials include PE, PP, and PVC, among others. Some shrink wrap is used for pallet-loads of goods, but that market is dominated by stretch film, due in part to its lower energy requirements. Most shrink-wrap is used for individual packages or small bundles of products. A non-packaging use of shrink film is window insulation, where the film is placed on the inside of a window and then shrunk using a hair dryer.

7.5   Film and Sheet Coextrusion

Multilayer structures are often used in packaging in order to take advantage of the properties of several polymers at once. Multilayer structures are made by laminations, coatings, and coextrusions. Because of the multiple handling of materials and the thickness of the layers, laminations (see Chapter 8) may be relatively expensive materials. Coextrusion can be used to reduce the cost of multilayer plastic film and sheet. Coextruded packaging materials are multilayer plastic sheet or film constructions produced from more than one plastic resin in a single step, using either the cast or blown film process. In coextrusion, the materials never exist as separate webs.

Each type of resin is melted in an individual extruder, and the melts are carefully brought together prior to or in the die, in a manner that keeps them in homogeneous layers, without mixing. The process used to combine the polymers is usually different in the cast and blown film processes. In cast processes, as illustrated in Figure 7.12, the polymers are typically combined in an adapter, called a feed block, before they enter the coat hanger die. This permits a simpler design for the die itself. Multi-manifold dies are used when plastics with widely different flow properties are to be combined, as such systems provide a shorter multilayer flow path before solidification, and thus minimize distortion of the interface.

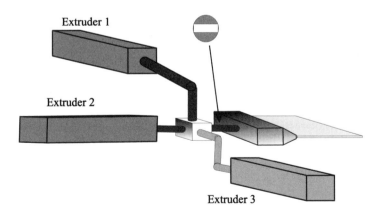

Figure 7.12   Cast film coextruders and die.

For blown film operations, a multilayer die is most often used, with cavities, including manifolds, for each polymer cut within the die. The reason for this is that the spiral channel design normally used in the die cannot accommodate pre-combined flows and result in a layered structure. Figure 7.13 shows a schematic of a multichannel die. In such systems, the layers are combined just before they exit the die, or immediately after they exit.

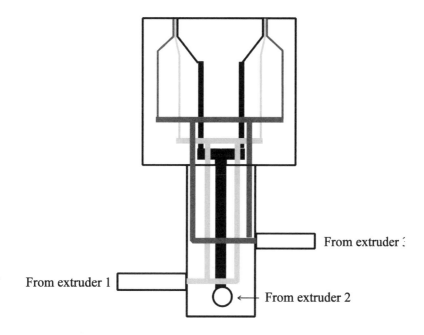

From extruder ?

From extruder 1

From extruder 2

Figure 7.13 Multichannel blown film die.

Processing of a coextruded material downstream of the die is identical to that of single layer materials. However, in design of the coextrusion, careful attention to the rheology of the polymers, especially flow patterns, melt viscosity, and elasticity, is required in order to prevent layers from intermingling or moving to a position in the structure where they are not supposed to be. Figure 7.14 illustrates some of the difficulties encountered during the flow of multiple materials in a pipe. If two polymers are flowing in a tube, as in Figure 7.14(a), the stability of the interface is dependent on the viscosity differences of the two materials. In order to minimize the energy required for flow, the lower viscosity material will try to cover as much of the wall area as possible, as shown in Figure 7.14(b). A viscosity mismatch of <3 is necessary for stability in piping and in dies for cast coextrusions. Since viscosity of a polymer is dependent on temperature and flow rate, these must be taken into account. The layer thickness affects the shear rate to which the polymer is exposed. As a layer gets very thin, the stress on the melt increases, and may exceed its elongational strength, resulting in discontinuities in the layer.

Figure 7.14 Viscous instability.

For many combinations of materials, adhesion between the two desired polymers is weak, and the layers tend to separate under stress. An example is HDPE and nylon, which will not bond to one another. In such cases, it is necessary to add a third component, known as a *tie layer*, which functions as a thermoplastic adhesive, bonding strongly with both components and providing integrity to the multilayer material. Tie layers must be melted in a separate extruder and fed into the die as an additional distinct layer. Thus, many coextruded materials are three-layer structures, one layer for each of the desired plastics, and an additional tie layer. Structures with four layers or more are also common. For example, structures containing EVOH as an oxygen barrier layer generally require it to be sandwiched between other polymers to protect it from moisture. With the addition of tie layers, this results in a five-layer structure. Each resin in the structure requires a separate extruder, although the melted resin from a single extruder may be split into two or more layers in the coextruded product.

7.6   Surface Treatment

Surface treatments can be used to provide special properties to plastic film or sheet, and are especially important in improving the ability of polyolefins to adhere to inks or adhesives in subsequent printing and converting operations. Flame treatment, corona discharge treatment, and ozone treatment all work by imparting some degree of oxidation to the surface of the polymer. The polar groups left by these processes provide stronger secondary bonding characteristics, improving the adhesion of inks and adhesives.

Corona discharge is used more often than flame treatment for films, and flame treatment is used more often than corona discharge for containers. In corona discharge, electric energy flows from a high-voltage conductor, the treater bar, through ionized air and through the film, which is passing across a grounded conductive roller with a nonconductive coating. In the ionization, some of the oxygen in the air is converted to ozone, which can oxidize the surface of the film, and the film surface is slightly ionized. The "corona" in the name of the procedure comes from the faint glow that occurs, along with a crackling sound, during the process.

In this process, the effect of the treatment is to increase the surface energy of the film. This energy is typically measured in dynes/cm^2, using solutions of known surface energy and finding the point at which they wet the film, rather than beading up. Untreated LDPE film typically has a surface energy of about 34-36 dynes and HDPE about 38 dynes. For printing and laminating, the surface energy is usually increased to 40-50 dynes. ASTM D2570 provides a standard procedure for measuring surface tension.

The effectiveness of corona discharge treatment dissipates somewhat over time. Most of the loss occurs immediately on film winding, when the treated side of the film contacts the untreated side. In addition, rubbing of the treated surface as it passes metal rollers or drums can produce substantial losses. High levels of treatment can cause excessive blocking of the film in the roll (layers sticking together, interfering with unwinding), and high treatment levels are associated with high loss levels, as well. Migration of slip additives to the film surface after the film is wound also generally reduces the effectiveness of the treatment. For all these reasons, it can be very difficult to deliver a film to the converter with the required degree of surface modification. Therefore, it is not uncommon for either the initial corona treatment, or an additional treatment, to be carried out on the conversion line, just prior to printing or laminating.

7.7 Yield of Film

The yield of film is a useful parameter in determining how much film, by weight, is required for a given packaging application, where area is the controlling variable. Yield provides a measure of the area of film of a specified thickness that is produced from a given weight of polymer resin. The relationship is as follows:

By definition:

$$Yield = area/weight$$

In addition, we know:

$$density = weight/volume$$
$$volume = area \times thickness$$

Therefore,

$$\text{weight} = \text{density x volume} = \text{density} \times \text{area} \times \text{thickness}$$

So

$$\text{Yield} = \text{area/(density} \times \text{area} \times \text{thickness}) = 1/(\text{density} \times \text{thickness})$$

Thus, yield, Y, can be calculated as the inverse of the density times the film thickness:

$$Y = \frac{1}{\rho \ell} \qquad (7.1)$$

where ρ is density and ℓ is thickness.

Example: What is the yield of a 25.4 µm (1 mil) low density polyethylene film (specific gravity 0.92)?

$$Y = \frac{1}{\rho\, t} = \frac{11\, cm^3}{0.92\ g\ 25.4\ \mu\ m} \times \frac{1 \times 10^4\ \mu\ m}{cm} = 428\ cm^2\ /\ g$$

or

$$Y = \frac{1}{\rho\, t} = \frac{1\, cm^3}{0.92\ g\ 0.001\ in} \times \frac{454\ g}{lb} \times \frac{in^3}{(2.54)^3\ cm^3} = 30,000\ in^2\ /\ lb$$

7.8   Testing and Evaluation of Films

A variety of tests are available for determining film quality. Many are based on ASTM standards. For example, tests for permeation of oxygen, water vapor, or other substances may be carried out, as discussed in Chapter 14. Tests for impact strength, tensile strength, tear strength, and other mechanical properties are important for many applications. Tests particularly related to quality control in the film production process include those that relate in some fashion to the appearance of the film.

ASTM D3351 describes the gel count test. Gels are imperfections found in the film, often related to thermally-induced cross-linking. In the standard test, a 640 in^2 area of film is examined and the number and size of gels are counted.

ASTM D2457 is a standard method for determining the specular gloss of films, using measurement of reflected light from a standardized source. Haze can be measured by following ASTM D1003 and using a special hazemeter.

Thermal shrinkage of films can be determined by following ASTM D2732 and immersing a sample in a temperature-controlled liquid for a defined time, or by making measurements at a series of temperatures in the range of interest.

Blocking of films can be measured using ASTM D3354. Standard D1894 is designed to measure the coefficient of friction. Use of ASTM D2570 in determining film surface tension and its relationship to printing and converting were discussed in Section 7.6.

References

1.   Knittel, R., *J. Polym. Film Sheet*, Vol. 3, p. 23, 1988.
2.   Loonsbury, D. , *Proc. ANTEC '87*, Brookfield, CT: Society of Plastics Engineers, p. 158, 1987.

Study Questions

1.   Name and describe the functions of the three main sections of an extruder screw.

2.   Describe the cast film process.

3.   Describe the blown film process.

4.   Discuss the differences in properties you might expect between cast film and blown film produced from the same resin. What are the major reasons for these differences?

5.   How is a film oriented? What changes in properties does this produce? Why might we want to use an oriented film rather than one that has not been oriented?

6.   Explain the similarities and differences between stretch and shrink wrap.

7.   What is the yield of 200 gauge (2 mil) PVC film, if its specific gravity is 1.35? If the film sells for $0.60 per pound, what will be the film cost for a package that uses 100 in^2 of the material? (Answer: 146 cm^2/g, 0.58¢)

8.   Why might you wish to use a coextruded film for a packaging application?

9.   Calculate the yield of 1.5 mil HDPE film, if its density is 0.94 g/cm^3. State your answer in English units. (Answer: 19,600 in^2/lb)

10.   What rheological characteristics are necessary for a stable blown film bubble? What molecular characteristics contribute to those rheological properties?

11. Polypropylene must be made into a film by a process that cools the films rapidly, or it will be brittle. What molecular property causes this issue?

12. Why do polymers with a broad molecular weight distribution have better processing characteristics?

8 Converting, Lamination and Coating

Lamination and coating are operations resulting in the production of multilayer flexible structures. The multilayer structures are actually composite materials, which may contain plastic, paper, and/or metal (usually aluminum foil), held together by adhesives. Operators fabricating these flexible structures, called converters, also typically incorporate a printing process. The group of converters makes up the converting industry, which is economically very important. In this chapter, we describe extrusion coating and laminating, hot melt coating, adhesive lamination, thermal lamination, metallization, and silicon oxide coating.

8.1 Extrusion Coating and Laminating

Extrusion coating and extrusion laminating operations are normally described together because the process equipment for accomplishing them is essentially the same. Extrusion coating is the operation by which a coating of a thermoplastic material is applied to a substrate web, such as paper. In extrusion lamination, a second web is incorporated and both webs are combined by the adhesive action of the extruded thermoplastic material. The processes are depicted in Figure 8.1. In operations for converting packaging materials, these two processes are often combined in the same line to create a web such as:

paper/acid copolymer/foil/heat sealant layer.

In the first operation, the paper and the foil are extrusion laminated using an acid copolymer as the adhesive. Next, the sealant layer is applied as an extrusion coating. The operation can be accomplished in two passes, or in a single pass in a tandem extrusion line as shown in Figure 8.2. In such a line, both extrusion heads are usually capable of coextrusion, as well. To understand the process of extrusion coating and laminating, it is necessary to start with the process in its simplest form, which, by the way, is how it is still practiced today by some converters.

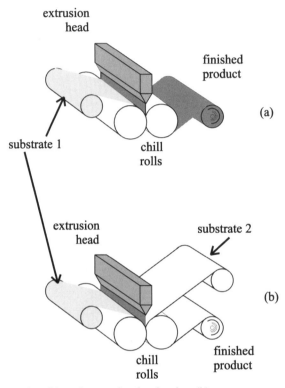

Figure 8.1 Extrusion coating (a) and extrusion laminating (b).

Extrusion laminating and coating represent early applications of polymers in packaging. Polyethylene coated paper was substituted for waxed paper as a bread overwrap in the 1950s. Since polyethylene is non-polar and paper is polar, there is little or no adhesion between the materials unless the PE can flow into the paper pores, or some modification is done to the PE. A solution to this problem was the discovery that allowing the web of molten PE to drop through the air for a certain distance causes the hot surface to oxidize slightly, creating carboxyl (-COOH) groups on the surface that are polar enough to bond to the paper fibers. To get the PE to oxidize quickly enough for the desired line speeds, the process is often run at a melt temperature of 316°C (600°F) or higher. One of the drawbacks to this approach is that the oxidation is not always confined to the surface, but can continue into the bulk of the coating. If too much oxidation occurs, off-odors may develop which can affect the product.

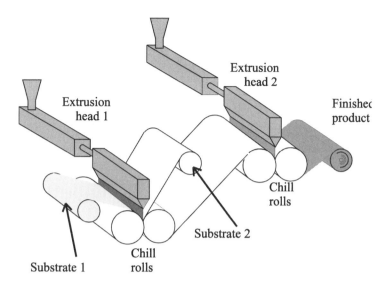

Figure 8.2 Tandem extrusion laminating line.

To reduce the need for such high temperatures, primers are used to improve adhesion. These primers are organic compounds that are coated onto the paper in-line with the extrusion coating to provide non-polar groups to which the PE can adhere, as well as polar groups which can adhere to the paper. Polyethylene imine (PEI) is one such compound. These primers allow the melt temperature to be reduced to 288°C (550 °F), so less oxidation occurs and there is less potential for off-odors.

Primers can be used on foil also, but another approach often used with foil is to use a polymer that has polar groups on the chain, such as an acid copolymer or an ionomer, for the coating. Polyethylene acrylic acid is a typical acid copolymer. Ionomers, as discussed in Section 4.1.4, are partially neutralized acid copolymers, of which Surlyn™ is the best known. Acid copolymers and ionomers can also be used as sealant layers.

Because the polymer melt is expected to drop though an open space before contacting the web, the polymer must have sufficient melt strength to support its weight; also the polymer *curtain* should not contract too much in the cross-direction (neck-in) during the operation (Fig. 8.3). Melt strength requirements are greater and neck-in is more of a problem in extrusion coating than in film extrusion because of the greater distance the polymer curtain travels without support. The distance between the die and the chill roll (the air gap) is typically 6 to 8 inches in extrusion coating, while in film or sheet extrusion it is much less.

These special requirements must be provided by the polymer molecular architecture. A polymer used for extrusion coating and laminating needs to have reasonably high elongational viscosity, and the melt should exhibit strain hardening at high elongation rates. Strain hardening means the resistance to deformation increases at a more rapid rate as deformation continues (the slope of the stress-strain curve increases). Both of these characteristics keep the melt from pulling apart in the MD, and from necking-in too much in the CD. Not all polymers that one might want to use for this operation have these characteristics. However, if a polymer that has certain desirable properties does not have the necessary melt characteristics, using coextrusion it can be carried on another polymer layer that does have the required characteristics. LLDPE's are examples of polymers with poor elongational melt characteristics, but they are desirable as sealant layers in many applications because of their hot tack. They can be used for extrusion coating in a coextruded structure where they are carried on a regular extrusion-coating-grade LDPE layer.

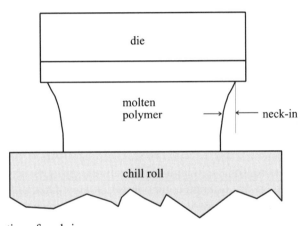

Figure 8.3 Illustration of neck-in.

The tandem coating line seen in Figure 8.2 can be used to reduce the cost of producing products that require both laminating and coating, such as a paper/polyethylene/foil/sealant type of structure. If this product is made in two passes through the machinery, the cost is higher because of the increased cost of handling the materials twice and the increased start-up and shut-down waste.

8.2   Hot Melt Lamination or Coating

Hot melt laminating is a close relative of extrusion laminating. This process uses a lower molecular weight adhesive polymer that does not need an extruder to melt and pump it. The material's viscosity in the molten state can be low enough for gravity delivery to the die, but most of the time a pump is used for more accurate control. The process beyond the die is similar to extrusion lamination; there is a chill roll nip and a take-up winder. The most frequent use of this process is for coating a sealant material on a web. It allows the use of a material with a much lower melting point, which can be advantageous in achieving high production rates.

8.3   Adhesive Lamination

The process where a solution or emulsion of a low molecular weight polymer adhesive material is coated onto the surface of one web, before the joining of the second web, is typically called *adhesive lamination*. If the adhesive layer is dried before the joining of the second web, this is called *dry lamination* or *dry bonding*. If the webs are joined while the adhesive is still wet, it is called *wet lamination* or *wet bonding*. For wet laminating, one of the webs must be porous to allow the evaporation of the water or solvent from the adhesive layer. A form of wet laminating can be done by using reactive adhesives which dry by cross-linking, and thus give off no volatiles, but this is usually called solventless adhesive lamination.

The process equipment for wet or dry laminating is rather similar as shown in Figures 8.4 and 8.5. In both cases, the adhesive is applied by a roll applicator system. In dry laminating, the oven for drying comes in front of the nip for laminating, and in wet laminating it comes afterwards.

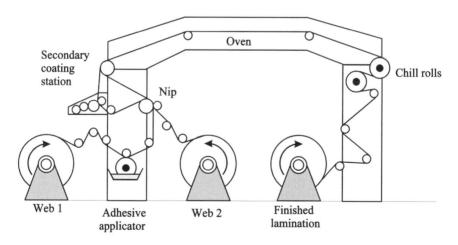

Figure 8.4 Wet laminating.

As mentioned above, the adhesive is applied to the web by a roll applicator system. These systems require some method of metering the amount of adhesive applied. This can be done by a blade on a gravure roll, a Mayer rod, or a doctor blade, as shown in

Figure 8.6. The gravure roll is a metal roll covered with engraved cells. The coating is picked up from the coating pan; the excess is doctored from the roll by a metal blade running in contact with it. A Mayer rod is a stainless steel rod, wrapped with a wire that is rotated while in contact with the web after the adhesive has been applied. The excess adhesive is forced onto the rod, and flows back to the coating pan. A doctor blade is set at a certain distance from the web and removes any excess coating as the web runs by. The gravure roll applicator requires a low coating viscosity; the other systems are less viscosity sensitive.

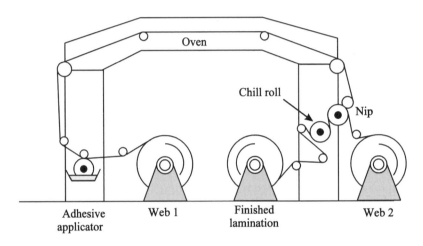

Figure 8.5 Dry laminating.

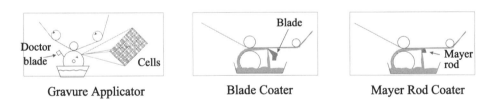

Gravure Applicator   Blade Coater   Mayer Rod Coater

Figure 8.6 Various roll coater set-ups.

Adhesive lamination can also be done as a part of a printing press operation, if the press is properly equipped. This is the way many OPP/ink/adhesive/OPP/cold seal type structures are made. In this case the laminating station, with its secondary unwind and nip rolls, is inserted in line with the press. An additional printing station following the laminating section applies the cold seal adhesive.

8.4   Thermal Laminating

In thermal laminating, heat energy is used to join two substrates together, in a process analogous to heat sealing. One or both of the materials are heated, and then the two substrates are pressed together and cooled, as shown in Figure 8.7. At least one of the layers must contain a plastic on the joining surface that can be heat-sealed to the joining surface of the other substrate. The materials that can be used, then, are similar to those used in extrusion laminating. The main difference is that in thermal laminating, all the layers of the two substrates are already in place; no additional adhesive material is being added during the laminating step.

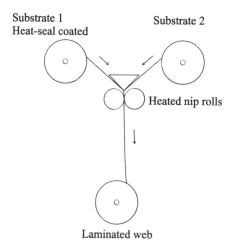

Figure 8.7 Thermal laminating (reprinted with permission from [1]).

8.5   Metallized Film

Metallized film contains an extremely thin layer of aluminum, which acts to greatly increase the barrier characteristics of the material, as well as to provide a shiny metallic appearance. Because the layer is so thin, it does not appreciably affect the mechanical properties of the film, such as strength and flexibility.

Metallization is done in a special chamber under a very high vacuum, usually as a batch process. As shown in Figure 8.8, aluminum particles from a wire which is vaporized in a crucible, or boat, at about 1700°C (3100°F) condense on the film being metallized as it runs past a cool roll. The layer of aluminum is typically 400 to 500 Å in thickness, and is commonly specified in terms of light transmission, optical density, or electrical resistance. Thickness is controlled by the speed of the film, number of plating stations, and metal temperature, with a higher temperature resulting in a thicker coating. The film can travel at speeds up to 500 m/min. It passes through rollers that adjust film tension and provide cooling, and it is then wound up on a take-up roll. Roll widths are typically 24 to 84 in. Coating an average sized roll takes about 30 minutes.

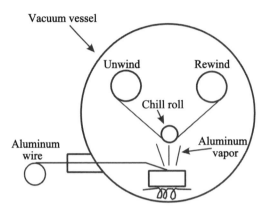

Figure 8.8 Metallization of film.

The barrier properties of metallized film approach those of aluminum foil. When new unflexed materials are compared, foil is a better barrier, but flexing of the foil can cause flex-cracking and significantly decrease its barrier. Metallized film is not as subject to flex-cracking as foil, so it is generally somewhat better at maintaining its barrier. Consequently, after flexing, the barrier provided by a metallized film can in some cases be superior to that of a foil lamination. Coupled with the lower cost of metallized film, this has resulted in considerable replacement of relatively heavy foil laminations with thin metallized film packages.   A variety of plastics can be successfully metallized. The structures used most often include biaxially oriented nylon, PET, and PP. Metallized films are opaque to visible and microwave radiation, which is a disadvantage when transparency or microwavability is desired. There are techniques for metallizing only certain areas of the film, or for removing the metallization from certain areas, which can provide visibility while also, of course, reducing barrier.

8.6 Silicon Oxide Films

Silicon oxide films contain a thin surface layer of silicon oxide, a glass-like material that acts to improve barrier in much the same way as the aluminum in metallized film. Because the layer is extremely thin, the mechanical properties of the resulting material are essentially the same as the uncoated film. The SiO_x layer may impart a slight yellowish color to the film, but, in contrast to metallized materials, preserves its transparency. This is particularly significant for microwavable packaging. The base film is usually 12.5-25 μm (0.5-1.0 mil) PET, though polypropylene, polystyrene, and polyamides can be used.

There are three major processes for applying the silicon oxide coating: evaporation, sputtering, and chemical plasma.

The evaporation process is very similar to vacuum metallizing, though it requires a much better vacuum. A silicon source is vaporized in a vacuum chamber, using a high-energy electron beam. Only about 20% of the SiO_x vapor is deposited on the film; the rest is deposited on the inside of the chamber. The glassy SiO_x layer usually has a thickness of 400 to 1000 Å, depending on process conditions. The chemical formula in the layer is not exactly SiO_2. The actual average number of oxygen atoms per silicon atom lies somewhere between 1.0 and 2.0. The yellowing comes from the SiO structure, so the higher the amount of this material, the greater the color. Very good barrier can be obtained, with exact results depending on the base film as well as on the thickness of the SiO_x layer.

The sputtering method uses bombardment by argon ions to dislodge silicon atoms from a target rather than heating and vaporization. This method is slower than evaporation, and is generally regarded as not cost-effective.

Chemical plasma deposition is very different in concept. It uses a chemical plasma created in a silicon-containing gas such as tetramethyldisiloxane or hexamethyl-disiloxane, plus oxygen and helium, to produce the SiO_x coating. Because there is only minimal heat, this process can be used with heat sensitive materials such as LDPE and oriented PP. The degree of vacuum required is less than that for evaporation, and it produces a thinner and clearer coating, with less yellowing. A key advantage of chemical plasma deposition is that it acts on all surfaces, so it can be used to coat containers as well as film.

8.7 Other Inorganic Barrier Coatings

Aluminum oxides can be used in place of silicon oxides for coating films, and combinations of SiO and MgO have also been used, resulting in transparent films with enhanced barrier properties. Evaporation using electron beams is used to deposit the aluminum oxide (or other metal oxide) on the base resin. These materials are still in relatively early stages of development.

Clay nanocomposites are also being developed as barrier coatings for film and for containers. The nanocomposite is deposited on the film from a solution of PVOH/EVOH copolymer in a mix of water and isopropyl alcohol which has been used in a supersonic dispersion system to nano-disperse 7 nm diameter silica and titanium dioxide particles. The ratio of polymer to silica depends on the barrier properties required. Typical microgravure equipment can be used to coat the solution onto a plastic substrate. The result reportedly is a transparent barrier coating which is superior to silica- and alumina-coated films, and is comparable to aluminum-coated materials. Oxygen permeability at a coating thickness of 2 μm is less than 1 cc/m^2 d atm, and moisture permeation less than 1 g/m^2 d. Costs are reported to be competitive with ceramic coatings [2].

8.8   Building Multilayer Structures

Now that we have discussed the processes that can be used to create multilayer structures, coating, coextrusion, and lamination, the question arises, which process should be used, and when? Liquid adhesive systems usually run faster than extrusion laminating and coating systems. If one is using a solvent-based system, there is the issue of whether or not any residual odors that might affect the product to be packaged remain. Water-based systems do not have solvent odors, but they are not necessarily odor-free.

For sealant layers, it is difficult, using an adhesive, to deposit and quickly dry enough polymer to create a sealant layer thick enough to seal through dust and to caulk corners or folds on a pouch. Therefore, extrusion coating is the preferred method for heat seal coatings. If the sealant film needs to have more capabilities than just sealing, e.g. barrier or carrying additives, then lamination may be needed to produce a film having all the desired properties.

Extrusion laminations are obviously used to create structures containing paper and foil, but extrusion laminating can be done to join two polymer films, also. One reason for choosing to extrusion laminate rather than adhesive laminate films is to obtain more bulk in the structure to give it more stiffness. As indicated above, extrusion lamination of a multi-layer film having a sealant layer as its outer surface may be necessary to achieve the desired performance in the packaging structure.

Obviously, the equipment that a supplier has available will affect the type of materials offered. Decisions should be based on the needs of the product and the packaging machinery available. Once one has set these goals, compromises because of business issues, such as limited supplier base, economics, etc., may be necessary.

References

1. Selke, Susan, *Understanding Plastics Packaging Technology*, Carl Hanser Verlag: Munich, 1997.
2. Moore, Stephen, *Modern Plastics*, Feb. 1999, pp. 31-32.

Study Questions

1. Discuss the similarities and differences between extrusion laminating and hot melt laminating.

2. Describe the process for making metallized film. Give two examples of packages that use metallized film. What are the major advantages and disadvantages of metallized film, compared to laminations containing aluminum foil?

3. Coextrusion, coating, and lamination can all be used to produce multilayer structures. What are the major advantages and disadvantages of each?

4. Discuss the differences between wet laminating and dry laminating. Include differences, if any, in both the processes themselves and in the characteristics of suitable materials.

5. Discuss the probable functions of each layer in the following packaging material (5 layers): LDPE/paper/LDPE/foil/LDPE (outside to inside).

6. Discuss advantages and disadvantages of metallization compared to use of silicon oxide coating.

9 Flexible Packaging

9.1 Characteristics of Flexible Packaging

The major advantage of flexible packaging is economy. Flexible packaging makes very efficient use of both materials and space. The ratio of delivered product to package material is high, and use of cube is efficient, so distribution packaging can be smaller. Storage of unfilled packages occupies very little space, especially if they are stored as webstock. Forming packages is generally rapid and simple. All of this contributes to lower cost.

The primary disadvantages of flexible packaging are its lack of convenience for the user, and lack of strength. Flexible packaging has no appreciable ability to support a load, so secondary packaging must provide any strength that is required. Flexible packages tend to be difficult to open, and they are often impossible to reseal effectively.

A number of developments have added to the consumer appeal of flexible packaging. One feature that is increasingly being used is zipper closures to provide reseal capability. Easy-peel seals have been developed to facilitate opening, and some of these provide for reclosure. Spouts have been added to some flexible packaging for dispensing, and some of these contain caps. While these features may add substantially to the package cost, they can also have a significant impact on sales, especially in markets where flexible packages compete against rigid containers.

Historically, the way to compensate for the lack of rigidity and load-bearing capacity of flexible packaging has been to combine pouches with paperboard cartons or with corrugated fiberboard boxes in bag-in-box packages. Such packages are commonly used for dry products, such as breakfast cereal, as well as for liquid products, such as inexpensive wines. Stand-up pouch designs provide enough rigidity that a product, as the name indicates, can stand on the store shelf. Use of pouches has been growing very rapidly in recent years, as new structures provide improved barrier and retortability, allowing these pouches to substitute for rigid packages such as cans, as well as for paperboard cartons. Demand for pouches in the U.S. is expected to reach nearly 80 billion units by 2006, for a total value of $4.6 billion [1]. Flexible packaging, for a number of years, has been the most rapidly growing segment of the packaging industry, accounting for about half of all use of plastics in packaging.

The simplest form of flexible packaging is a wrap, a flat piece of material designed to the folded around the product in some way. Stretch wrap and shrink wrap are the most commonly used, and were discussed in Chapter 6. In this chapter, we will consider packages that have some type of form, where some assembly of the plastic film into a package has occurred. These packages can be categorized as bags, envelopes, sacks, and pouches. All are made by folding and sealing the plastic together in some way. These terms are not clearly defined, and there is, consequently, quite a bit of overlap and confusion in the way the terms are used. Rather than attempting to order the flexible packaging universe into these neat categories, we will, instead, refer to such packages generically as pouches, when talking about smaller packages, and bags when talking about large bulk packages.

9.2 Pouch Styles

Pouches can be made in a variety of styles. However, the majority can be categorized into one of four groups: pillow pouches, three-side seal pouches, four-side seal pouches, and stand-up pouches.

9.2.1 Pillow Pouches

Pillow pouches (Fig. 9.1) are constructed with seams at the back, top, and bottom. They tend to billow in the middle, giving these pouches their characteristic "pillow" shape, as seen in potato chip bags, for example. The seam or seams in a pouch can be made in a variety of ways, but the most common, by far, is heat sealing, which was discussed in Chapter 6. The back seam of a pillow pouch is most often a *fin* seal, where the two inner layers of the material are brought together and sealed, resulting in a seal which protrudes from the back of the finished pouch. Sometimes a *lap* seal is used, where the inner layer of one side is sealed to the outer layer of the other side, producing a flat seal. The fin seal produces a stronger seal and requires only the inner layer of the material to be heat-sealable. The lap seal uses slightly less material, but is weaker and requires both the inner and outer layers of the pouch to be heat-sealable. The top and bottom seals are made between inner layers of the pouch material, analogous to the fin seal. If desired, gussets can be incorporated into the pouch design, permitting the pouch to expand in cross-section to a larger degree.

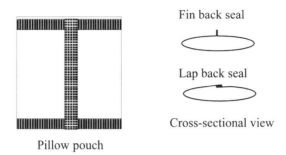

Pillow pouch

Figure 9.1 Pillow pouches and fin and lap seals (adapted with permission from [2]).

9.2.2   Three-side seal pouches

Three-side seal pouches (Fig. 9.2a) contain two side seams and a top (or bottom) seam. The remaining side contains a fold. All seals are made between inner layers of the pouch material. If the pouch design incorporates gussets, these pouches can be capable of standing erect. In some applications, three-side seal pouches are used, but the side with the fold is also sealed, thus blurring the lines between three- and four-side seal pouches.

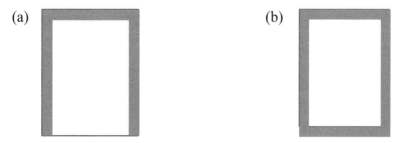

Figure 9.2 Pouches, (a) three-side seal, (b) four-side seal (reprinted with permission from [2].

9.2.3   Four-side Seal Pouches

Four-side seal pouches (Fig. 9.2b) are made from two webs of material, which are sealed together at the top, bottom, and both sides of the pouch. Unlike the pouches previously discussed, the front and back sides of the four-side seal pouch can be made from different materials. A variation on this idea is production of pouches in non-rectangular shapes. These pouches may have a variety of decorative shapes, but are alike in being made from two pieces of flexible materials, sealed all the way around.

9.2.4   Stand-up Pouches

The term stand-up pouch refers to a pouch that is able to support itself in an upright position when it is filled with product. Such pouches contain gussets that expand when the product is added, providing a relatively flat base to the pouch. Such pouches are increasingly replacing more rigid structures such as cans and bottles, for both dry products and liquids.

Design of the gusset, usually a horizontal bottom gusset, is critical in pouch performance. It can be made from a separate piece of material, sealed to the rest of the pouch, or the pouch material can be folded into a W shape to produce the base. Sometimes side folds or gussets are used, with an overlapping flat sheet base. Such pouches tend not to stand up as securely as do those containing a bottom gusset, so they are used more frequently for dry products than for liquids. In both cases, the design usually depends on the weight of the product to cause the gusset to spread out, so the pouch will stand up only when it contains enough product. If a pouch that can stand up even when empty is desired, a design which provides a rectangular bottom shape, or which can be folded into such a shape, can be used.

Some experts predict the North American market for stand-up pouches, which was only about 2 billion units in 1998 but reached over 4 billion by 2001, will exceed 8 billion units by 2006. Use in Western Europe has been greater than in North America, at about 2.3 billion units in 1998, and is projected to exceed 11 billion units in 2006 [3,4]. Pouches are also very popular in Japan for food, personal products, and household chemicals. Much of this growth has been made possible by recent design innovations in flexible packaging. These pouches are now available with snap caps and threaded closures, nozzles, and zippers for resealability. One company has even introduced a low cost hook-and-loop ventilating closure for flexible packaging. These innovations are permitting pouches to invade markets customarily held by rigid containers.

While changing from a rigid to a flexible package often offers savings in material and hence often in cost, as well as resulting in less waste for disposal, some experts see the better display graphics offered by the pouches compared to competitive packaging as the true driving force for the rapid growth in their use. The pouches are printed as rollstock, facilitating the use of high-quality multicolor images. The upright presentation makes the product readily visible to the consumer. Several shaped standup pouches have been introduced where the non-rectangular design is a significant advantage in catching the eye of the consumer and appealing to them, particularly in products designed for children.

Technological innovations in production of high barrier materials have also been important in ability to use pouches for sensitive products. Many such pouches fall into the dual category of standup retort pouches, and will be discussed in section 9.4.

A remaining drawback to the use of pouches is their slow line speeds, which for beverage packaging is often only about half the speed used with bottles of the same capacity.

9.3   Forming Pouches

The most common way to make pouches (and to package products in pouches) is to use a *form-fill-seal* (FFS) machine, in which pre-printed roll stock is formed into a package and the package is filled and sealed with product, all in a continuous operation within one piece of equipment. Cutting the pouches apart is usually accomplished within the FFS machine, as well.

Two configurations, vertical and horizontal, are defined by the direction of travel of the package through the machine (Figs. 9.3 and 9.4). The pouches are always produced and filled vertically in a vertical FFS machine, and can be produced and filled either vertically or horizontally in a horizontal FFS machine. A variety of pouch types can be made on either type of equipment. The sealing and cutting apart can be done simultaneously, or the pouches can first be sealed, and then cut apart at a subsequent station.

An alternative to form-fill-seal equipment is to use preformed pouches. In this case, the preprinted pouch is supplied ready to be filled with product, and then after it is filled the top seal is made. In such cases, filling and sealing are most often done on two separate pieces of equipment.

Both form-fill-seal and preformed pouches have advantages and disadvantages. For large operations using materials which seal readily, form-fill-seal operations are usually the most economical. However, use of preformed pouches requires less capital investment, since the equipment is simpler and less expensive. It also requires less quality control, since only one seal must be monitored. Therefore, for low volume operations or materials that are difficult to seal correctly, use of preformed pouches can be advantageous. Consequently, most moderate-to-high volume packaging pouch operations use form-fill-seal technology, but operations using retort pouches or stand-up pouches are an exception, most often using preformed pouches.

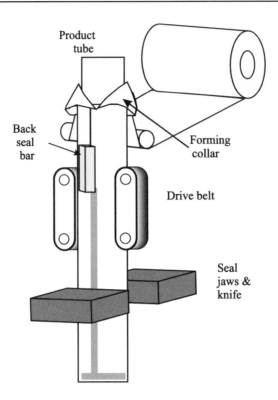

Figure 9.3 Vertical form-fill-seal machine.

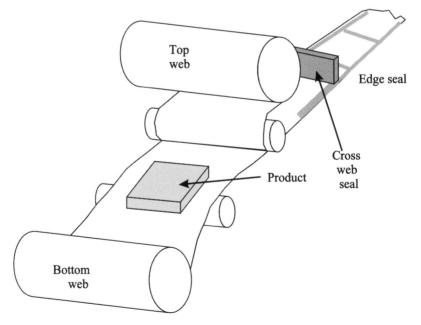

Figure 9.5 Horizontal form-fill-seal machine.

9.4   Retort Pouches

Retort pouches are pouches that are designed to be filled, usually with a food product, and then retorted (heat-sterilized in a procedure analogous to canning) to produce a shelf stable product, one that does not require refrigeration. Retort packages some time ago replaced cans in the U.S. military MRE (meals ready to eat) program. Their flexibility, smaller volume, and much lighter weight than cans are a significant advantage. In the consumer segment of the market, retort pouches have, until recently, been much less successful. They were introduced by a number of companies, and generally failed to win consumer acceptance. The major consumer packaging use for many years remained a small market for foods targeted at backpackers and other campers.

The initial design for retort pouches, and the one still used by the military, was a multilayer lamination containing an outside layer of polyester, a layer of aluminum foil, and an inside layer of polypropylene. The polyester provides strength and puncture resistance, the aluminum provides barrier, and the polypropylene provides the sealant and product contact layer. A significant disadvantage of this structure is that the food cannot be heated within the pouch by microwaving.

There are obvious trade-offs between choosing a material which is easy to seal, and choosing a material whose seal will remain strong at the elevated temperatures reached during retorting. Consequently, the retort pouch is not easy to seal. In addition to the difficulty in working with polypropylene as the sealant layer, to ensure sterility, any wrinkling in the seal area must be eliminated. Therefore efficient manufacture of these pouches is difficult. Nearly all operations using retort pouches buy preformed pouches rather than using form-fill-seal systems, letting the experts deal with producing all but the final seal.

In the last few years, the retort pouch, especially in its stand-up variations, has started to take off, replacing cans or bottles in a number of significant applications. In addition to the advantages associated with flexible packaging in general, retort pouches provide an additional advantage. Because of their thin profile and high ratio of surface area to volume, food products can be sterilized in less time, typically 30% to 50% less than is required for canning, and sometimes even more. This results in greater retention of product quality. Simply put, products in retort pouches taste better than equivalent products processed in cans. The products also look better, and have greater nutritional value.

Development of improved sealing layers has facilitated sealing of retort packages. Developments in filling equipment permit preheating of the package, injection of steam or nitrogen into the headspace to minimized the amount of oxygen in the pouch and increase shelf life, and more rapid line speeds. Some retort pouches now even incorporate zippers for reclosure. A variety of complex ultra-high barrier laminate structures are now available as alternatives to the old aluminum foil structures. Retort pouches have been even more successful in a market few consumers see - replacing the large institutional size cans used by food service operations such as cafeterias and restaurants.

While retort pouches continue to be more successful in Asia and Europe than in the U.S., there are clear signs that U.S. consumers at long last are embracing the advantages that retort pouches can bring. The success of StarKist™ tuna in pouches was one of the early signs. Now it is increasingly common to find pet food and a variety of other products appearing in pouches as an alternative to cans. Some have gone so far as to predict that cans will soon be on the "endangered species list" [5].

9.5   Bulk and Heavy-Duty Bags

Bulk bags and heavy-duty bags are designed for packaging large quantities of solid or liquid product. They can contain as much as 5,000 kg (11,000 lbs) and, therefore, must have high tensile strength. Woven PP fabric is usually the material of choice, although HDPE, PVC, and polyester fabric are also used. Some bags, especially in smaller sizes, are made of film, usually LLDPE or HDPE, rather than fabric. Other bags include a plastic film layer in a construction that is mostly paper. An alternative is to use the plastic as a coating on the paper. HDPE, LDPE, PVDC, PP, and combinations are all used in such applications.

Bags made of woven fabric do not generally provide sufficient containment for liquid products, so such bags often incorporate a film liner. Polyethylene is used most often, but aluminum foil or PVDC copolymers can be used to provide improved barrier ability. The liner may be a single material, or have a multilayer construction. Alternatively, the bags can be coated with PVC or latex to make them waterproof. Liners can also be used on bags for solid products. Often the liner is designed to be disposable to facilitate reuse of the bag.

The bag seams can be heat-sealed or sewn, depending on the material and the application. Some bags are made from tubular material, so sealing is required only at the top and bottom. Others have seams at the back, bottom, and top; or they can be made like three-side seal pouches, with a fold forming the bottom of the bag.

Large bulk bags are most often designed to be filled from the top and emptied from the bottom. Filling and discharge openings may be sealed, or may simply be tied shut. Spouts or other dispensing devices may be included, or the bags may simply be cut to open. Handling devices, such as loops of fabric, are often incorporated. Ultraviolet light stabilizer must be incorporated into the resin formulation if the bags will be stored and used outdoors.

9.6 Bag-in-Box

Bag-in-box structures of various kinds are very common. As discussed in Section 9.1, one of the common ways to add rigidity to a flexible package, for stacking and other purposes, is to combine it with a folding carton. The most familiar example is breakfast cereal. The product is contained in a pouch, but that pouch is itself contained in a folding carton. In addition to rigidity that facilitates shipping, stacking, and display in the store, and storage in the consumer's kitchen cabinet, the paperboard provides an excellent surface for printing, adding to the sales appeal of the package.

Bags can also be contained in corrugated boxes, when additional rigidity is desired. This is the approach taken for flexible pouches containing wine. Bag-in-box wine pouches are multilayer structures, typically containing a foil or metallized film layer for oxygen protection. The pouch has a dispensing orifice and closure attached, which in use protrudes through a special opening in the corrugated box, allowing for ease of use. Such packages provide much greater efficiency in terms of weight and use of space than glass bottles. They also provide excellent product quality over an extended period after opening, since they prevent oxygen from reaching the product and causing deterioration. However, they do not have the same consumer appeal as glass bottles. They are used almost exclusively for inexpensive brands in large quantities, or for institutional use.

References

1. Freedonia, Pouches, Study #1614, November 2002, http://www.freedoniagroup.com/pdf/1614web.pdf.
2. Selke, Susan, *Understanding Plastics Packaging Technology*, Carl Hanser Verlag: Munich, 1997.
3. Modern Plastics, March 1999, p. 28.
4. Packaging Strategies, Stand-up Pouches: Continuing Global Markets, Economics and Technologies: 2001-2006, www.packstrat.com.
5. Mykytink, Andres, Retort Flexible Packaging: The Revolution has Begun, Flexible Packaging, October 2002, pp. 18-25.

Study Questions

1. What are the major advantages and disadvantages of flexible packaging compared to rigid packaging?

2. Find an example of a pillow pouch. Describe it. Does it have a fin or lap seal? Why do you think that type of seal was chosen for that application?

3. Find an example of a stand-up pouch. Describe it.

4. Find examples of three-side seal and four-side seal pouches. Describe them.

5. Potato chip pouches often have gas inside, giving them a "blown-up" appearance like a balloon. Why do you think this is done?

6. What is an ffs operation? Describe it, including both major variations.

7. You are packaging a dry powder in three-side-seal pouches. Protection of the product from moisture is important. Selling price of the product is low, so package cost is an important consideration. Production volumes are quite high. Recommend a packaging material and justify your choice. Sketch the package. Would you recommend buying formed pouches or making your own? Why?

10   Thermoforming

10.1   Introduction

Thermoforming is a method of forming plastic sheet (or film) into useful packaging by shaping it into or around a mold. It is widely used to produce *blister packages*, in which the product is encapsulated between a molded plastic blister, usually transparent, and a backing, often paperboard. In production, thermoforming may be highly automated with the various parts of the cycle being sequenced by a computer control system. It is usually considered to be one of the least expensive ways to form three-dimensional packages from plastic. It competes in many applications with blow molding and injection molding. Since thermoforming is a low pressure forming technique, less expensive materials are required for the molds and less robust equipment is needed to form parts. In general, thermoforming is much less expensive than injection and blow molding for forming large parts. Since most packages or components made by thermoforming are not large, the process lends itself to the formation of large numbers of packages per cycle. The main drawbacks of thermoforming are the inability to form parts to very tight tolerances, and difficulty in making significant undercuts. For most packaging applications, these are not serious drawbacks. This chapter will cover the basics of thermoforming, including process steps and part design.

The thermoforming process has three basic steps: heating the sheet, forming the sheet, and trimming the part. Each of these steps must be performed correctly or the parts will not be formed properly. Generally, one determines the proper heats, cycle times, mold designs, etc., through experimentation; however, there is some fundamental science underlying these steps.

10.2   Heating the Sheet

10.2.1   Temperature Selection

The polymer sheet must be heated to get it to the proper state for forming. The optimum temperature is dependent on the polymer and on the mold design that is being used to make the part. Therefore, even if one knows the correct temperature for production of a 4-inch diameter, 1.5-inch deep polystyrene clamshell from 0.016-inch thick sheet, that might not be the correct temperature for forming the same sheet into a 9 x 4-inch, 10 compartment cookie tray. The process parameters depend on the part being manufactured, as well as the starting material.

Amorphous polymers generally have a fairly wide range of forming temperatures that can be used. The forming range generally starts about 20-30°C above the polymer's glass transition temperature (T_g), with typical forming temperatures 70-100°C above T_g. Parts must be cooled to 10-20°C below T_g before they are removed from the mold. For crystalline polymers, the range of forming temperatures is much narrower, confined to within a few degrees of the melt temperature. Special techniques such as solid phase pressure forming (SPPF) can be used to thermoform plastics at temperatures lower than their normal forming temperatures.

10.2.2   Radiative Heating

Heat can be transferred to the sheet in several ways. The most common for thermoforming is radiation from heater elements. The sheet is moved into, or through, an oven with radiant heating elements. In some thermoform/fill/seal equipment, heating is by conduction. In this case, the sheet is pulled up against a heated plate, and heated by contact.

Selection of the proper temperature for the heater requires understanding of the way in which plastics absorb radiated energy. A higher temperature setting for the heater will not necessarily mean a more rapid heating rate of the plastic. A short review of physics will aid our understanding. Temperature is a measure of molecular motion in a material. All objects in the world have radiation of various wavelengths impinging on their surfaces. This radiation is either reflected, absorbed, or transmitted. The chemical nature of the material determines the percentage of each of these at any given wavelength. A body that is a perfect emitter and absorber is called a "black body". The amount of energy emitted by a black body radiating at a given temperature is:

$$E_b = \sigma \ T^4 \tag{10.1}$$

where σ is the Stephan-Boltzmann constant, σ = 0.5674 x10^{-10} kW/m^2 °C^4. This energy will be radiated at a range of wavelengths. The amount of energy produced at any specific wavelength is given by the following equation:

$$E_{b,\lambda} = C_1\lambda^{-5}/\left[\exp(C_2/\lambda T)-1\right] \qquad (10.2)$$

where C_1= 3.734 x 10^{5}kW μm^4/ m^2, C_2= 1.439 x 10^4 °K μm, and λ is the wavelength in μm. Thus the radiation emitted by a black body depends on its temperature and spans a range of wavelengths, as shown in Fig. 10.1.

Black Body Radiation Spectrum as a Function of Temperature

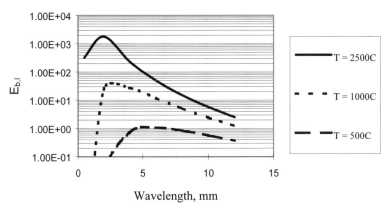

Figure 10.1 Black body radiation spectrum.

Plastic materials, even if they appear transparent in the visible spectrum, absorb energy in the infrared region. The amount of absorbance at any wavelength can be predicted from the IR spectrum of the material. Although the material may absorb over a broad range, each material has an optimum absorbance, a frequency at which it absorbs most strongly.

The ideal temperature for a radiant heater, then, is that temperature which produces the most energy at the wavelengths that the plastic material most absorbs. Table 10.1 gives the ideal wavelengths for absorbance and the matching temperature range for black body radiating heaters, calculated from Eq. 10.2, for several plastics. In some cases, plastic materials have two primary absorbance regions, leading to the calculation of two different heater temperatures. Usually operation at the lower temperature is preferred. The practical upper temperature limit on most thermoformers is 900°C, since above this temperature the materials of construction of the thermoformer begin to fail.

In reality, no material is a perfect black body, but rather emits and absorbs energy at some fraction of the ideal. This proportionality constant is called "emissivity". A heater with higher emissivity, therefore, will deliver a larger amount of energy at a given temperature than one with a lower emissivity.

In practice, the thermoformer oven temperature is set to match the theoretical best temperature for absorption of the particular polymer, and heating time is determined by trial-and-error. Obviously, one should discuss the heating temperatures with the equipment manufacturer and the plastic supplier for recommendations. The heaters usually used in thermoformers are either calrod type or quartz lamps. Less frequently, other types of heaters are used.

Table 10.1 Ideal Radiant Heater Temperatures for Various Plastics

Plastic Material	Ideal Wave Length μm	Temperature Range °C
LDPE	3.2 - 3.9	470-630
HDPE	3.2 - 3.7	510-630
PS	3.2 - 3.7	510-630
	(6.4 -7.4)	120-180
PVC	1.65 - 1.8	1340-1480
	(2.2.- 2.5)	885-1045
PMMA	1.4 - 2.	1910-3265
Nylon 66	11- 2.8	765-1250
	(3.4 - 5)	310-580
Cellulose Acetate	21- 3.6	990-1910
	(5.2 - 6)	440-545

10.3   Forming the Sheet

10.3.1   Basic Methods

There are many methods for molding the sheet of plastic, once it has been softened by heat. The simplest are drape forming, vacuum forming, and pressure forming. Other thermoforming techniques are modifications of these basic methods.

10.3.1.1 Drape Forming

In drape forming (Fig. 10.2), the main forming force is the influence of gravity on the hot plastic sheet, Typically, a positive, or male mold of the part, with a convex shape, is used. The forming is assisted by the application of vacuum to pull the material down around the shape.

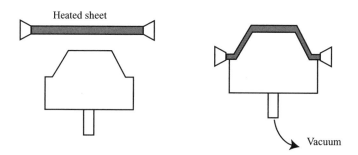

Heated sheet

Vacuum

Figure 10.2 Drape forming.

10.3.1.2 Vacuum Forming

In vacuum forming (Fig. 10.3), the main forming force is air pressure, caused by normal atmospheric pressure on one side of the sheet and a near vacuum on the other side. Typically a negative, or female mold, with a concave shape is used. The heated sheet is clamped onto the mold, and vacuum applied, so that the sheet is forced into the shape.

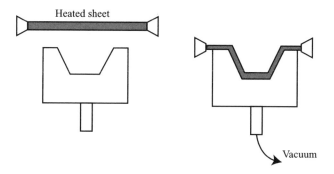

Heated sheet

Vacuum

Figure 10.3 Vacuum forming.

10.3.1.3 Pressure Forming

Pressure forming (Fig. 10.4) uses additional pressure to form the sheet. Either a positive or a negative mold can be used. Most often air pressure greater than one atmosphere is supplied to one side of the sheet, while a vacuum is drawn on the other side. Pressures of 20-80 psi are common, but pressures as high as 500 psi are sometimes used. Pressure forming can achieve better detail and allow more rapid forming than forming methods that have a maximum driving force of one atmosphere.

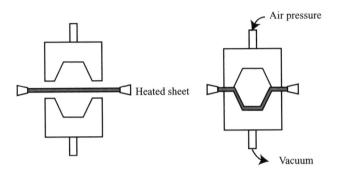

Figure 10.4 Pressure forming.

10.3.2   Sheet Deformation

As the sheet is stretched into the shape of the part, it thins and cools. As soon as the sheet touches the mold surface, dramatic cooling takes place and the change in dimensions of that portion of the part is nearly stopped. Drawing of the sheet involves three-dimensional deformation, because as the planar dimensions increase, the thickness decreases, as shown in Figure 10.5. Thus, the choice between use of male and female molds has implications for the characteristics of the final part. Imagine a bowl shape as the final package. If this shape is produced with a male mold, the first contact between the mold and the sheet will occur at the base of the bowl, so this will be the first to stop deforming, and consequently the thickest section. If that same part is produced with a female mold, the first contact is at the rim of the bowl, and the last at the base. Consequently, the thickest section of the bowl will be at the rim.

The polymer at forming temperature can behave as an elastic solid, a viscoelastic fluid, or a combination of the two. Modeling of thermoforming has been done using all of these models. On a molecular scale, things are just as complex. For an amorphous material, such as polystyrene or PMMA, the forming temperature determines the chain mobility and the ease of flow. For semi-crystalline materials, such as polyolefins, nylons, and polyesters, the chains that are not crystalline will deform more readily; but if enough heat is present, and enough energy for deformation, then crystallites can also be unraveled and rearranged. Because during the deformation and cooling, strains are frozen into the material, a thermoformed part will want to return to the flat sheet form if it is reheated to a temperature near the forming temperature. For materials such as CPET that are used for packages intended to be heated in an oven, it is necessary to heat-set the material. Heat setting involves heating the part, under constraint, in the mold to a temperature above that which it will be exposed to during use, to allow the strains to relax.

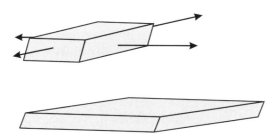

Figure 10.5 3-D dimensional change.

Some polymeric sheet work hardens during deformation. Deformation of the sheet involves the flow of molecules. As the deformation proceeds, the amorphous regions become fully extended, and unraveling of the entanglements cannot occur rapidly enough to allow the deformation to continue at the previous rate. The result is that the force necessary for continued deformation increases; this is called work hardening. Semi-crystalline polymers, such as polyolefins, nylons, and polyesters, work harden more than amorphous polymers because the crystalline regions have great difficulty in unraveling during the stretching, and act as pseudo-crosslinks. Heating the sheet to a higher temperature avoids work hardening. However, at some temperature the sheet begins to sag and becomes impossible to handle through the equipment. If work hardening is a problem, the usual solution is to reduce the deformation rate.

There are methods for estimating the pressures necessary for forming and the optimum temperatures. However, these techniques are based on constant strain rates and constant temperatures, neither of which exists in real thermoforming applications. For that reason, processing parameters are usually determined by experimentation.

10.3.3 Thermoforming Variations

One of the most common problems in thermoforming is non-uniform walls. If a draw is too deep, a corner too sharp, etc., non-uniform thinning may result. A number of thermoforming variations are directed towards alleviating problems associated with nonuniformity of wall thickness. Other thermoforming variations are designed primarily to increase production speeds. Some do both.

10.3.3.1 Plug-Assist Thermoforming

In plug-assist thermoforming, a mechanical device is used to help stretch the sheet into the mold, as shown in Figure 10.6. Where the plug touches the material, it is cooled, and the material does not stretch much more. The optimal shape for the plug is difficult to predict. Again a trial-and-error approach is usually employed, starting with a plug that may be bigger than needed. One can always cut material away from a plug, but adding material back is not usually feasible. Plugs can be made from thermally insulating material or thermally conducting material, or the plug can be temperature-controlled. Most often, they are simply made from hard wood. The plug typically occupies 60-90% of the cavity volume of the mold. Often plug assist is combined with pressure forming, in plug-assist pressure forming, with the air pressure being supplied through the plug.

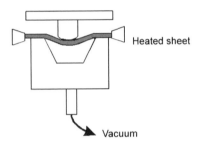

Figure 10.6 Plug assist thermoforming.

10.3.3.2 Solid Phase Pressure Forming

Solid phase pressure forming (SPPF) combines the use of pressure forming and plug-assist to form crystalline plastics at temperatures below their normal forming range, 5-8% below the crystalline melting temperature, in order to produce stiffer parts with less material.

10.3.3.3 Bubble or Billow Forming

Another method of compensating for uneven draws is to pre-stretch the sheet above the cavity, in bubble or billow forming, as shown in Figure. 10.7. This may be sufficient by itself, or it may be used in combination with plug assist. The primary advantage of bubble forming is that during the initial stretching of the plastic, the main body of material is not in contact with the mold, so stretching occurs evenly. Thus, deeper draws are possible, with more uniform wall thickness.

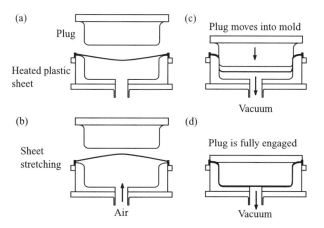

(a) Plug

Heated plastic sheet

(b) Sheet stretching

Air

(c) Plug moves into mold

Vacuum

(d) Plug is fully engaged

Vacuum

Figure 10.7 Billow forming with plug assist (reprinted with permission from [1]).

10.3.3.4 Vacuum Snap-Back Thermoforming

Vacuum snap-back thermoforming (Fig. 10.8) is similar in appear to plug assist. However, the negative cavity is not the mold, but simply a vacuum box for pre-stretching of the plastic. The actual mold is positive, taking the place of the plug. The heated plastic is drawn partially into the vacuum box, using a low vacuum, as the mold is lowered. Next, a high vacuum is drawn through the mold, as the vacuum in the vacuum box is stopped. This causes the plastic to "snap back" onto the positive mold. In a variation, bubble forming can be used before the initial draw into the vacuum box.

10.3.3.5 Matched Mold Forming

In matched mold thermoforming (Fig. 10.9), matched sets of negative and positive molds are used to form the part from the softened sheet. Vacuum is generally applied through the negative cavity, as well. The result is excellent dimensional control, and the ability to form very complex shapes. This method is most often used in thermoforming of foams, due to their tendency to deform if not restrained.

10.3.3.6 Scrapless Thermoforming

The scrapless thermoforming method (Fig. 10.10) was invented to allow the economical use of expensive high barrier coextruded materials in thermoforming containers where the high scrap rate would otherwise be prohibitive. The sheet is cut into rectangular sections, which can be done with little scrap. These rectangles are then formed into thinner disks using matched mold solid-phase forming. Finally, solid phase plug-assist pressure forming is used to make the final container.

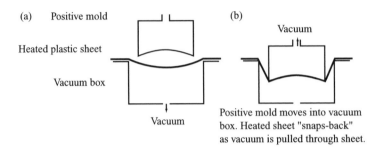

(a)   Positive mold

Heated plastic sheet

Vacuum box

Vacuum

(b)

Vacuum

Positive mold moves into vacuum box. Heated sheet "snaps-back" as vacuum is pulled through sheet.

(c)   Positive mold moves out of box

(d)

Molded product removed, ready for trimming.

Figure 10.8 Vacuum snap-back thermoforming (reprinted with permission from [1]).

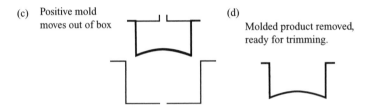

Top half of matched mold

Heat-softened sheet

Bottom half of matched mold

Mold closes.
Vacuum map be applied through bottom.

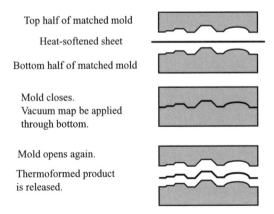

Mold opens again.

Thermoformed product is released.

Figure 10.9 Matched mold thermoforming (reprinted with permission from [1]).

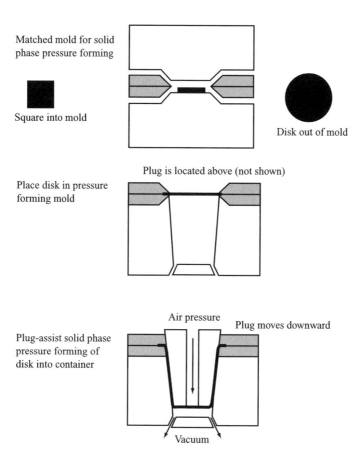

Figure 10.10 Scrapless thermoforming (reprinted with permission from [1]).

10.3.3.7 In-Line Thermoforming and Melt-to-Mold Thermoforming

In-line thermoforming differs from the thermoforming variations discussed previously in that it is not a specific type of thermoforming, but instead reflects the relationship between the thermoforming operation and the upstream production of the sheet. In in-line thermoforming, placement of the thermoforming operation immediately after sheet production is used to decrease energy use and handling requirements. It allows residual heat in the sheet to be used to provide part of the heat-softening required for the thermoforming operation.

Melt-to-mold thermoforming takes this to the ultimate step by extruding the melted plastic directly onto a chilled chrome cylinder that contains the thermoforming dies. In addition to energy savings, this process reduces thermal stress in the finished containers.

10.3.3.8 Twin-Sheet Thermoforming

Hollow objects can be produced using twin-sheet thermoforming, in which two sheets of plastic are thermoformed and sealed together. This can be done sequentially, with one sheet formed first and then the other added and formed, or simultaneously, with both sheets formed and sealed together in a single step (Fig. 10.11).

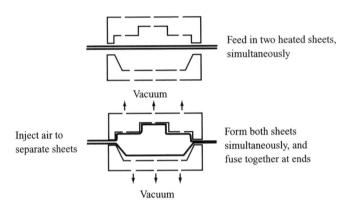

Figure 10.11 Twin-sheet thermoforming, simultaneous forming (reprinted with permission from [1]).

10.3.3.9 Skin Packaging

The skin packaging method (Fig. 10.12) takes quite a different approach to thermoforming. It uses plastic film instead of sheet, and the product itself becomes the mold. The product is placed on a backing material, which is typically a heat-seal coated paperboard. The heated film is lowered over the product and the backing, and drawn down tightly by a vacuum pulled through the backing. The film then heat-seals to the backing, with the product held tightly in place. This process is sometimes called *pneumatic sealing* and is frequently used for retail goods as an alternative to blister packaging.

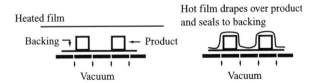

Figure 10.12 Skin packaging (reprinted with permission from [1]).

10.3.4 Selection of Thermoforming Method

The type of thermoforming that should be chosen depends on the product performance required, as well as on the material characteristics. In medical device packaging, for example, very complex cavities are often the norm. In order to form these complex shapes uniformly, a combination of plugs, clamps, and inter-cavity vacuum ports may be necessary to make uniform parts. For parts that are to be heat sealed on a flange, uniformity of the flange is essential for good seals. Some mold designs utilize a flange cooling station to decrease the stretch of material in that area. If this is not done, then plug assist will usually be a requirement in order to correctly draw the material in the flange area. Obviously from the preceding discussion, part design is critical to the successful implementation of thermoforming. In Section 10.5, part design as well as mold design will be discussed.

10.4   Trimming the Sheet

Once the part is formed, the excess material must be removed from the part. This can be done in a wide variety of ways, but is most successful when the stress-strain characteristics of the plastic at the trimming temperature are considered. Thin gauge parts in a roll-fed process are usually trimmed in the mold. Heavy gauge materials, which are usually sheet fed, are often trimmed in an external process step. Many packaging applications are thin gauge, roll-fed parts and are therefore trimmed in the mold. Cutting of the polymer involves tensile fracture, so the choice of trimming process depends on the thickness of the material, and on temperature, which affects the modulus of the material.

For most packaging applications, compression cutting or shear cutting are used. LDPE is most often compression cut because its low modulus leads to necking in a shear die, and therefore a ragged edge is formed. For HDPE, PP, PMMA, PS, nylons, and PET, shear cutting can be used very effectively. However, the same die will not work equally well for each material because of the differences in modulus.

10.5   Part and Mold Design

In designing a new thermoformed part, it is necessary to consider the application, the function of the part, and its lifetime. Each of these areas has many sub-parts that must be considered also, in order to select the material and refine the design concept. Along with considering thermoforming for part manufacture, it is necessary to consider alternative processes and determine if these alternatives would yield better functionality and/or cost effectiveness.

The most critical design issue is *draw ratio*. As mentioned earlier, most applications require reasonably uniform wall thickness in the part. The depth of draw divided by the areal extent of the draw defines the areal draw ratio. Areal draw ratios for various shapes are given in Table 10.2. A more meaningful measure of draw is the ratio of the depth of the draw to the mold opening at the rim, H:d. Many materials have limiting values of H:d that cannot be exceeded without excessive thinning of the sheet. For example, polystyrene foam cannot be drawn to H:d > 1:1; crosslinked polyethylene foam cannot exceed 0.6-0.8:1. Study of the literature and discussion with suppliers can guide the packager in determining the maximum draw ratio that can be used with a given material. Thinning of material in corners is a particular problem. If the radius of a corner is not sufficiently large, the draw into the corner causes excessive thinning. Sharp corners on parts also act as stress concentrators that lead to fractures on impact. Therefore, most thermoformed parts have tapered profiles and corner radii of at least 8 mm. The tapered profile also facilitates nesting of parts for shipping.

10.5.1   Prototype Molds

Once the part has been designed, it must be molded. The normal procedure is to make a prototype mold to determine the process parameters and any part modifications that are necessary before making a more costly production tool. Prototype molds can be made rather simply. Often they are constructed from wood or plaster. Thermoset plastics can also be used. None of these materials are good heat conductors, but the rate of part manufacture is slow during the prototyping stage, so this is not a problem. Prototype molds can be made by hand, but are most often machined. This allows the tool shop to refine the tool paths and techniques that they will be using for manufacturing the production mold. Part drawings created in most CAD software programs can be electronically translated into tool paths for CNC milling machines by other types of software. Generally, a sanded surface on the mold is satisfactory, since the low pressure does not cause surface replication on the part to be an issue.

The mold begins with machining of the basic cavity or the basic male form. In choosing between male and female molds, there are several important considerations. One is the draft angle (slope of the sides of the cavity or male form). This needs to be large enough to allow easy removal of the part from the mold. When the plastic shrinks, it will pull away from a female mold, but adhere more tightly to a male mold. Therefore draft angles on negative molds are typically at least 2-3°, and on positive molds 5-7°. Another consideration is cost. Positive molds are generally cheaper to produce, but more likely to be damaged. Additionally, the location of any molded-in surface features must

Table 10.2 Areal Draw Ratios for Regular Shapes

Shape	Figure	Area	Areal Draw Ratio, Ra
Hemisphere		$2\pi R^2$	2
Right Cylinder		$\pi R^2 + 2\pi Rh$	$1+2(h/R)$
Right Cone		$\pi R(R^2 + h^2)$	$[1+(h/R)^2]^{1/2}$
Truncated Cone		$\pi R_2^2 +\pi(R_1+R_2) *$ $(R_2/R_1)^2+(1+R_2/R_1)$	$[(R_1-R_2)^2+h^2]^{1/2}$ $*[(1/R_2/R_1)^2+(h/R_1)^2]^{1/2}$
Cube (side = a)		$5a$	5
Parallelepiped $a * b * h$		$2ah + 2ab + 2bh$	$2(1+h/b+h/a)$
Wedge $a * b * h$		$ah + 2b(h^2 + (a/2)^2)^{1/2}$	$(h/b)+[1+(2h/a)^2]^{1/2}$
Pyramid $a * b * h$		$(ab/2)[1+(2h/b)^2]^{1/2} +$ $(1/2)(ab/2)[1+(2h/a)^2]^{1/2}$	$[1+(2h/b)^2]^{1/2} +$ $(1/2)[1+(2h/a)^2]^{1/2}$
Frustrum of Pyramid $a_1 * b_1 * h$ and $a_2 *b_2$		$a_2b_2+(a_1-a_2)[(b_1/2-$ $b_2/2)^2+h^2]^{1/2} + (b_1-b_2)*$ $[(a_1/2-a_2/2)^2+h^2]^{1/2}$	$(a_2b_2/a_1b_1)+(\frac{1}{2})(1-a_2/a_1)$ $* (1-b_2/b_1)(\{1+[2h/(1-$ $b_2/b_1)]^2\}^{1/2}+ 1+$ $\{[2h/(1-a_2/a_1)]^2\}^{1/2})$

be considered. Surface finishes, identification symbols, etc., which are intended to be transferred to the finished part will appear on the surface which is in contact with the mold.

Next, vent holes must be added to the mold. The air in the mold must removed by vacuum, and it is best if this can be done rapidly. The vent holes are usually placed in the last part of the mold form to be contacted by the plastic.

The diameter of the vent holes is limited by the acceptable height of the nipples that will form on the mold-side of the part. The drawing of the plastic into the vent holes is dependent on the diameter of the hole, the modulus and Poisson's ratio of the polymer, and the sheet thickness. The problem is similar to a beam suspended across an opening, as shown in Figure 10.13.

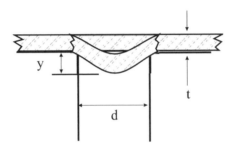

Figure 10.13 Vent hole.

The deflection, y, can be described by the equation

$$y = \alpha\, t = \left(\frac{3q\,d^4}{16\,E\,t^3}\right)\frac{(5+v)}{(1-v)} \tag{10.3}$$

where q = applied pressure   d = diameter of vent hole
       E = modulus of elasticity   t = thickness of sheet
       v = Poisson's ratio.

The maximum nipple height is established by the designer, so we can establish an upper limit on α in terms of the physical constants of the material, the applied pressure, and the thickness of the sheet. The value of v is between 0.35 and 0.5 for most plastics, so we will take the upper limit and substitute into Eq. (10.3). Rearranging the equation, we get

$$\frac{d}{t} = 1.18\left(\frac{\alpha\,E}{q}\right)^{1/4} \tag{10.4}$$

Therefore, the engineer needs to determine the maximum nipple height, the temperature-dependent modulus, and the molding vacuum to solve for the permissible diameter of the vent holes. After determining the correct vent hole size, the value is rounded down to the next smallest drill size.

After determining the diameter of the vent holes, the task is to determine how many should be used. The volume of air in the cavity to be formed must be removed relatively quickly, in some short time θ. For air flow, the limiting velocity is the sonic velocity, C. C is about 335m/s (1100 ft/s). We can write an equation representing the velocity of the air through our N-number of holes in terms of some fraction of the sonic velocity that we feel safe in approaching.

$$\frac{C N \pi d^2}{4} = \frac{\beta V}{\theta} \tag{10.5}$$

β is the safety factor, or fraction of the sonic velocity, and V is the volume of the cavity. Solving this equation for N, the number of holes

$$N = \frac{4\beta V}{\theta C \pi d^2} \tag{10.6}$$

For most packaging applications, β=10 and θ=0.1 s will give satisfactory results.

Example:

      Assume V=1000 cm³, d= 1.59 mm,
      then N = 4 x 10 x 1000/(0.1 x 335 x 3.1416 x 1.59²) = 150 holes

---

As mentioned, vent holes are placed in the mold in the area that is the last to be contacted by the plastic. If the shape is very complicated, then there may be a need for vent holes in other areas, such as bottoms of steps in the mold cavity. As mentioned previously, use of plug assist or other techniques may be necessary to help in achieving uniform walls in complex shapes.

If undercuts are required, a split mold design or some retracting feature in the mold is likely to be required, unless the plastic article has the ability to flex enough to release from the mold.

10.5.2 Production Molds

The difference between prototype and production molds is mainly durability and the addition of cooling to a production mold to decrease cycle times. Often a production mold is a multiple cavity tool that produces many replicates of the same part at once. Production molds usually are made from aluminum, steel, or other durable materials with good heat transfer characteristics. Tool steel is usually used for long run items.

Mold cooling is typically provided by water channels in the mold, when metal molds are used. A system of fans blowing air over the top of the mold can aid in cooling, and this can be used alone for wood or plastic molds, which do not generally contain internal cooling.

10.6 Thermoform-Fill-Seal Systems

Systems which thermoform packages, fill them with product, and seal the package are called thermoform-fill-seal operations. They are analogous to the form-fill-seal operations commonly used for flexible packaging. Applications range from vacuum packaging of meat to modified atmosphere packaging of fresh vegetables. Most systems run in a horizontal configuration. The bottom web is thermoformed, the product filled into the cavities, a lidding web indexed on, air evacuated, a modified atmosphere added if desired, and the lidding sealed on. Modifications include thermoforming both webs, or thermoforming only the top web. A variety of thermoforming techniques can be used. These systems can also be adapted to aseptic packaging.

References

1. Selke, Susan, *Understanding Plastics Packaging Technology*, Carl Hanser Verlag: Munich, 1997.

Study Questions

1. What is the relationship between the discussion of the optimum temperature for radiative heaters for thermoforming and plastics identification using IR spectrophotometry?

2. Compare and contrast drape forming and vacuum forming.

3. Describe plug-assist thermoforming. Why is it used?

4. What are the major advantages and disadvantages of skin packaging compared to blister packaging?

5. Why might you choose to use matched-mold forming for a packaging application?

6. If billow-forming is used, it is usually combined with plug-assist. Why is this the case?

7. Give an example of a package where you would want the surface texture or marking to be within the cavity, and an example where you would want such markings on the convex surface. Which basic thermoforming method would you use in each case?

8. What are the major advantages and disadvantages of thermofom-fill-seal systems compared to buying preformed packages? Give an example of the type of situation in which thermoform-fill-seal would be most advantageous, and one where filling preformed packages would be advantageous.

9. Positive and negative forming result in differences in the wall thickness of the finished containers. Explain why this is the case. Give examples of cases where positive forming and negative forming would be most beneficial for final package performance.

11 Injection Molding, Closures, Rotational Molding, Compression Molding, and Tubes

11.1 Injection Molding

Injection molding is the process in which melted plastic is forced into a mold where it is shaped and cooled. In packaging, it is the usual method for manufacturing closures (caps and lids) of various types. Injection molding is also used for producing margarine tubs, 5-gal pails, and other containers and package components. A very important application is production of preforms for injection blow molded bottles.

11.1.1 Injection Molding Machines

The injection molding process begins with delivery of the plastic, usually in the form of pellets, to the injection molding machine (Fig. 11.1). This is in essence an extruder much like those used for producing cast or blown film, as described in Chapter 7, with one addition. The screw has the capability to move backwards and forwards in the barrel; this is known as a reciprocating screw. As the screw turns, the melted plastic is contained within the barrel and accumulates ahead of the screw, and behind a valve at the end of the barrel. The pressure created by the melt forces the screw backwards in the barrel. When the desired volume of melt is accumulated, the screw stops rotating. It is then driven forward, mechanically or hydraulically. This forces the melted plastic through a nozzle into the mold. Typically the pressure is raised to a high level and held there after the mold is filled. This is called packing the mold. After a period of time (determined by the processing necessary for the particular parts being made), the screw is retracted, screw rotation begins, and the process starts over again.

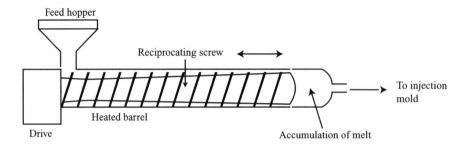

Figure 11.1 Injection molding machine (reprinted with permission from [1]).

Some injection molding machines are designed with separate plasticizing and injection functions, particularly those designed for production of large parts or those used for processing of materials that are temperature sensitive. In one such design, a non-reciprocating extruder feeds the plastic into a separate chamber, known as the shooting pot. When the desired amount of melt is accumulated in the shooting pot, an injection piston forces the melt into the mold. This design allows the extruder screw to rotate continuously, and also provides greater accuracy in shot volume. A third design involves a plunger, instead of a screw. The plunger forces the plastic over a heated torpedo, melting it, and then the plastic is forced into the mold. The major disadvantage of this process is poor homogeneity of the melted plastic due to lack of mixing. Consequently, the torpedo design has been almost totally abandoned.

The size of injection molding machines is usually specified in terms of injection capacity and clamp force. These factors are not directly related. Other important variables are L:D (length to diameter) ratio, barrel size, plasticizing rate, injection rate, and injection pressure. Major components of injection molding machines include the injection unit, the clamp unit, and the machine base which contains the power and control units.

11.1.2 Injection Mold Units

Injection molds have two main parts, or platens. Typically the half containing the hollow cavity side is attached to the injection molding machine, and is referred to as the fixed platen. Because the melted plastic enters through this side, it is also called the hot side, or the hot half (Fig. 11.2).

The section of the mold containing the concave core is usually the piece that moves back and forth to open and close the mold, and is referred to as the moving platen. In some designs, especially in stack molds, the mold cavity is mounted on an intermediate section called the floating platen, rather than directly on the injection molding machine.

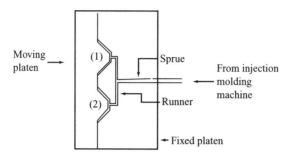

Figure 11.2 Injection mold, with mold cavities (1) and (2) (reprinted with permission from [1]).

The clamp unit platens hold the two halves of the mold together, providing the force needed to keep the mold closed when the plastic is injected. This clamping action is provided using hydraulics or by mechanical toggles. Because of the very high pressures that are common in injection molding, the clamping units must be able to apply high counter-pressures to keep the molds from being forced open. Injection pressures of up to 200 MPa (29,000 psi) are not unusual. The clamping pressure required is determined by multiplying the pressure in the mold by the projected area of the part(s).

To achieve accurate dimensions in the finished part, the mold cores must be aligned accurately with the cavities. Mold halves have large guide pins in the corners to assure alignment.

The mold is cooled, usually by water, to solidify the plastic. When the plastic object has been cooled sufficiently to maintain its shape, the mold is opened, along the mold parting line, and the object is ejected. The cycle time in injection molding depends on the size of the molded part, as well as on the molding conditions and on the thermal and mechanical characteristics of the plastic being molded. In some case, the cycle time can be as short as two seconds.

Injection molds are usually made from high quality tool-grade steel. They may include beryllium copper inserts to improve heat transfer rates for more rapid cooling. Molds are quite expensive, and often contain wear plates which can easily be replaced, so that the molds can be more easily refurbished and have a longer life.

11.1.3   Polymer Flow

The melted plastic enters the mold block through the sprue, then typically flows through a system of runners, and finally enters the mold cavity through the gate (Fig. 11.2). Most often, the mold block will contain several mold cavities, sometimes as many as 64, so several objects are molded at the same time.

When a package or component is produced by injection molding or injection blow molding, it can usually be recognized by the characteristic mark left at the gate where the plastic flows into the mold. This mark appears as a little nub, often surrounded by a slight depression called a sink mark. Usually the gate is designed to be thin in cross section, so the plastic solidifies there quickly, sealing off the mold. This allows the screw to retract without causing backflow of the plastic out of the mold.

The gate is most often located at the center of the molded object, though it can be located elsewhere. Its location can have a significant impact on the properties of the molded object, since it determines the evenness of flow of the polymer into the mold, the polymer orientation in the finished part, and the presence or absence of weld lines and stresses. Weld lines are places where the flow of plastic, initially separate, unites. For example, suppose we are molding the cup shown in Figure 11.3(a). When the polymer enters the mold, small temperature differences in the mold walls will cause the polymer front to advance unevenly, as shown in Fig. 11.3(b). Eventually, when the polymer fronts meet in the flange area, they must join together. At these points, a weld line is formed. This has not quite occurred in Fig. 11.3(c). Proper selection of gate location is important in determining the type and amount of stresses in the molded object, and consequently its susceptibility to distortion.

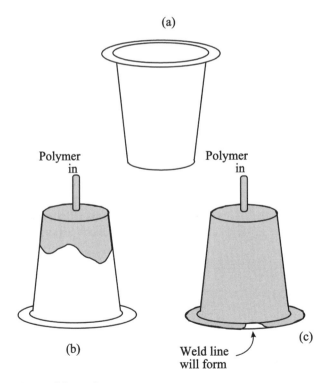

Figure 11.3 Injection molding of a cup.

For some parts, multiple gates are needed. For instance, if the cup shown in Figure 11.3 had very thin walls, it might be necessary to add gates in the flange area to avoid short shots, incomplete filling of the mold due to freezing (solidifying) of the melt before filling is complete. Also, if the part is an oblong tray or similar part with a large aspect ratio (ratio of diameter to height), one gate will usually not be sufficient to allow proper filling without frozen-in strains or short shots. It is usual to gate such parts in the center and also around the periphery of the tray lip, as shown in Figure 11.4.

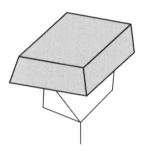

Figure 11.4 Gate placement in trays.

Obviously, proper filling of the mold is essential in producing good parts. Insufficient filling, known as *short shots*, will produce objects with gaps or thin spots. Overfilling, on the other hand, results in stresses, and may also result in the presence of flash seeping into the mold parting line. The problem is further complicated by thermal expansion and contraction, since the mold can be overfilled initially yet be underfilled when the polymer has cooled. Therefore, a certain amount of overfilling, or packing, is generally desired. This helps to reduce the presence of sink marks, which result from thermal contraction and usually surround the gate. Locating the gate in an inconspicuous spot, if the product requirements permit, can also help.

When the mold contains more than one identical mold cavity, it is important that the cavities fill equally. The usual approach to accomplishing this is to balance the flow paths for the plastic, so that distances and geometry, and thus pressure and flow, are equalized. Where non-identical objects are being produced, this job is even more complex, but that seldom applies in packaging applications. It is also important to design the runner geometry to avoid dead spots, where plastic can accumulate and be subjected to an excessive heat history.

A key to efficient production of components by injection molding is the ability to rapidly fill the mold. Therefore, polymers used for injection molding usually have relatively low viscosities (high melt flow rates). Injection rates of up to 2 kg/s (4.4 lb/s) are common.

Temperature of the melt is an important variable. Higher temperatures produce lower viscosities, and hence more rapid mold filling. On the other hand, high temperatures also produce slower cooling, and increased chances of polymer degradation. Therefore selection of an appropriate melt temperature requires balancing these attributes. Of course, the molecular weight of the polymer also affects viscosity.

The temperature also affects the orientation of the polymer in the finished part. Higher injection and mold temperatures, along with slower cooling and thicker parts, decrease orientation by providing less stress on the polymer molecules during flow, or greater ability for them to relax into a more random coil conformation once they have entered the mold. Orientation produces increased strength in the flow direction and decreased strength in the perpendicular direction, as well as greater shrinkage on cooling in the flow direction than in the perpendicular direction. Such uneven dimensional changes can produce warping of the finished parts. Warpage of parts can occur any time after molding, whenever the temperature is raised high enough for molecular rearrangement and relaxation of the molded-in stress. This can even occur slowly at room temperature, taking perhaps 24 hours before the changes begin to be visible. This warping could occur during shipping, or during filling on a packaging line. Therefore, the molder must take great care not to have a problematic degree of molded-in stresses.

11.1.4   Removal of Molded Parts

Parts can be ejected from the mold either by using air pressure or with a mechanical ejection system. A common system for mechanical removal uses stripper rings surrounding the core, which physically push off the parts. Pin-type devices can also be used. In air ejection, blasts of air loosen and blow the parts off the cores. Air ejection is used more often than mechanical ejection, since it requires fewer moving parts, and thus less maintenance, as well as permitting the molds to be more compact in design. The choice of ejection method depends on both the resin and the part design. Mechanical injection, for example, is preferred for PS, since PS has a tendency to crack during the higher stress typical of air ejection. Air ejection works better for thinner parts than for thick ones. Air injection usually allows parts to be ejected earlier in the molding cycle, because it requires less sidewall strength in the component than mechanical injection.

Successful removal of the part from the mold is a major factor in mold design. Many packaging components, such as closures, contain undercuts, portions of the molded object that protrude into the side wall of the mold (Fig. 11.5). When these are present, the object being molded cannot simply slide out of the mold, since it is restrained by the protruding area. Therefore, some mechanism for releasing the object must be provided. Sometimes the molded object is flexible enough, and the undercut small enough, that it can be blown or popped loose from the mold. If this is not the case, a more complex mold design is required. One approach is to use a "collapsing core" where the withdrawal of a central part of the core allows other parts of the core to retract, freeing the molded object. An alternative often used for threaded closures is to provide a mechanism that unscrews the closure from the core. For relatively soft materials, such as polyolefins, closures can be stripped off the core without damage to either the part or the core.

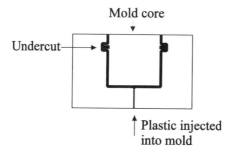

Figure 11.5 An undercut in an injection mold (reprinted with permission from [1]).

11.1.5   Hot Runner Molds

When the molded object is ejected, in simple mold designs the solidified plastic in the runners is also ejected and must be trimmed off. This can be avoided if heat is kept in, or applied to, the runners to prevent the plastic from solidifying. If the runner channels are buried in a mold half and insulated, the core of the runner will not solidify during a normal injection cycle. In a more sophisticated mold, there is a heated runner system, or "hot runner," used to keep the plastic molten. In these systems, the plastic that is in the runners becomes the first plastic delivered to the mold in the next cycle. This shortens the cycle time, as well as significantly reducing the generation of scrap. The main drawback is that mold design is more complex, and therefore more costly. Hot runner molding has become routine in packaging applications.

11.1.6   Venting

As is the case in molds for thermoforming, the design of injection molds must provide for venting to remove trapped air, which otherwise will mar the finish of the molded object, and may, in extreme cases, prevent accurate mold filling. The simplest way to accomplish this is to build vent holes into the mold parting line, where the core and the cavity come together. For large molded objects, this may be insufficient, and additional vents may be needed.

11.1.7   Applications of Injection Molding

The shapes of packages and components which can be obtained by injection molding are somewhat limited by the requirement of being able to free the molded article from the mold. On the other hand, injection molding permits the dimensions of the molded objects to be controlled with a great degree of accuracy. For some articles, such as margarine tubs, which can be produced by either injection molding or thermoforming, the ability to control dimensions can permit light-weighting, which can result in better overall economics in spite of the higher processing costs. In other applications, the lower costs and high productivity of thermoforming outweigh this advantage. Some package components, such as threaded closures, must usually be injection molded, since it is difficult to achieve the required shape through any other method. (Owens-Illinois recently introduced a compression-molded continuous-thread closure made from high density polyethylene.)

The combination of injection molding and blow molding, injection blow molding, is one of the major methods for producing bottles, and will be discussed in Chapter 12.

11.2   Closures

The term closures refers to caps and lids. Because they provide the seal (and often re-seal as well) in many packaging systems, they are often critical components of packages. All forming processes result in some variation in the dimensions of the final product. When sealing is critical, packagers often depend on a closure to compensate for the manufacturing inaccuracies not only in the closure, but also in the container. To achieve a tight hermetic seal that is capable of protecting the contents of the container, closure systems typically rely on good dimensional control of the closure itself, as well as provision of some resilient sealing surface which can deform to compensate for variations in the dimensions of both the package and the closure. While a number of materials can be successfully used, thermoplastics are increasingly the material of choice for closures, and injection molding the method of production. The plastic most often used for closures is PP, followed by PS, HDPE, LDPE, and PVC, in that order. Thermoset phenolics and urea used to be commonly used for closures, but have increasingly been replaced by thermoplastics. Thermoplastics have also replaced a large share of the metal (steel and aluminum) closure market. Three major categories for plastic closures can be identified: friction, snap-on, and threaded.

11.2.1 Friction Closures

Friction closures have the simplest design. The closure is designed to deform as it is pushed into the mouth of the container. The resilience of the material causes it to attempt to return to its original dimensions, providing a seal by the tight fit between the closure material and the container. Friction between the outside of the closure and the inside of the mouth of the container provides the force that resists removal of the closure. A cork is the classical example of a friction closure; a rubber stopper is another. One of the oldest designs of plastic friction closures (Fig. 11.6) is used for inexpensive bottles of champagne. It is hollow inside, and has a ridged outer surface.

Plastic cork

Figure 11.6 Example of a friction closure (reprinted with permission from [1]).

In the last several years, an increasing problem of cork taint, estimated to affect 2 to 10 percent of bottles of wine, is causing bottlers to be more receptive to the use of alternatives to natural cork. Bottlers are increasingly turning to metal screw caps or to plastic corks for lower-end wines, and are even sometimes using them for higher-end wines. In contrast to the plastic cork pictured above, these corks are shaped like solid cylinders, essentially identical in shape to the natural cork they replace. Some are produced by molding, and others by extrusion. The plastic corks do not generally provide as tight a seal as screw caps, but, surprisingly, this may provide an advantage. An absolutely hermetic seal evidently may lead to development of some off-flavor in wine as it ages, while the plastic corks more closely mimic the tight, but not absolutely hermetic, seal of natural cork. Another advantage of plastic corks is no more worry about cork crumbs falling into the wine!

11.2.2 Snap-Fit Closures

A snap-fit, or snap-on, closure is made of a resilient materials and is designed to deform as it passes over a protruding feature on the container (Fig. 11.7). Removal of the closure requires deformation again to "snap" the closure back over the protruding feature, which is typically a retaining ring. When the closure is in place, some resilient part of the closure system that is in contact with the container remains deformed, and provides a seal as it attempts to return to its original dimensions. This category of closure is widely used, and a variety of designs exist. Examples include plastic lids on coffee cans, as well as "line-up-the-arrows" child-resistant caps on medicine bottles.

One of the major advantages of snap-fit closures is that they can be applied very quickly. A disadvantage is that they cannot be used for containers with internal pressures exceeding one atmosphere, since the pressure may act to snap the closure back off of the container.

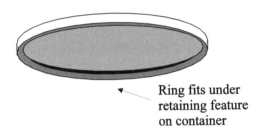

Ring fits under
retaining feature
on container

Figure 11.7 Snap-fit closure (reprinted with permission from [1]).

11.2.3   Threaded Closures

Threaded closures are applied by screwing them onto a container, and removed by screwing them off. They contain a set of threads that is designed to match the threads on the container. These threads are usually continuous, leading to the designation CT (for continuous thread) closure. Threaded closures are very versatile, and can be used for packages that contain internal pressure, such as carbonated beverages, as well as for vacuum packages and those at atmospheric pressure. Child-resistant designs of several types are available, as well.

The amount of force required for application and removal of the closure is determined by how far the closure is rotated on the container. Charts are available that provide recommended torques for various sizes of containers. Some examples are provided in Table 11.1. Removal torques are generally less than application torques, unless there has been some interaction between the liner and the contents, resulting in sealing the liner to the container. Removal torque typically declines with time for the first several days to a month or so after application, and then stabilizes. The change is caused by stress relaxation and creep in the liner, closure, and/or container. Recommended minimum removal torques are typically about half recommended minimum application torques. If an application torque is too low, the container may not be sealed adequately. On the other hand, if the application torque is too high, consumers may have difficulty removing the closure. It should also be noted that if application torques are too high, a variety of additional problems can result. The capping machinery may not be able to reliably deliver the torque that is set. The forces involved may result in permanent deformation of the cap or the container, resulting in poor sealing, or even leakers. There may be excessive wear of the machinery, etc.

The sealing action is provided by deformation of a resilient surface that is either built into the closure design, or provided by a liner used in conjunction with the closure. As is the case with the other closure designs, the attempt of the resilient material to return to its original dimensions exerts force against the container and provides the seal. The liner may contain plastic as the product contact layer, plastic foam or paperboard for resilience, and may also contain aluminum foil for barrier. Liners can be glued in place, but most often are held in place by being snapped behind a retaining ring built into the closure. Linerless closures are designed to provide a resilient sealing surface without requiring the use of a separate liner component. This resilient feature is molded into the closure itself. Usually the resilient feature is made of the same material as the rest of the closure and is produced during molding of the closure, providing resilience by a combination of its geometry and the nature of the plastic used for the closure. In some cases, it is a distinct plastic material, with superior resilience characteristics, produced by a technique such as coinjection molding.

Table 11.1 Recommended Application and Removal Torques for Rigid Closures on Plastic Containers (reprinted with permission from [1])

Cap Size mm	Application Torque N m (in lb)	Removal Torque N m (in lb)
15	0.7 (6)	0.3 (3)
24	1.1 (10)	0.6 (5)
33	1.7 (15)	0.8 (7)
48	2.3 (20)	1.1 (10)
70	3.2 (28)	1.6 (14)
120	5.4 (48)	2.7 (24)

The size of CT closures is specified by the nominal outside dimension of the container opening in millimeters, plus a number that represents the style of finish. Both container and closure finishes are standardized so that, at least in theory, a closure of a given size and style should fit any bottle of that same size and style, from any manufacturer. U.S. closure standards have been established by the Closure Manufacturers Association, standards for glass bottles by the Glass Packaging Institute, and for plastic bottles by the American Society for Testing and Materials (Fig. 11.8). Common closure diameters are 22-120 mm.

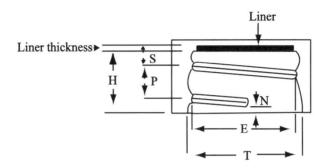

Figure 11.8 Standard closure dimensions (reprinted with permission from [1]).

Critical dimensions for closure performance include T, the diameter of the root of the thread (the deepest part) inside the closure; E, the inside diameter of the closure; H, the distance from the inside top of the closure to the bottom of the closure skirt; and S, the distance from the inside top of the closure to the starting point of the thread.

11.2.4   Specialty Closures

In addition to the basic designs described above, a wide variety of special closure designs are available. Many of these are designed to provide a dispensing function, such as pumps, sprays, flip-open caps, etc. They range in design from rather simple to extremely complex.

An important set of specialty closures are those that are designed to make packages child-resistant, difficult for young children to open. Child-resistant packaging, known as "special packaging" in the relevant regulations, is required on most prescription drugs, aspirin, and other over-the counter drugs, and on household chemicals that pose a serious risk to children if they are accidentally ingested. The regulations prescribe standards that must be met for packages to qualify, and include the ability to successfully prevent opening by young children, while at the same time permitting opening by adults, including the elderly.

Some closures also are designed to provide tamper-indicating features, typically, in closures, by incorporating some type of tear-off ring. Closure liners that are designed to release from the cap and seal to the bottle mouth can also be used to provide tamper-evidence. In the U.S., over-the-counter drug products are required by law to contain some tamper-evident feature, designed to alert consumers if a package has been opened and may have been tampered with. Such features are increasingly found on packages for food and other products as well, although they are not required in these applications. Closures and liners, of course, are not the only way of providing tamper evidence. Other common mechanisms involving plastic packaging include shrink bands on package necks and shrink wraps around containers.

11.2.5   Fitments and Overcaps

Fitments are another set of devices associated with closures. These are components that are used in conjunction with a closure to provide some added utility by regulating the flow of product out of the package. A common example is the shaker-top on a spice jar. When the dispensing device is built into the closure, instead of in a separate piece, the result is sometimes referred to as a fitment closure.

Overcaps, as the name indicates, are designed to be applied over the closure. They may serve purely a decorative function, but most often they are designed to offer some protection to a dispensing closure so that it is not activated prematurely.

11.3   Rotational Molding

In rotational molding, plastic in powdered form is placed inside a hollow mold. The mold is closed, and then spun on two perpendicular axes, simultaneously. Centrifugal force distributes the powdered resin throughout the mold. Next, the mold is heated, while still spinning, to melt the plastic. Spinning continues as the mold is cooled to solidify the plastic object. When it has cooled sufficiently, the mold is stopped, opened, and the item removed. For many objects, such as containers, cutting in openings or other features is the next step. Rotational molding has some use for forming large bulk containers and also for forming the round balls used in dispensing containers for roll-on deodorant, but otherwise is used very little in packaging.

11.4   Compression Molding

Compression molding is most often associated with forming of reinforced plastics. Fiberglass or carbon fiber reinforced thermoset polymers are often formed into useful shapes by compression molding. This process is also used to produce closures from thermoplastic polymers. Instead of forcing a molten polymer from an extruder into a die, as in injection molding, compression molding starts with solid polymer placed in the die cavity. The die cavity is heated to help melt the polymer, and pressure is applied by means of a ram, or piston, to force it to flow and fill the mold cavity. For the manufacture of closures from polypropylene, for instance, a puck of material is cut from a hot log of polypropylene that has been extruded and partially cooled. This puck is dropped into the mold cavity, and a ram, the mold core, applies pressure to force the material to fill the mold. This process is shown in Figure 11.9, where one-half of a mold cavity is represented. This process does not usually work well with multiple cavity molds, as are typically used for injection molding, but works rapidly with a single cavity mold. To achieve the production volume needed, multiple lanes of the same process are carried out.

One of the chief reasons for using the compression molding process is to gain uniformity of the closures. Defects in the molding process cannot be caused by short shots, or rheology changes in the polymer, as might be the result in a multi-cavity injection molding machine. Because the amount of polymer added to the cavity can be precisely controlled, only mold imperfections will cause part imperfections. These imperfections will be uniform, which makes them much easier to detect by quality control inspection procedures.

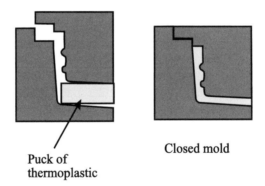

Puck of
thermoplastic

Closed mold

Figure 11.9 Compression molding of a closure; one-half of the mold is shown.

11.5 Plastic Tubes

Plastic tubes are most commonly produced by a process that combines extrusion with injection or compression molding. The tube body is usually produced by extrusion. As the plastic emerges from the extruder, it is corona treated for ink adhesion, and then drawn over a chilled, internal forming mandrel as cold water is applied to the outside of the tube. As the tube cools, it shrinks to the diameter of the forming mandrel. The tube body, called a sleeve, is then cut to length. Tubes produced in this way can be single-resin or coextruded. The next step is to produce the tube head and attach it to the tube body. This can be done either before or after the sleeve has been printed and coated. The head can be produced either by injection molding or by compression molding. In one common method, the top of the sleeve is captured in an injection mold, and plastic is injected, at relatively low pressures but high temperatures, to fill the cavity, forming the head. The head and the sleeve are bonded together in this step. Next, the sprue from the injection molding is cut off the head, and the cap is applied. In another common method, a disk is punched out of a continuous hot strip of LDPE, it is adhered to the sleeve, and then compression molding is used to form the final head shape. A third method uses injection of a ring of melted plastic in a female mold cavity, followed by compression molding of the head. A fourth method uses a premolded head that is spin-welded to the tube.

Rather than forming the tube sleeve and then subsequently forming the head, the head can be immediately injection-molded onto the extruded tube, in a continuous operation. Another option is to use blow molding to form the tube in one piece, really making a bottle, and then transform it into a tube simply by cutting off the bottom. This is a procedure that is being used increasingly as an alternative to the traditional methods of tube manufacture. Stretch blow molding of injection-molded preforms is used to produce PET and PP tubes with improved barrier and high clarity, in a reheat-stretch-blow process (see Sec. 12.4). The tube walls can be as thin as 150 microns, with very precise thickness distribution. An integrated closure can also be provided. The preform skirt (body) is typically stretched to 3 to 5 times its original length, while the preform head is unchanged. The bottom of the tube is then cut to the precise length desired.

Laminated tubes are made from multi-layer materials that usually contain paper as well as plastic, and often also include aluminum foil as a barrier layer. The preprinted tube bodies are sealed into a cylinder, with the edges of the sleeve overlapped and compressed, squeezing some of the plastic out around the raw edges of the foil and paper to make a good seal. Next the tube is cut to length, and finally the head is molded and assembled to the body. To improve the barrier in the head, a premolded insert of polybutylene terephthalate or of urea can be incorporated in the injection mold when the head is formed.

After the tube is formed, it is printed and coated, if this has not already been done, and the closure is attached. The end of the tube is left open for filling. Tubes are filled through the bottom and then sealed, using radiant heat, heated jaws, ultrasonic sealing, or other methods. High frequency sealing is often used for laminated tubes containing aluminum foil. Hot air sealing can be used for both laminated and all-plastic tubes. In this process, the seal area inside the tube is heated with hot air and the tube is then pressed shut and chilled. During the pressing operation, a code is commonly applied, indicating the batch number, expiration date, and other information.

References

1. Selke, S. *Understanding Plastics Packaging Technology*, Carl Hanser Verlag: Munich, 1997.

Study Questions

1. Describe the process of injection molding.

2. Why is removal torque of closures generally less than application torque? What are the consequences of a removal torque that is too high? too low?

3. Tubes are generally filled through the end, rather than through the mouth. Why?

4. Why isn't thermoforming used to manufacture continuous thread caps?

5. Why are molds used for thermoforming much less expensive than molds used for injection molding of similar products?

6. Which thermoforming process variation is closest to providing the benefits of injection molding? Why?

7. Describe friction closures. What basic property must a material have to be suitable for this application?

12 Blow Molding and Bottles

12.1 Blow Molding

The most desired shape for a rigid plastic container for many products is a bottle or jar. However, bottles and jars, even those with large necks, cannot effectively be produced by injection molding because of the difficulty of getting the solid core out of the molded object. As glass blowers learned centuries ago, an effective way to produce such containers is to use air pressure to shape the inside of the object. In blow molding, air pressure is used to shape the inside of a plastic object, while a mold shapes the outside. Blow molding is the only practical way to make plastic bottles and jars, and this method is also used for large plastic containers such as drums. There are two major types of blow molding, extrusion and injection blow molding.

Injection blow molding combines injection molding of a precisely formed parison with blow molding of the finished container. Therefore, it is able to give fairly accurate control over container dimensions, especially in the critical finish area of bottles. Injection blow molding is more expensive than injection molding alone, since it requires two sets of molds and two molding processes. However, it is capable of producing shapes that cannot be produced by injection molding. It also produces very little scrap. Injection blow molding can be used with resins without sufficient melt strength to be handled by extrusion blow molding. Injection blow molding is used for most PET bottles, as well as for most bottles used for pharmaceutical products. It generally cannot economically produce bottles with handles, although a new development in this area is discussed in Section 12.4.

Extrusion blow molding is the simplest and generally the most economical process for making plastic bottles. Control over wall thickness is not as good as in injection blow molding, but can be enhanced by techniques such as parison programming and die shaping. It is capable of producing a very wide variety of bottle shapes, including bottles with handles, offset necks, dual chambers, and more. Considerable scrap is produced, especially with complex designs. Usually scrap can be reused in the process, but not always. Extrusion blow molding is also not very economical for very small bottles, so these are usually produced by injection blow molding. The resins used must have sufficient melt strength for the parison to remain unsupported until it is captured in the mold. More than half of all plastic bottles are produced by extrusion blow molding.

Stretch blow molding can produce a bottle with biaxial orientation, thus improving its strength and barrier properties. It requires a two-step process, with very accurate temperature control of the parison, so the distortion during stretching happens as desired, rather than some parts remaining too thick or too thin. In principle, it is a variant on either extrusion or injection blow molding, but in practice it is associated almost exclusively with injection blow molding of PET bottles.

12.2   Extrusion Blow Molding

12.2.1   Basic Extrusion Blow Molding Process

Extrusion blow molding (Fig. 12.1) is the oldest, and still the most common, process for producing plastic bottles. The first step is extrusion of a hollow plastic tube, just as is done for blown film, except that the extrusion is usually in a downward direction for making bottles, and usually in an upward direction for film. The two halves of the mold next close on the tube, capturing it as it is cut off from the extruder. A blow pin or a needle is inserted and air is blown into the mold, expanding the tube, or parison. In some cases the blow pin, cooled by water, assists in forming the finish by compressing the finish section into the mold, rather than simply blowing it in. This results in a smooth interior in the finish region. In the needle blow case, the needle is inserted into a part of the molded object that is trimmed off in forming the final container shape, and the inside of the finish is formed only by air. The mold is cooled, usually with water, to solidify the plastic. When the container is cool enough to maintain its shape, it is ejected from the mold. The flash (excess material) is trimmed from the container neck and bottom, as well as from other areas that are pinched off, for instance to form handles or offset necks (Fig. 12.2).

The marks left from the removal of the flash serve as an easy means for identification of extrusion blow-molded containers. Usually this is easiest to see on the bottom of the container. It typically appears as a rough area along the mold parting line, centered in the middle of the bottom and running half or so of the distance to the heel of the bottle. It is also possible, on careful examination, to identify the roughness at the top of the finish, or on other areas where flash was trimmed.

The flash, after being trimmed, usually is immediately ground up and fed back into the extruder at a controlled rate, mixed with the virgin resin. The use of regrind can be problematic for heat-sensitive resins like PVC, especially if the proportion of flash is high.

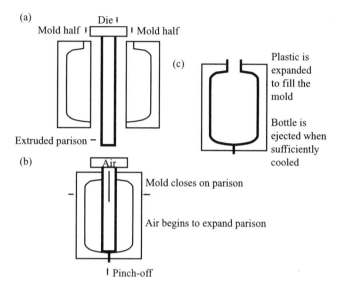

Figure 12.1 Extrusion blow molding (reprinted with permission from [1]).

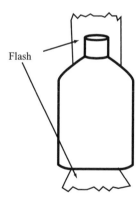

Figure 12.2 Flash from extrusion blow molding of a bottle.

The process of parison extrusion can be continuous or intermittent. For intermittent extrusion, the melt from the continuously rotating extruder may be fed into an accumulator, from which it is periodically ejected, or a reciprocating extruder like those used for injection molding may be used. Continuous extrusion is preferred for most packaging applications. It provides higher productivity and reduces thermal degradation, since the melt is not held up. Intermittent extrusion is commonly used for the production of very large blown containers, where a large parison must be produced in a very short time.

12.2.2   Parison Dimensions

Usually the diameter of the parison is nearly the same as the diameter of the container finish, although this depends strongly on bottle design. Containers with an offset neck or a handle require a wider parison. It is generally recommended that the ratio of the container diameter to the parison diameter not exceed four to one. The parison dimensions are not identical to the die dimensions, since the stress relaxation and elastic memory characteristics of plastics cause the parison to shrink in length and swell in diameter and wall thickness as it is being produced. The amount of such swelling is dependent on the polymer elasticity, so it is influenced not only by the type of resin, but also by the molecular weight distribution. Swell is also highly dependent on the design and operating conditions of extrusion, with important factors including the shear rate and the land length in the die. The land refers to the outer portion of the die where no further dimensional changes are being made in the flowing plastic.

Swelling of the extruded parison can be expressed by three parameters. Die swell refers to the percent change in the outer dimension of the parison, calculated by dividing the outer diameter of the parison by the outer diameter of the die opening, converted to percent. Tube wall swell, or parison swell, refers to the percent change in the wall thickness of the parison. Flair swell, or mandrel swell, refers to the percent change in the inner diameter of the parison (Fig. 12.3). Polyethylene is subject to a rather high degree of swell, with die swell values of 150-300% not uncommon. When designing a die, it is important to know the amount of die swell to expect so that material is not wasted as flash, and so that weaknesses are not formed in the bottle wall at the pinch points. Of course, knowledge about the extent of longitudinal shrinkage is also important, so that the cycle times and extrusion rate are matched to give parisons of the correct length when they are captured in the mold.

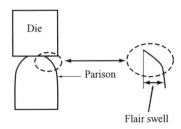

Figure 12.3 Die swell and flair swell.

12.2.3   Extrusion Blow Molding Variations

A number of variations on the basic extrusion blow-molding process are used in commercial operations. A single extruder can feed a multiple orifice blow molding die, producing multiple parisons simultaneously, which are then blown in multi-cavity molds. The blowing can be done at the same machine station as the extrusion, though it is most often done at a second station in the machine. Both rotary and shuttle arrangements for the molds are commonly used. Bottles are most often blown from the top, but in rising mold designs they are blown from the bottom, and produced upside down. Wheel machines are available for very high speed, high volume outputs. Such a machine, shown in Figure 12.4, has a series of molds mounted on a rotary wheel. The parison is usually extruded upward in this case, and the mold captures it on the fly. These types of bottles are always needle blown because of the need for self-contained operations on the moving wheel. At the end of the cycle the mold opens and deposits the blown bottle, with its associated flash, on a conveyor that transports it to the trim station. In rotary and shuttle machines, the trimming may be done in the machine, or in a subsequent operation.

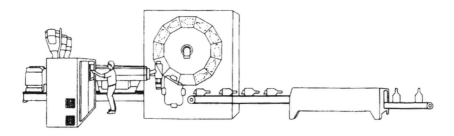

Figure 12.4 Wheel blow molder.

12.2.4   Container Designs

A very wide variety of container shapes and sizes can be produced by extrusion blow molding. This process is particularly useful if a bottle with a handle or with an offset neck is desired, since these cannot be effectively produced by injection blow molding. While it can be used to produce very large containers, it is not often used for small bottles less than 200 ml (7 fl oz) in volume. Injection blow molding is usually more economical for these small sizes.

Cylindrical bottles are the easiest to produce, but they require a lot of shelf and shipping space relative to the amount of product they carry, and also may be undesirable for marketing reasons. An additional drawback is that cylindrical containers tend to "panel-in" if any vacuum develops inside the container, as occurs if the container is hot-filled, or if there is any significant loss of product through the container walls. Paneling also occurs in some products without any obvious cause, and is a phenomenon that is still not well understood. Whatever the cause, these distorted containers are likely to be rejected by consumers. While a number of innovative designs have been developed to combat or disguise this distortion, often the simplest solution is to use an oval container. The relatively flat side panel on an oval container can move inward, often to an appreciable extent, without producing any readily visible sign of distortion. However, producing an oval bottle with uniform wall thickness is more difficult, as will be discussed below.

Another design style that presents challenges for wall uniformity is the F-style container (Fig. 12.5), often used for motor oil and chemicals. The offset neck of this bottle makes it easier for the consumer to accurately dispense the product in a restricted area, such as into the oil reservoir in an automobile.

Figure 12.5 F-style bottle.

12.2.5   Die Shaping

When the melted plastic in the parison is expanded by air pressure, it stretches and thins until it comes in contact with the mold. At this point it is immediately cooled enough to stop the stretching. Therefore, the parts of the parison that come in contact with the mold first are the thickest, and the parts that have a greater distance to travel are the thinnest. For a cylindrical bottle, this is not a concern. At any given bottle height, all parts of the bottle wall have stretched the same distance, so if the parison was uniform in thickness, the bottle will also be reasonably uniform. However, if the bottle cross section is oval, some parts of the bottle will have stretched significantly farther than others; so if the parison began with uniform thickness, the bottle will have some parts of the wall that are significantly thinner than other parts. The result is that to make the bottle strong enough in its thinnest sections, some parts of the bottle wall must be made thicker than would otherwise be necessary. This means excess plastic is used, increasing costs. Fortunately, there is a relatively simple solution that can permit significant light-weighting of such containers - die shaping.

The key to the concept of die shaping is the statement above: "if the parison began with uniform thickness." If the parison is not uniform to begin with, for example if it is thicker in the areas that will be stretched farther, then a uniform wall thickness can be produced even in an oval bottle. Die shaping produces such a non-uniform parison by permanently modifying the shape of the annular die. The opening for plastic flow in such a die is not symmetrical, so the parison produced does not have equal wall thickness on all sides (Fig. 12.6). The thickest part of the parison is aligned with the part of the bottle that is farthest from the midline. The result is a container with more even wall thickness throughout. Of course, the amount and location of the modification of the die shape must be matched with the demands of the container to be produced. The shaping of the die can be done on either the die mandrel (the central part, core pin, or die pin) or the die bushing (the outer part).

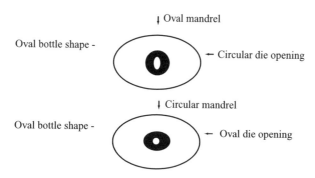

Figure 12.6 Die shaping (reprinted with permission from [1]).

12.2.6 Programmed Parison

A different problem occurs when the bottle has vertical rather than horizontal asymmetries, such as bottles with narrow waists or wide shoulders. In such cases, a one-time change in the die dimensions cannot produce even wall thickness. The solution is to provide the ability to change the die dimensions as the parison is being produced, so that a thicker profile is produced for the part of the parison that forms the wider part of the bottle, and a thinner profile where the bottle is thinner. This technique is called parison programming.

A programmed parison is produced by moving the die mandrel up and down inside the die bushing to increase or decrease the gap between the two, as the parison is extruded (Fig. 12.7). This process is also known as *dancing mandrel* blow molding. An alternative is to move the die bushing instead of the mandrel.

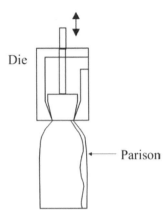

Figure 12.7 Parison programming.

Die shaping and programmed parison techniques can be combined. In the most sophisticated wall thickness control systems, the use of deformable die rings permits both the mandrel motion and the die shape to be controlled as the parison is extruded. This permits extremely accurate control of the thickness profile of the parison and hence of the blow molded container.

12.2.7 Coextruded Bottles

For some applications, no single polymer can best deliver the desired combination of price and performance, so there is a need or desire to combine one or more resin streams in production of the finished bottle. Use of such coextruded bottles has grown rapidly. Applications range from inclusion of a layer of recycled content in laundry detergent bottles to provision of oxygen barrier in mayonnaise containers.

A coextruded bottle is made by melting each type of resin in a separate extruder, feeding the melts into a die where they are formed into a multi-layer tubular parison, and then blowing the container as usual. As with coextruded film, the combining must be done carefully to avoid blending, with attention paid to melt viscosities and other factors. When dissimilar resins are used, a tie (adhesive) layer is often required to obtain adequate adhesion.

Figure 12.8 shows two examples of coextruded bottle structures. The PP/EVOH ketchup bottle was the first commercially successful bottle in the U.S. for an oxygen-sensitive food product. Because of poor adhesion between the chemically dissimilar PP and EVOH, a tie layer is needed to promote adhesion. The addition of a regrind layer benefits the overall process economics by permitting reuse of the flash from bottle production, avoiding waste. Since this regrind layer is predominantly PP, it adheres well to the virgin PP layer without requiring an additional tie layer.

The HDPE/regrind-recycle/HDPE bottle structure is widely used for liquid laundry detergent and similar products. It permits use of post-consumer recycled HDPE, blended with regrind, as the major bottle component, while providing layers of virgin HDPE to protect the product against contamination and also to preserve the desired surface characteristics of the bottle. Because all resin layers are HDPE, no tie layers are needed. Use of coextrusion also permits decreased use of the expensive pigments employed to give many of the bottles their bright colors.

A very different type of coextrusion is used to produce opaque bottles with clear viewing stripes, so that the consumer can see the product level in the container (see Fig. 12.5). In this case, the separate layer, rather than extending around the container, occupies the whole wall thickness in a defined portion of the container. Typically, unpigmented resin is used, which is processed in a small satellite extruder and introduced into the die as a separate vertical stripe in the parison. Since the resin is the same type as the pigmented resin found in the rest of the bottle, no tie layer is needed. Such containers are used for motor oil, liquid laundry detergent, and other products.

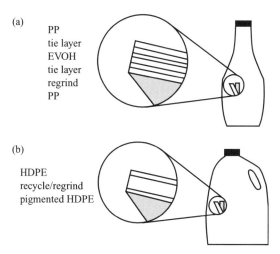

Figure 12.8 Typical coextruded bottle structures, (a) for oxygen sensitive foods, (b) for laundry products (reprinted with permission from [1]).

The use of coextrusion to minimize use of expensive colorants is also employed in cosmetics packaging, where expensive color enhancers such as pearlescents, irridescents, and metallic flakes can be confined to the outer layer, which typically accounts for only 10-15% of the total wall thickness, thus decreasing the cost of using these materials.

Multilayer containers can also be used with clear inner and outer layers and a colored inner layer, to get a smoky or misty appearance. In one application, the inner layer has a metallic look, enhancing light reflection and refraction so that the package appearance changes in different light and at different angles. Soft-touch bottles for lotion and other products can be produced using HDPE cores and a special outer layer.

Some cosmetics containers are using HDPE/EVOH/HDPE/EVOH/HDPE coextruded bottles for better barrier performance. Others are using nylon exterior layers on HDPE containers for added gloss.

12.3 Injection Blow Molding

In previous sections, we have covered extrusion blow molding. One can also use injection molding to make preforms for bottles. Injection molding allows much tighter tolerances on the finish of the bottle. This is particularly important for small bottles used in pharmaceutical applications. Small bottles made by injection blow molding can have tighter wall thickness tolerances, if the preform is carefully designed. The smaller stretching distances also contribute to tighter wall tolerances.

12.3.1 IBM Process

Injection blow molded bottles (IBM) are always made in a two-step process. First, a preform, or parison, is produced by injection molding (Fig. 12.9). Next, this preform is placed in a second mold and blown to produce the final container shape. The injection molding of the preform is identical to the injection molding processes described in Chapter 11, except that the mold has three parts, the two halves of the cavity and a core rod. Hot runner molding is used almost universally, so no trimming is required. The bottle finish is formed in the injection molding step, and the parison typically has a test tube-like shape. Usually the preform is cooled only enough maintain its shape with the support of the core rod, before being transferred, still on the core rod, to the blow mold. In this step, air is introduced at high pressure through the core rod, to blow the bottle. Blowing pressures are higher than for extrusion blow molding because of the cooler parison temperature, and hence the higher viscosity of the parison. As in extrusion blow molding, the mold is cooled, usually with water. When the container is cool enough to maintain its shape without distortion, it is ejected from the mold. Bottles produced by this process can be identified by the mark left by the gate on the bottom of the bottle at the initial injection molding stage.

Injection blow molded bottles are generally blown on the same machine as the one making the preforms. It is a multi-station machine, where the first station does the injection molding, the second station blows the bottle, and the third station ejects the finished bottle from the machine. The process is often arranged on a horizontal table, as shown in Figure 12.10. Multiple cavity preform molds and bottle molds can be used in the process. However, the cavitation (the number of cavities) is limited by the size of the rotary table.

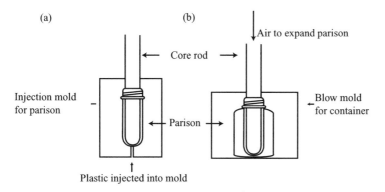

Figure 12.9 Injection blow molding, (a) injection molding of parison, (b) blow molding of container (reprinted with permission from [1]).

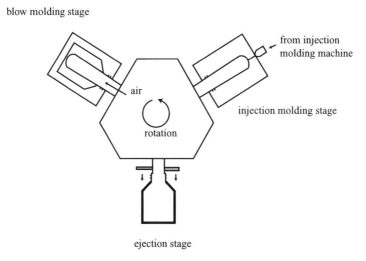

Figure 12.10 Rotary injection blow molding.

An alternative approach is to cool the parison completely after the injection molding step, and remove it from the core rod. The parison can then be stored or shipped elsewhere before blow molding. In this case, it is necessary to reheat the parison to the desired temperature before the blow-molding step. This approach is seldom used for ordinary injection blow molding, but is not uncommon for stretch blow molding, which is discussed in Section 12.4.

12.3.2   Preform Design Process

Early injection molded preforms, like extrusion molded preforms, were straight tubes. There are general design guidelines for such preform shapes. For best results, it is recommended that the diameter of the finished bottle not be greater than three times the preform diameter, and the length of the preform not be greater than ten times its diameter. The ovality of the container, defined as the ratio of the major diameter to the minor diameter, should not be more than two. The injection molded preforms usually have a wall thickness between 1 and 5 mm (40-200 mil). The ratio of the inside diameter of the parison to the inside diameter of the blown container is known as the *hoop ratio*, and can be used to calculate the ratio of the wall thickness of the finished container to the wall thickness of the parison. The length of the parison is typically about 1 mm (40 mil) less than the inside length of the container mold, so there is no significant stretching in the vertical direction when the bottle is blown.

To get even wall thickness throughout the container, just like for extrusion blow molded containers, the preform must be "programmed" to put additional polymer where the most stretch will occur. Since the core and the cavity of the injection mold are fairly permanent, one cannot make changes on the fly as one can in extrusion blow molding. Therefore, the designing must be done upfront.

The most efficient method is to have the supplier use finite element analysis (FEA) to do a simulation. One can use trial and error approaches, but since we are cutting metal on a mold, it is faster and less expensive to use a computer simulation. This is not a simple task because the program must be capable of non-linear calculations. Not all FEA programs are capable of this type of analysis. Therefore, one must determine whether the program being used can handle non-linear material properties.

The first step is to create a mesh model of the preform using estimates of the wall thickness requirements needed for the bottle. A computerized model of the bottle is also required to model the blow cavity. The program can then simulate the stretching of the polymer from the preform shape into the cavity shape. By using the Poisson's ratio of the material, one can then calculate the thickness at any point by knowing the amount of stretching. The assumption is made during the calculations that the preform is at a uniform temperature. If this is not the case in real life, significant differences may arise between the predictions and the final bottle. Jerry Dees described modeling of the process in a 2003 ANTEC paper [2]. Figure 12.11 shows two figures from that paper, which show the models and the wall thickness simulation.

The author (Culter) then made bottles from the preform design and checked the resulting wall thickness distribution, which is shown in Figures 12.12. The two plots are for a bottle corner and for the middle of the side panel. Reasonably close agreement was found between the model and actual practice.

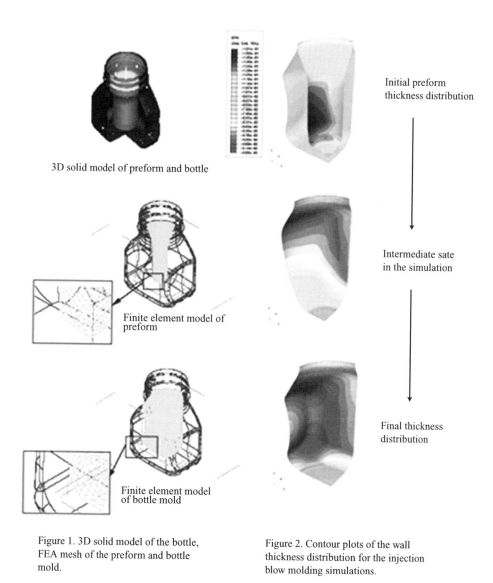

Figure 1. 3D solid model of the bottle, FEA mesh of the preform and bottle mold.

Figure 2. Contour plots of the wall thickness distribution for the injection blow molding simulations.

Figure 12.11 Simulation process (reprinted with permission from SPE [2]).

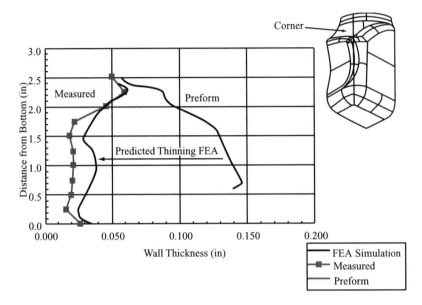

Figure 3. Comparison of simulation results with measured data at the corners.

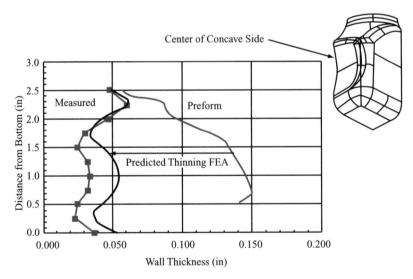

Figure 4. Comparison of simulation results with measured data at the concave side.

Figure 12.12 Predicted versus actual thickness measurements (reprinted with permission from SPE [2]).

12.3.3   Comparison of Injection and Extrusion Blow Molding

As discussed in Section 12.2.4, injection blow molding is generally preferred over extrusion blow molding for very small bottles. Since two sets of molds are required for injection blow molding, one for the preform and one for the final container, mold costs are higher. A significant disadvantage of injection blow molding is the extremely limited ability to produce handles. Processes to produce small solid handles exist, though they are seldom used, but hollow handles cannot yet be made. Injection blow molding produces higher quality containers than extrusion blow molding, especially in terms of finish tolerances. Extrusion can produce a wider variety of sizes and shapes, has higher production rates, and is more economical except for small bottles. Extrusion blow molding is preferred for heat sensitive materials such as PVC, while injection blow molding is preferred for materials with limited melt strength such as PS and PET. HDPE and PP can be readily molded by either process.

In either extrusion or injection blow molding, the polymers in the containers are oriented because of the radial stretching which takes place in the blowing step. If desired, the mechanical or barrier properties of the material can be maximized by producing biaxial orientation, stretching the preform vertically as well as horizontally, using stretch blow molding.

12.4   Stretch Blow Molding

12.4.1   Plastic Soft Drink Bottles

Back in the early 1970s, the use of plastic bottles as a substitute for one-way glass carbonated soft drink bottles (CSD) was being explored. Part of the driving force was the danger posed by exploding packages in the grocery stores - either exploding spontaneously or through impact with the floor. Both acrylonitrile copolymers and PET were being developed as alternatives to glass. Interestingly, the earliest preforms were not made using injection molding. They were made from pipe segments, which had the thread section and the bottle closing done by thermoforming. Injection molding eventually won out because of the better control of the preform dimensions that it provided.

Acrylonitrile copolymers were, in fact, the first bottles introduced by Coca Cola. A suspected issue with monomer extractables developed with the PAN, and it was pulled from the market. PET moved into the business, and dominates today even though the extractables problem was eventually shown to be a non-issue. The damage was done, and PAN never became a player in the market for CSD.

The major idea behind stretch blow molding is to use the stress-induced crystallization (SIC) to improve the barrier and increase the modulus of the polymer. The blow molding is done at a temperature that is advantageous for creating SIC. The temperature must not be too cold and result in whitening of the bottle, and it must not be too hot and result in relaxation of the polymer during cooling. Since we are going to do a "cold" drawing of the polymer, we make a preform by some other means than just extruding it (and in any case PET does not have sufficient melt strength for parison extruding). Since it is going to be cooled down before being reheated to the blowing temperature, one can use the parison making as an opportunity to put a really accurate finish on the bottle. Injection molding is used today for CSD to produce the preforms; in the process, the preform shape is varied to result in the proper programming to minimize the amount of wasted polymer in any portion of the bottle.

A CSD bottle is not heat stable, so it cannot be used for hot-fill products. If one puts hot, 82°C (180°F) or hotter liquid in a CSD bottle, the bottle will try to return to the preform shape. To make hot-fill bottles, we must "heat-set" the bottles. Heat setting causes thermally induced crystallization to occur. It is done by holding the bottle in the mold at an elevated temperature for a period of time. The amorphous regions undergo molecular rearrangements that relax thermal stress, and thermally induced crystallites form. Some of these crystallites are of the straight chain form from the aligned molecules in the amorphous regions.

12.4.2   Overview of Stretch Blow Molding

Stretch blow molding begins with the production of a preform, either by injection molding or extrusion blow molding (or, rarely, extrusion alone). The container finish is formed in this step. The preform is shorter than the final height of the container. Next, the body of the preform, on a stretch rod, is conditioned to an accurate, consistent temperature, usually just above its T_g, while the bottle finish is kept cool to avoid distortion. Then, the preform is placed in the container mold and is stretched by a vertical movement of the stretch rod while air is blown through the rod to expand the bottle into its final shape, stretching it axially (Fig. 12.13).

Stretch blow molding produces a biaxially oriented container with improved impact strength, gas barrier, stiffness, clarity, and surface gloss. Consequently, containers can have thinner walls, saving money through light-weighting.

For successful orientation, the preform must be brought to a temperature within its orientation temperature range, below its crystalline melting point but above its T_g. For PET this temperature range is 88-116 °C, for PVC 99-116 °C, and for PP 104-127 °C . The temperature and amount of stretching affect the degree of orientation obtained. The *axial ratio* is the ratio of the length of the stretched part of the bottle to the length of the stretched part of the parison. The finish is subtracted from both lengths in calculating the axial ratio since it is not stretched. The *blow-up ratio* is the product of the hoop ratio and the axial ratio. Maximum creep resistance, burst resistance and barrier properties require a blow-up ratio of at least 10:1 and a hoop ratio of at least 4.8:1.

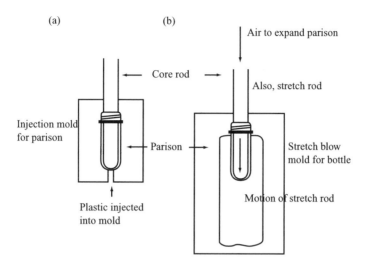

Figure 12.13 Injection stretch blow molding, (a) injection molding of parison, (b) stretch blow molding of container (reprinted with permission from [1])

Extrusion stretch blow molding begins with an extrusion blow molded preform; injection stretch blow molding with an injection molded preform. Injection stretch blow molding is the method of choice for most PET bottles, including all soft drink bottles. It is also used for PEN bottles and some PP bottles. Extrusion stretch blow molding is used for some PVC bottles.

Stretch blow molding can be a one-step or two-step process. In one-step stretch blow molding (also known as one-stage or in-line), the preform production, stretching, and blowing are all done in the same machine. The preform, after molding, is rapidly cooled to the stretch temperature and then immediately stretched and blown. In the two-step (two-stage or reheat-blow) process, the preform is completely cooled and then reheated to the stretch temperature before stretching and blowing. A cost saving through reduced energy use is the primary advantage of the in-line process. The primary advantage of the reheat-blow process is that the preform and the container can be molded at different times, in different places. For example, parisons for beverage bottles can be produced at a central location and shipped in a relatively compact form to the filler, where they are blown. This avoids the high cost of shipping a lot of air in the fully blown bottles, while the bottler need not have the expertise and equipment necessary for the injection molding part of the process.

Stretch blow molding has become much more common-place with increased use of PET bottles and widemouth jars, most of which are produced in this manner. Reheat stretch blow molding has been successfully used for 5-gallon PET water bottles.

Preferential heating of PET preforms, in which some sections are heated more than others, can be used to optimize wall thickness distribution in oval, square, and rectangular PET bottles. The stress relaxation permitted by this method allows higher molding precision in large flat areas of bottles. This technique used for cosmetics bottles was reported to permit material savings of up to 7 g/bottle compared to extrusion blow molded containers [3].

One of the newest developments in injection stretch blow molding is the development of a stretch blow molded bottle with an integral handle. The first commercial use of such a container was reported for a vegetable oil product in New Zealand. The bottle, made by Carter Holt Harvey Plastic Products in Hamilton, New Zealand, is produced on a single-stage injection stretch blow molding machine. The 12-g handle is 15% of the bottle weight, and is molded as an integral part of the preform during injection. According to the manufacturer, the preform tooling is a little more complex, but does not slow down production rates nor add substantial tooling costs. During the blowing stage, the lower section of the handle is encapsulated by the bottle material. Reportedly, the distribution of material in this part of the container is aided by a proprietary process enhancement, which is critical to the success of the technology [4].

12.4.3   Manufacture of PET Preforms

PET preform manufacture today is the result of years of development to optimize the design and speed of manufacture. CSD, water, and other PET beverage bottles are commodity products today, so they must be made inexpensively. The injection molding machines and molds are specialized to the job of making preforms. Since bottles are a commodity product, the molding operation must be able to do the operation at high output rates. Current molds are designed for 144 cavities. To speed things up even more, multiple sets of cores are used, so the preform can be removed from the cavity before it is completely cooled; cooling is completed on the core during the next shot cycle before the preform is removed from the core.

Figure 12.14 shows a Husky molding machine, with the cores on the left. Two sets are used and rotated out of position with the mold, which is not visible in the photograph. The device on the right is a cooling device added by Husky to their newest machine. In Figure 12.15, the mold section is visible on the left, and the cooling collets are more visible. The preforms are transferred to the collets, which are then raised to allow air to be blown into each preform by the needles shown at the top right of Figure 12.15. All of these features are added to speed up the molding process.

Preforms must be designed properly for the package they are expected to make. This means that they come in many different shapes, as shown in Figure 12.16. The one dimension that controls the cycle time the most is the wall thickness of the preform, because cooling is the limiting factor. If the wall thickness is held constant, the change in volume from a 19 g to a 46 g preform would only cost 2.5 s, whereas a change in wall thickness from 2.2 mm to 4 mm would cost 6 s in cycle time.

Figure 12.14 Husky PET preform molding machine, view 1, reprinted by permission of Husky.

Figure 12.15 Husky PET preform molding machine, view 2, reprinted by permission of Husky.

Figure 12.16 PET preforms and bottles, reprinted by permission of Husky.

Another compromise is made when blowing multiple bottle sizes from the same preform. This may not be optimum from the perspective of polymer utilization, but it may save lots of money on inventories and mold costs. As shown Section 12.3.2, one can design the perform for optimum wall thickness distribution by using finite element analysis. Sometimes polymer is less expensive than molds.

The desired end result of PET preform manufacture is a preform with excellent clarity, good color, low acetaldehyde (AA) concentration, no defects or imperfections, minimum stresses, and proper weight and dimensions. Good clarity implies no crystallinity, which would result in whitening of the preform. Hydrolytic degradation can reduce the IV (intrinsic viscosity) of the preform, resulting in bottle blowing problems such as ovality in the finish, poor strength, easily crystallized bottle, etc. Moisture in very small amounts has a large effect on the IV of PET, as shown in Figure 12.17. The curve shown is for 0.84 IV, but the same pattern is found regardless of the beginning IV.

Thermal degradation can also cause IV loss, color generation, and acetaldehyde production. Since we want colorless bottles, any color is bad. AA has a fruity flavor that can affect product quality because it is very potent at small concentrations. Figure 12.18 shows the effect of temperature on haze and AA. From the figure, one can see that the two attributes are affected differently by changes in temperature, with haze decreasing and acetaldehyde concentration increasing as temperature goes up.

Theoretical Effect of Moisture

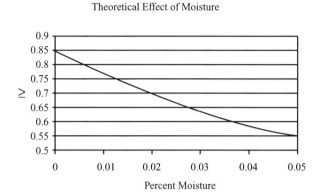

Figure 12.17 Effect of moisture on viscosity of PET.

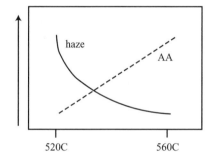

Figure 12.18 Effect of temperature on haze and acetaldehyde.

12.4.4 Bottle Blowing

Preforms, especially for soft drink bottles, are often produced in one location and then shipped to the bottle blowing location. PET preforms must be protected from moisture until they are blown, as PET will pick up moisture in the preform stage as well as in the pellet stage. If the moisture content of PET rises too high, the reheating operation will lower the IV and increase the AA content. If the ISBM process is a single stage operation, then this is not an issue since the preform is only in that stage for a short period of time. However, doing all of the operations in a single machine tends to limit the speeds and the versatility of the use of the equipment. In addition, it is much more economical to ship empty preforms than empty bottles, if the bottles are not filled near their point of manufacture.

Most high volume stretch blow molding systems use two stage blow molding. Blow molding of hot fill bottles differs from that of soft drink bottles in requiring a heat setting step. In either case, the preform is loaded into the machine automatically from a preform hopper, transferred into the heating section, and then to the blowing station. Operations on the blow molder may be linear, or may use wheels for higher outputs. There are machines with linear ovens and wheel blowing stations, and there are machines with wheels for both the heating and the blowing sections. Wheels offer higher outputs for a smaller footprint.

12.4.4.1 Preform Heating

The first step in the process is to reheat the preform to the proper temperature for blowing. The preform, mounted on a carrier (it can be oriented in many different manners), is passed in front of a series of IR lamps (Fig. 12.19). Each lamp is usually controlled individually for temperature, so the proper temperature gradient can be achieved in the preform to result in the plastic moving to where it is desired in the finished bottle. A schematic is shown in the drawing. Note that there is a heat shield for the finish portion of the preform, since we do not want this area to distort during the blowing process. The heating process is not necessarily continuous. The walls of the preform are thick and the heat conduction is slow, so in order to not overheat the outer wall, the heating may involve a "soak" time where the preform is not in front of the heaters.

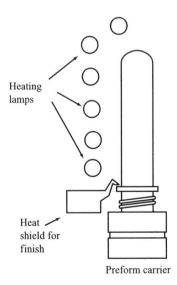

Fig. 12-19 Preform in the oven.

12.4.4.2 Blowing the Bottles

After exiting the oven, the preform is moved to the blow mold area. Here, the mold is closed around the preform. The stretch rod begins to stretch the preform toward the base of the mold and the preblow air comes on. Next, the high pressure air comes on to finish the blowing to the bottle to its final shape. The timing of all of these functions is critical to the final bottle properties. Obviously, material used and the size and shape of the bottle have an effect on the choice of the timing variables. During the blowing process, the material is following the stress-strain behavior inherent to the polymer. First it deforms elastically. Then the material yields and stretches to the "natural stretch ratio" under the pressure of the preblow air. In the region beyond the natural stretch ratio, the material starts to strain harden, and the air pressure must be increased dramatically to finish the blowing (Fig. 12.20). We can see similar behavior in the lab using test strips in a tensile tester, but the level of stress is much higher in the blowing process because the strain rates are much higher.

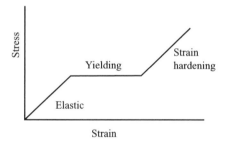

Figure 12.20 Stress-strain during blow molding.

It is in the strain-hardening region that the orientation of the bottle becomes the most pronounced. The orientation causes the amorphous portion of the polymer to form regions of aligned molecules that are very close together. The close alignment of the molecules in any one of the regions causes the entropy to be lowered locally, which in effect raises the crystallization temperature. Therefore, the material forms small regions of straight-chain crystals when it is cooled. These regions are as effective in reducing the permeation as folded chain crystals, but because of their small size, they do not scatter light, so they do not reduce the transparency of the plastic.

Sometimes, there are imperfections due either to crystallinity in the preform or due to crystallinity allowed to form during the reheat process. These imperfections show up as haze - a pearlescence. This is due to micro-cracks in the PET induced by the stresses during blowing. This is different from the white crystallinity around the gate area of the preform, which is due to thermal crystallization.

The finished bottle has an amorphous neck finish, an oriented sidewall, and a poorly oriented base. The bottle has a large amount of frozen in stresses. The crystallinity is not high compared to polyolefin bottles, but it is sufficient to produce a reasonable barrier to oxygen and water vapor, permitting the shelf life required for a carbonated beverage. Water bottles are usually the same design as a CSD bottle, but of a lighter weight because they do not have to withstand the high internal pressure. Neither bottle is stable to temperatures near or above the T_g of PET (about 70°C). If you have never tried the experiment, fill a CSD bottle with boiling water (preferable done in a sink) and watch it shrink back to a shape looking like the original parison. It is for this reason that hot fill bottles have a different design.

12.5   Hot Fill Bottles

Hot fill bottles are used for many juices, other beverages, and food products that use the process temperature to minimize the bacterial load in the bottle and the cap to protect the product from spoilage. Filling temperatures are 82-95°C. While some bottles, such as PP, can be filled at these temperatures without much distortion, this is above the T_g for regular bottle grade PET. This means that for PET, something has to be done to the bottle to keep it from shrinking during filling. At the fill temperatures, the polymer would start to crystallize and shrink; the bottle would therefore shrink in volume and distort. In addition, during the cooling of the product after capping, the pressure decreases in the package due to the shrinkage of the liquid and the headspace gas. This can lead to paneling of the bottle, if design compensations are not made.

    To minimize shrinkage and moisture absorption in hot fill bottles, the first step is to "heat-set" the bottle. This is a process in which the bottle crystallinity, mainly in the sidewalls, is increased to about 30% under stress to maintain the bottle shape. The crystallization that is desired is strain-induced crystallization, which results in smaller crystallites that do not scatter light. The preform weight for hot fill bottles is higher also to help stabilize the finish area. Illustrations of the different preform weights commonly used in industry are in Table 12.1.

Table 12.1 Common Weights for PET Bottles

Container Type 500 ml (20 oz)	Weight (g)
Water	15-19
Carbonated soft drink	23-27
Hot fill	33-38

If one can afford the added cost, one can use a slightly higher T_g PET for hot fill bottles. This higher T_g is particularly useful in minimizing the dimensional changes in the finish.

The process for blowing hot fill bottles is different from that used for cold fill. In the most common process, preforms are heated to 10-15°C higher than for cold fill. The molds are heated to 120-140°C to cause the heat setting. The base of the bottle mold is usually cooled to keep thermally induced crystals from forming because the orientation in the base is minimal. The bottles are "cooled" by circulating air into the bottle after the molding for 0.5 to 1 sec. to stop the crystallization process before the bottles are de-molded. An alternative process uses two sets of molds, with the bottle initially blown to a larger size, shrunk somewhat in an oven and then reblown to achieve its final and more stable shape.

Most suppliers have their own, patented vacuum panel designs to compensate for pressure changes in the bottle during cooling. However, a cursory look at the bottles does not necessarily reveal a large difference in these designs - particularly for generic shapes. The photos in Figure 12.21 show two designs for hot fill. The Twister bottle on the left uses the oblong shape and the label panels to allow for volume change, whereas the Snapple bottle on the right has built the logo into the vacuum panels.

Figure 12.21 Hot-fill bottles.

The design of vacuum panels for a bottle is predicated on a certain liquid and gas shrinkage. Therefore, if the volume of the headspace increases, or the temperatures change, one may end up with a paneled bottle. Paneling may not affect the performance of the bottle as a package, but it is esthetically unpleasing and it makes the bottle hard to label.

Example:

Assume we have a bottle with an overflow volume of 283 ml. It is filled with 263 ml of product and capped at 82°C. After cooling to 25°C, what is the pressure in the bottle?

From the data, we have a headspace of 20 ml. If we assume the headspace gases act as an ideal gas,

$$\frac{P_1 V_1}{T_1} = \frac{P_2 V_2}{T_2}$$

We need to figure the change in liquid volume first, so we will know what V_2 is. To do this, if we assume the product behaves the same way as water does (since it is mostly water), we can use a set of steam tables, which will list specific volumes for liquid water at various temperatures, or another source of this information. If you do not have such a table, go to your library, ask a chemical engineer, go online, . . .

We find that at 82°C, water has a specific volume of 0.00103033 m³/kg (density of 970.564 kg/m³). At 25°C, the specific volume decreases to 0.00100295 m³/kg (density of 997.062 kg/m³). Therefore, the liquid volume of 263 ml at 82°C will be:

$$263 \, ml \, \frac{0.00100295}{0.00103033} = 256 \, ml$$

Therefore, the headspace volume will increase from 20 ml to 31 ml.

The pressure in the bottle, using the PVT equation, is:

$$\frac{1 \, atm \, 20 \, ml}{355.2 \, K} = \frac{x \, 31 \, ml}{298.2 \, K}$$

$$x = 0.54 \, atm$$

So, the pressure has been reduced to only 0.54 atm, or a vacuum of 0.46 atm.

In hot-filled PET bottles, distortion of the finish may cause problems with successful sealing of the bottle. If the finish of the PET preform is heated while the body of the preform is kept cool, spherulitic crystals result which turn the finish an opaque white and greatly increase its rigidity. Such crystallized finishes have less tendency to deform during hot-filling of the bottle.

12.6   Coinjection Blow Molded Bottles

Technology to successfully produce multi-layer injection blow molded bottles is much newer than that to produce multi-layer extrusion blow molded bottles. Heinz, Inc.'s ketchup bottle was the first U.S. example (Fig. 12.22). The use of PET bottles made by coinjection blow molding is growing rapidly. As is the case for coextrusion blow-molding, the key is production of the parison. Once the multilayer parison is produced, the remainder of the process is the same as for single layer materials. Even stretch blow molding can be used.

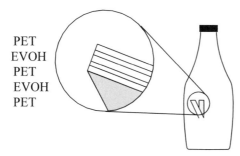

PET
EVOH
PET
EVOH
PET

Figure 12.22 Coinjection stretch blow molded bottle (reprinted with permission from [1]).

Two techniques are used for producing the multilayer injection molded preforms. The technology used for the Heinz PET/EVOH ketchup bottle involves simultaneous injection of two or more resins into the mold cavity in such a way that they remain in distinct layers. The second process involves a multiple set of preform molds, with a single resin injected in each step.

As illustrated in Figure 12.22, the coinjection stretch blow molded PET bottles actually have a five layer structure, containing two layers of EVOH and three of PET. Presumably, the reason is that two thin EVOH layers provide better oxygen barrier than one thick EVOH layer, because the effects of any defects in one layer are diminished by the presence of the second. No tie layers are used in the structure, to facilitate recycling of the containers in existing PET recycling facilities. When the bottle is granulated and washed, most of the EVOH is removed. The small amount that remains does not significantly impact the quality of the recyclate.

As was the case for ketchup bottles, most uses of coinjection blow-molding have the aim of improving the barrier characteristics of the container, to extend the shelf life and better preserve the flavor, aroma, or other characteristics of the product. Reheat stretch blow molding is the most common process, with three to five layers of resin forming the parison. The skin layers, both inside and outside, are typically virgin PET, with inner layers of barrier resins, most often EVOH or special nylons. The designs for the preform tooling are closely guarded, with major patents held by American National Can Co. for simultaneous coinjection tools, and by Continental PET Technologies Inc. for sequential coinjection tooling. The rapidly growing market for these containers is fueling new developments in tool designs which are expected to bring cost and performance more in line with the market demand.

One characteristics of the latest coinjection equipment is separate hot-runner temperature control systems for the resins, allowing a core layer like EVOH to be processed at a temperature 70°C lower than the PET skin layers. Computer simulation is used to develop tooling which manipulates the barrier resin distribution to fortify critical areas such as wall sections. Both wide mouth and narrow neck bottles are being produced.

Many of the applications of coinjection blow molding are for oxygen-sensitive products. In this area, PET beer bottles are a major target. While PET by itself has been used successfully for some beers in Europe, Japan, and Canada, the applications have generally been for heavier body beers with a robust flavor that is less sensitive to oxygen off-taste, and shelf life is limited. The light beers favored in the U.S. market require a better oxygen barrier than can generally be achieved with PET alone. Protection from light is also important for many beers. As discussed in Chapter 4, one solution is to use PEN or PEN/PET blends or copolymers, but these are costly. Multilayer PET bottles with ethylene vinyl alcohol as an oxygen barrier, or with an oxygen scavenger, are an alternative. Two polymers being used as oxygen scavengers are a copolyester being produced by Amoco Chemicals, and a nylon with a cobalt catalyst. In both cases, the oxygen that penetrates through the PET reacts with the scavenger. The first targets for beer in plastic bottles were sporting events, beaches, and other facilities where glass is not permitted.

In 1997, Coca Cola introduced a high-barrier 12-oz PET/EVOH coinjection stretch blow molded bottle that was reported to double the shelf life of the beverage in hot and humid conditions. It also permits the use of post-consumer PET content in the center layer of the container.

12.7   Foam Blow Molding

Foam blow molding was pioneered in about 1998 in Germany and Austria for shampoo bottles. The bottle was developed by Wella AG in Darmstadt, Germany, in partnership with Alpla Technik of Hard, Austria, and Boehringer Ingelheim KG of Ingelheim, Germany, the formulator of the blowing agent. The HDPE bottles are reportedly the first foamed blow molded product made commercially, as well as being by far the lowest-density bottles ever produced for the household and industrial chemical market. The density of the walls is around 0.75 g/cm^3. The overall bottle density is 0.78-0.80 g/cm^3.

The total bottle weight is about 25% less than the equivalent in HDPE homopolymer and 15% lighter than PP, but the rate of diffusion of perfume through the bottle walls is reportedly unchanged. For some compounds such as peroxide, the cells provide increased barrier, allowing the replacement of multilayer bottles by single layer bottles in some applications. Savings come primarily from being able to make a stronger part with less resin and less pigment, sometimes in shorter cycles. In Germany and Austria, this saves not only in initial costs, but also in the Green Dot recycling fees, which are based on packaging weight.

Alpla has produced bottles with a density as low as 0.68 g/cm^3 in developmental work. Addition of pigment slightly increases the density. The foam cells cannot be seen with the naked eye, but are visible with a microscope. The cell diameter is between 50 and 100 microns in the bottle neck and base, and about 30 microns in the walls where the polymer is stretched. The surface layer is solid HDPE, with microbubbles in the core. Two-layer versions of the bottle have a smoother bottle surface but are somewhat heavier than monolayer bottles.

The endothermic nucleating foaming agent uses hydrocerol agents reacting with baking soda and citric acid to release carbon dioxide with limited moisture. It is formulated in a PE-based master batch, and added at 1.5 to 2% levels. One grade has FDA food-contact approval.

The foam adds iridescence and brightness to the bottles because light reflects off the micro bubbles, and therefore less pigment is required. Use of titanium dioxide can be eliminated because the foam structure produces a whitening effect in the HDPE.

Wall stiffness and strength of the bottles is reported to be equivalent to solid HDPE bottles, while squeezing the bottles requires only about one-third of the hand pressure needed for solid bottles. The matte texture of the outer surface also provides a better grip of wet bottles. The foam bottles have a good memory, allowing them to return to their original shape after pressing or squeezing [5].

12.8   Blow Molds

Aluminum and beryllium copper are often used for blow molds for plastic bottles. Beryllium copper has better heat transfer and resistance to wear, but is more expensive. Aluminum is used most because it is significantly less expensive and provides good heat transfer, although it does wear more quickly. Because blow molds are not subject to pressures nearly as high as the pressures in injection molding, the molds can be less rugged. Beryllium copper or steel inserts are often used in the parts of the mold most subject to wear, such as the pinch-off in an extrusion blow mold. A typical blow mold can produce about 12 million containers before it must be discarded, provided some refurbishing is done periodically.

Removal of trapped air must be provided in blow molding, just as in injection molding. Venting is usually provided along the mold parting lines. For large containers especially, additional venting may be required. For bottles that do not have high clarity, such as HDPE and some PP containers, the molds are often sandblasted to help reduce air entrapment and reduce requirements for venting. This cannot be used for clear bottles such as PVC, PS, PET, and some PP bottles, which require highly polished molds.

The determining factor in cycle time for blow molding is most often the time required for sufficient cooling of the bottle. Therefore, it is important to design the cooling channels properly. In addition to cooling by heat transfer to the mold, bottles may be cooled internally using carbon dioxide, air, or a combination of air and water vapor. These methods have been reported to increase production rates by as much as 50%, but they are not often used. In many cases, the extruder output then becomes controlling, and cannot support the increase.

12.9   In-Mold Labeling

In-mold labeling means that the label is placed on the container during the molding cycle, so that the label is already in place when the container is first produced. This process is applied to well less than half of all plastic bottles, but its use continues to increase. To accomplish in-mold labeling, the labels are placed in the blow mold while it is empty. They are held in place by a slight suction applied through the mold, and form a heat-seal to the bottle when the hot plastic contacts the label as the container is blown. The first generation of labels were paper, backed with a heat-seal coating. Current labels are increasingly likely to be plastic, often a plastic compatible with the bottle, to facilitate recycling. The appearance of plastic labels is generally preferable to that of paper labels.

The most important advantage of in-mold labeling is that it eliminates the need for a subsequent labeling operation. Since labeling is typically the part of a packaging line associated with the most down-time, this is significant in achieving higher productivity. In-mold labeling also eliminates the need to flame-treat the bottles prior to labeling in order to achieve adequate adhesion. Moreover, since the labels are nearly flush with the bottle wall, there is less scuffing of the print, enhancing container appearance. Claims have been made of improved strength and bulge resistance in the container, but these are disputed. When in-mold labeling is coupled with in-case filling, containers can leave the manufacturing facility in corrugated boxes and be filled and capped without de-casing.

The major disadvantage of in-mold labeling is the increased process complexity, and the required investment in capital equipment. It is not well suited to small production runs or where label copy changes frequently, even if bottles do not. Some experts claim in-mold labeling increases overall cycle time by 1.5 to 3.3 s, but others claim cycle time is not increased.

In-mold labeling is used most often with extrusion blow molded HDPE bottles. It has also been used on PET and PP stretch blow molded bottles, as well as on injection molded tubs.

12.10 Aseptic Blow Molding

Aseptic packaging involves a sterile product that is sealed into a sterile package under sterile conditions, thus eliminating the need to sterilize the product-package system after filling. This concept can be applied to blow-molded packages, along with other package forms. The heat of extrusion effectively sterilizes the plastic, if the residence time is long enough. However, in ordinary bottle manufacturing the container is exposed to airborne microorganisms, so this sterility is lost. In aseptic blow molding, the bottle is produced and sealed under sterile conditions.

Two process variants exist. In both, the first step is to provide sterility for the blowing operation, including the air used in forming the container. In the blow and hold process, the bottle is sealed in the mold to preserve its sterility, and can then be released into a non-sterile environment. In a subsequent filling operation, it is cut open under sterile conditions, filled, and sealed again. In the blow-fill-seal process, the bottle is blown, filled with product (often while still in the mold), and sealed, all within the sterile environment of the blow molding machine, either at a single station or at multiple stations.

12.11 Surface Treatment

Surface treatment of plastic containers is sometimes used to improve barrier properties, or to enhance printability or label adhesion.

12.11.1 Flame Treatment

Flame treatment is the most common surface treatment for containers. It is used to improve the bottle's adhesion to printing inks or to adhesives on labels. Printing and labeling polyolefin bottles, in particular, is difficult because their non-polar surfaces do not adhere well to most inks or adhesives. Passing the bottle near a flame causes some oxidation of the surface, leaving polar groups that improve bonding ability. The flame treatment can also remove surface contaminants, such as mold release agents, which may interfere with bonding. Of course, the heat exposure must not be enough to cause melting and distortion of the plastic. Typically, the bottle is passed rapidly over a flame, for a contact time of less than one second.

12.11.2 Coatings

Coatings of various types can be used to improve containers' barrier properties. PVDC coatings can increase barrier to hydrocarbons, water vapor, odors and flavors, and oxygen. They can be applied by a variety of methods, including spray and dip coating.

Silicon oxide-based coatings, of particular value as oxygen barriers, can be applied using chemical vapor deposition, most often to PET bottles. The process forms an SiO_x film less than 2000 Å in thickness, which can more than triple the oxygen barrier of the base container, and also improves water vapor barrier by 2 to 3 times.

Sidel, a French company, has developed the Actis plasma process which coats the inside of PET bottles with a 0.15 micron thick layer of amorphous carbon to improve oxygen and carbon dioxide barrier. Actis stands for Amorphous Carbon Treatment on Internal Surface. The carbon is deposited from acetylene gas, using a microwave-assisted process to excite the gas into plasma. The bottles are clear, and the process is reported to increase the carbon dioxide barrier of beer bottles by up to seven times, while not interfering with recycling. Actis has been approved by FDA, and is reported to cost 20 to 25 percent less than multilayer PET bottles with comparable barrier properties [6]. Sidel has also developed "Actis Lite" - a lower level of treatment for carbonated soft drink bottles, sparkling waters, juices, teas, and sauces, which do not require as good a barrier as beers and ciders.

12.11.3 Fluorination

Fluorination of polyolefin bottles, usually HDPE, is designed to improve their barrier capability, especially to hydrocarbons. In fluorination, the container is exposed to fluorine gas, which reacts on the surface of the container, replacing some C-H bonds with C-F bonds. The increased surface polarity decreases the container's permeability to nonpolar penetrants such as hydrocarbons. It also increases adhesion to inks, coatings, and label adhesives.

Two basic methods are used for fluorination. In the first, extrusion blow molded containers are blown with nitrogen containing a low percentage of fluorine, instead of with air. At extrusion temperatures, the fluorine reacts with the polyolefin, creating a fluoropolymer layer on the inside of the container. In the second method, containers are produced using normal methods and then post-fluorinated by placing them in a chamber where they are heated and exposed to a mixture of fluorine gas in nitrogen. In this case, the fluorinated layer is formed on both the inside and the outside of the container. The extent of fluorination, in both cases, is determined by the combination of fluorine concentration, temperature, and time. Fluorine gas is highly corrosive, and must be carefully purged and collected. Precautions are also needed to ensure adequate protection for operators of the equipment.

12.11.4 Sulfonation

Sulfonation can also be used to improve the barrier properties of HDPE bottles to hydrocarbons. After molding, the bottles are treated with sulfur trioxide in an inert gas, producing sulfonic acid groups on the bottle's surface, which are then neutralized with ammonia or sodium hydroxide. As with fluorination, proper procedures to ensure worker safety and avoid release of the sulfur trioxide and neutralization chemicals into the environment are necessary. Sulfonation also improves the adhesion of inks and coatings to the containers.

12.12 Dimensions and Tolerances for Plastic Bottles

The design of plastic bottles is highly standardized, as is the design for closures, so that in theory any plastic closure of a designated size should fit any plastic bottle of matching size. A number of standards organizations issue design specifications for dimensions and tolerances of plastic bottles, including ASTM. Bottle finishes are sized by the nominal outside diameter in mm, as indicated in Figure 12.23 and Table 12.2. The series designation reflects the finish style. Volume is specified by nominal overflow capacity. Thread profiles are of two main styles, the "L" or all-purpose thread, designed for plastic or metal closures, and the "M" or modified buttress thread, designed for plastic closures. Major dimensions include T, the diameter of the finish, including the threads; E, the outside diameter excluding the threads; and I, the inside diameter of the finish. These dimensions are measured across both the major and minor axis and then averaged, with the ratio of the measurements giving the ovality of the finish. Excessive ovality interferes with proper sealing of the container. S is the distance from the top of the finish to the top edge of the leading thread; and H from the top of the finish to the first feature on the bottle that exceeds the T dimension and therefore stops the downward motion of the cap. This is the transfer bead in the 400 and 444 series, and the shoulder in the 410 and 415 series. When the T dimension is to the shoulder, the L dimension is the distance from the top of the finish to the top of the transfer bead.

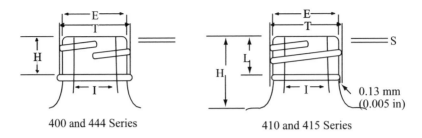

400 and 444 Series   410 and 415 Series

Figure 12.23 Standard finish dimension nomenclature for bottles (reprinted with permission from [1]).

Table 12.2 Standard Finish Dimension Tolerances for Plastic Bottles (reprinted with permission from [1])

Series	Millimeter size	T (in) min max	E (in) min max	H (in) min max	S (in) min max	I (in) min
SP-400	18	0.688-0.704	0.604-0.620	0.356-0.386	0.022-0.052	0.325
SP-400	28	1.068-1.088	0.974-0.994	0.385-0.415	0.031-0.061	0.614
SP-410	18	0.688-0.704	0.604-0.620	0.508-0.538	0.022-0.052	0.325
SP-410	28	1.068-1.088	0.974-0.994	0.693-0.723	0.031-0.061	0.614
SP-415	18	0.688-0.704	0.604-0.620	0.602-0.632	0.022-0.052	0.325
SP-415	28	1.068-1.088	0.974-0.994	1.067-1.097	0.031-0.061	0.614
SP-444	28	1.068-1.088	0.974-0.994	0.572-0.602	0.160-0.190	0.614

References

1. Selke, Susan, *Understanding Plastics Packaging Technology*, Carl Hanser Verlag: Munich, 1997.
2. Dees, Jerry, "Preform Optimization Using Non-Linear Finite Element Simulations," *ANTEC 2003*, Vo. XLIX, Nashville, TN, Society of Plastics Engineers, 2003, pp. 842-849.
3. Defosse, M. T., *Modern Plastics*, May 1999, pp. 51-54.
4. *Modern Plastics*, Nov. 1998, pp. 41-42.
5. Schut, J., *Modern Plastics*, March 1998, pp. 28-30.
6. Demetrakakes, Pan, *Food & Drug Packaging*, "Barrier coatings help bottlers overcome shelf-life obstacles," *Food & Drug Packaging*, March, 2003.

Study Questions

1. Describe extrusion and injection blow molding, identifying their differences and similarities.

2. Why might you want to use parison programming? die shaping? Can both be used together? Why would you want to do so?

3. Describe stretch blow molding. What is its major purpose? Why might you want to use it?

4. Explain what a tie layer is, and why and when it is used.

5. Describe in-mold labeling. What are its major advantages and disadvantages?

6. Why are HDPE bottles often flame-treated? What does flame-treating do?

7. Someone hands you a bottle and asks whether it was produced by extrusion or injection blow molding. How can you tell? What if they do not have the bottle with them - what can you ask about the bottle or the product that would let you make an intelligent guess about which process was used? Think of at least 3 different questions to ask.

8. Fluorinated bottles are available in various grades. What relationship would you expect between fluorination conditions and barrier performance?

9. PET has a much higher melting temperature than PP, yet distortion of PET bottles during hot filling is often more of a problem than distortion of PP bottles during hot filling. Why is this so?

10. It used to be true that considerably more than half of all bottles used in the U.S. were made of HDPE. Now, more PET bottles are made than HDPE bottles. What are some reasons for this change?

11. You are in charge of selecting a bottle for an oxygen-sensitive food product. Marketing wants a transparent container, with a one-year shelf life. What plastic materials would be reasonable to consider using? Why?

12. What bottle production processes would you recommend for the materials you decided to consider in question 11? Why?

13. Would you expect there to be differences in thread configuration (shape and size) between glass bottles and plastic bottles? Why or why not?

14. Would you expect there to be differences in finish dimension tolerances between glass bottles and plastic bottles? Why or why not?

15. Why is extrusion blow molding preferred to injection blow molding for heat-sensitive resins?

16. A "soak" time is used in preform heating for some stretch blow molding processes. Would you expect this to more likely be needed for thick or for thin preforms? Explain.

13 Foams, Cushioning and Distribution Packaging

13.1 Foams

Plastic foams are used in packaging both as cushioning materials and as containers. Polystyrene is used most often, but there is also considerable use of polyethylene, polypropylene, and urethane foams, along with some other materials. Foams are characterized by very light weight, good insulating capacity, and the ability to absorb shocks and protect the enclosed product. The most common use of packaging foams is as cushioning materials, to protect products during the distribution cycle.

Foam cushioning materials can be used as small shapes piled loosely around the object, loosefill, or as molded pieces designed for a particular application. Loosefill is used most often for light weight objects, and molded pieces are used most often for large heavy objects. A variant of molded foam is foam-in-place systems, where the foam is formed within the shipping container, using the container and the product as molds, and producing a foam that fits tightly around the product.

Foam containers are most often produced by thermoforming extruded foam sheet. They can also be produced by molding.

Foams can be divided into two groups. *Open cell* foams have communicating channels between adjacent cells. In *closed cell* foams, the cells are completely surrounded by the polymer. In an open cell foam, liquids or gases can travel through the foam by traversing the channels between the cells. Therefore these foams tend to be absorbent; a sponge is a common example. In a closed cell foam, a liquid or gas can travel through the foam only by diffusing through the polymer at the boundaries between the cells. Therefore these foams provide much better barrier and are nonabsorbent. The cells sometimes contain substantial amounts of residual blowing agent. Most packaging foams are closed cell. A closed cell foam can be converted to an open cell foam by subjecting it to enough force to rupture polymer membranes between the cells, providing channels between them. The percentage of open cells can be used to characterize where a foam is in the continuum between 100% open cell and 100% closed cell.

13.1.1 Polystyrene Foam

Polystyrene, as mentioned above, is the most common packaging foam. EPS is the material of choice if it can perform acceptably, since it is typically the least expensive packaging foam available. It is used extensively for containers as well as for cushioning material, in molded shapes and in extruded form. Molded shapes are commonly termed expanded PS, while the extruded material is called simply extruded PS foam. The term "styrofoam" is often incorrectly used for these materials, but Styrofoam is a Dow Chemical Company trademarked building insulation.

PS foam is relatively chemically inert, and is acceptable for use in food packaging. PS foam trays are widely used for meat and produce. The ubiquitous coffee cup utilizes the light weight of PS foam combined with its insulating ability to provide an inexpensive container for hot drinks. Fast food outlets often use PS foam in the shape of bowls or clamshells for products which need to be kept either hot or cold. Shippers of sensitive products ranging from fish to pharmaceuticals often rely on PS foam containers to keep products cold during distribution, especially if the products are shipped by air freight.

When insulation is not required, PS foam is still frequently used as part of distribution packaging systems due to its excellent shock-absorbing capabilities. In a very different application, the cushioning ability of PS foam is used in the form of labels on single-serving glass bottles for carbonated beverages to reduce surface abrasion of the glass, permitting significant light-weighting that makes the glass bottles more competitive with aluminum cans and PET bottles.

13.1.1.1 Expanded Polystyrene Foam (EPS)

To make expanded polystyrene foam, PS granules or beads are impregnated with a hydrocarbon blowing agent. The blowing agent produces bubbles in the plastic, forming the cells, as it vaporizes. Pentane is the most common choice, and is used in amounts up to 8% of the polymer by weight. Both the amount of blowing agent and the processing conditions influence the properties of the foam. The PS beads are pre-expanded by heating them to 85 to 96°C (185 to 205°F) to vaporize the pentane. Typically the beads reach 25 to 40 times their original size, resulting in a foam density of 16-26 lb/m³ (1 to 1.6 lb/ft³). These pre-expanded beads are aged so that they reach equilibrium. Next, they are packed into a mold, where they are held under several tons of pressure while steam is introduced directly into the mold. The heat and pressure cause the beads to fuse together, producing a semirigid closed cell foam. The mold is over-filled at the beginning of the cycle, since space is occupied by the voids between the beads, as well as by the foam itself. If the mold were not over-filled, gaps in the finished product would result. The product is held in the mold during a cooling cycle, and then released when it is dimensionally stable.

The mold can directly produce the final shape desired, or it can simply produce a block of material that is subsequently cut into the final shape. Additional aging and curing steps may be employed to further reduce the density of the material.

Expandable PS beads are manufactured in three sizes, small, medium, and large. The bead size required depends on the wall thickness of the molded part. Large beads can be used for thick walls, while small beads are necessary for products with thin walls.

Adhesive-coated expanded PS shapes have been used as an alternative to foam-in-place polyurethane systems (see Section 13.1.3). The shapes are deposited into a corrugated box, covered with a release film, the product added, another sheet of release film put in place, and the carton filled with the adhesive-coated shapes. When the water-based adhesive dries, a pair of resilient cushions surrounds the product.

13.1.1.2 Extruded Polystyrene Foam

Extruded PS foam is produced using extrusion, rather than molding. The PS resin is melted in an extruder, the blowing agent and a nucleator mixed in, and the blend extruded. As with molded foam, the blowing agent creates the cells as it vaporizes. The purpose of the nucleator is to aid in obtaining the desired cell size and uniformity by providing sites for bubble growth. Talc, citric acid, and blends of citric acid with sodium bicarbonate are often used as nucleators.

The blowing agent is usually a hydrocarbon or hydrocarbon blend, which is injected into the melt as a liquid, or sometimes as a pressurized gas. In recent years, use of carbon dioxide as a blowing agent, either in place of or blended with hydrocarbons, has increased substantially. Chlorofluorocarbons (CFCs) were common blowing agents in the past, but are no longer allowed due to their ozone depleting activity; hydrochlorofluorocarbons (HCFCs), which are less active in ozone depletion, cannot be used as blowing agents in the U.S., and are also being phased out worldwide.

The melt is kept under pressure until it leaves the die, preventing vaporization of the blowing agent. When the melt exits the die, the pressure is released and the blowing agent immediately vaporizes, expanding the melt. If the melt strength of the polymer is not high enough, this abrupt expansion will fracture the melt. It must be able to withstand the pressure exerted by the blowing agent, forming a uniform network of cells. For polystyrene, this requires that the melt be cooled after the blowing agent has been introduced and before the pressure is released. Two methods are commonly used for cooling the melt to the required temperature. One method is to incorporate a cooling zone in the extruder, after the blowing agent has been blended in. The more common approach is to use a two-extruder system, or tandem system. The polymer is melted and the blowing agent and nucleator added in the first extruder. The melt is then introduced, still under pressure, to the second extruder, where it is cooled and then released through the die.

The foam can be produced as a flat sheet, using a slit die as is done for cast film. Most often, however, a tubular film is produced, using an annular die. In this process, a compressed air ring around the circumference of the die forms a thin skin on the foam surface as it exits the die and begins to foam. The tube of foamed PS passes over a water-cooled mandrel, which provides additional cooling. Next the foam is slit, flattened, and wound into rolls.

The amount of blowing agent introduced is the primary determinant of the final density of the foam. The size and number of cells are controlled by the amount of nucleating agent. Extruded PS foams used for packaging typically have densities of 64-96 kg/m^3 (4-6 lb/ft^3).

As mentioned previously, the sheet is commonly thermoformed, often using matched molds, to produce the desired package shapes. Matched mold forming is used to minimize the distortion that would otherwise likely result when the sheet is heated. For best results, the sheet should be aged for 3 to 5 days before thermoforming, so the gas pressure in the cells can equilibrate. The scrap produced during thermoforming can be ground and densified in an extruder for recycling. Used foams can also be recycled.

13.1.1.3 Styrene Copolymer Foams

Copolymers of styrene and other monomers are also used in packaging. The most common is styrene-acrylonitrile (SAN) foam, which is a semirigid foam and offers better performance than EPS in repeated drops and for heavy products (high static loads). The density of SAN foam is usually about 16 lb/m^3 (1 lb/ft^3).

13.1.2 Polyolefin Foams

Polyolefin foams are somewhat higher in cost than PS. They are also more flexible and better able to provide protection from multiple impacts. Typical densities are 16-32 kg/m^3 (1-2 lb/ft^3). Polypropylene foams have somewhat greater rigidity than polyethylene foams. Polyolefin foams, like polystyrene, are available in two varieties, expanded (generally termed moldable) and extruded. PE/PS copolymer foams are also available, with characteristics generally intermediate between PS and PE, but with outstanding toughness.

The manufacture of polyolefin foams is very similar to manufacture of extruded PS foam. Hydrocarbons or blends of hydrocarbons and carbon dioxide are used as blowing agents. An annular die and forming mandrel is used to produce extruded sheet products. Extrusion through rectangular slit dies onto a conveyor belt is used to produce planks.

PE foams are also available in cross-linked grades. These are manufactured in a much different manner. Rather than physical blowing agents like hydrocarbons and carbon dioxide, chemical blowing agents such as azodicarbonamide are used. These agents produce a gas as a result of a chemical reaction. The resin, additives, crosslinking agents, and blowing agents are mixed together at temperatures below the activation temperature of the blowing agent, and then extruded into a flat sheet or other profile. Next, the material is crosslinked, either chemically or by radiation. Radiation is typically used for thin materials, and chemical cross-linking for thick profiles. The final step is to expand the foam by exposing it to hot air (about 200°C) to activate the blowing agent.

13.1.3   Polyurethane Foams and Foam-in-Place Systems

Polyurethane foams are not widely used in packaging, but they do have significance in systems that rely on foam-in place operations, rather than on molded cushions or loosefill. In these systems, a mixture of a polyol and an isocyanate, with other ingredients, is injected into the erected shipping container (usually a corrugated box). As the foaming reaction begins, a PE film is placed over the top of the mixture, the product is added, and a second layer of film placed over the product. Next another shot of the polyol/isocyanate mixture is added, and the case is quickly sealed. As the foaming action continues, the product, protected by the polyethylene sheets, is tightly encapsulated within the foam (Fig. 13.1).

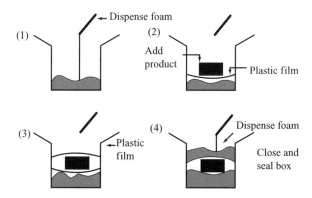

Figure 13.1   Foam-in-place packaging system (reprinted with permission from [1]).

Foam-in-place systems are not usually as cost-effective as molded cushioning or loosefill in high volume packaging operations. However, in low volume applications where relatively heavy products of many different shapes and sizes are packaged, they can be advantageous. In these situations, molded cushions to fit all the various packages and products are not economical, both in terms of initial investment and in terms of the logistics involved in getting the correct cushion for the correct product. Loosefill works well in such situations for light products, but for heavy products it tends to become displaced within the package, allowing the product to settle to the bottom, and perhaps become damaged.

A process that somewhat bridges the gap between molded packaging and foam-in-place uses a fixture, rather than the actual product, to produce polyurethane cushions, encapsulated in polyethylene sheet, of the desired size and shape. Foams can also be produced in bags rather than directly in the box, so that the product does not need to be surrounded by polyethylene. Polyurethane foams can be produced by extrusion methods, as well, but these are rarely used for packaging applications.

13.1.4   Starch-Based Foams

Starch-based foam loosefill packaging shapes are also available. While starch was not traditionally considered a plastic, the technique for producing these foams involves plasticizing the starch with water (and sometimes additional plasticizing agents) in an extruder and molding it. Starch foam shapes and cushions compete with loosefill PS and molded PS cushions.

The major advantage of starch-based foams is that they are considered by many to be more environmentally friendly, in part due to their water solubility. A related disadvantage is their tendency to be susceptible to moisture sorption, which may even result in collapse in environments with very high humidity. There have also been concerns, for some starch-based cushioning materials, about rodent or insect infestation.

13.2   Non-Foam Plastic Cushioning Systems

Not all plastic cushioning systems rely on the use of foam. Two major alternatives use air to provide resiliency, but are not foams. In bubble wrap, air bubbles of a defined size and pressure are sealed between two plastic sheets. These materials can be used as wrapping material, or as bags or envelopes. Several different sizes of bubbles are available. The larger sizes are intended for heavier duty applications, and the smaller ones for lighter duty. Bubble wrap is an exceptionally light weight packaging material, with a density as low as 11 kg/m^3 (0.7 lb/ft^3).

Another alternative is the use of transparent inflatable bags to provide a cushion of air around the product. A major advantage of these systems is that they occupy minimal space during shipping and storage prior to use, since they are inflated at the point of use. Some are designed to be collapsible and reinflatable, permitting reuse.

13.3 Cushioning

To select a cushioning system that will protect a product during transportation, it is necessary to know how fragile the product is and the distribution hazards it is likely to be exposed to, as well as the performance characteristics of the cushioning material.

Product fragility is generally determined by actual testing of the product. It is usually reported in "G-levels," the number of multiples of normal gravitational force that results in product damage. For analysis of distribution hazards, it is common to use characteristic drop heights that depend on the product weight and the type of handling system used. Then a calculation can be made, determined by the characteristics of the cushioning material, of the G-level a product would be subjected to in falling from the designated drop height, protected by the cushion under consideration. If this G-level were lower than that which causes product damage, the cushion would be expected to successfully protect the product. If it is higher, another cushion needs to be chosen.

The information about cushion performance is most often transmitted in the form of characteristic "cushion curves" that relate the average deceleration to the static loading as a function of drop height, cushion thickness, temperature, and number of impacts (see Fig. 13.2). Computer programs that contain mathematical models of cushion curves for various types of foams, sometimes along with cost information, are also available. Relatively simple calculations then permit selection of an appropriate cushion thickness and load-bearing area that can reasonably be expected to provide adequate protection for a given product.

In addition to impact forces, transmission of vibrations to the product can result in damage. Characteristic vibration transmissibility plots for foams are also available, which can be matched with the product's susceptibility to damage. Of course, the design process should be followed by the preparation of a model package and suitable testing of the product/package system, or, in case of a very expensive product, of a mock-up, using accelerometers, to determine actual performance.

It must be kept in mind that the fragility of the product may be different in different parts. For example, if the product falls on its side as opposed to its bottom, the result may be either greater or less damage. In general, at least for products contained in corrugated fiberboard boxes, flat drops are more damaging than corner drops, because when the box deforms during a corner drop, it absorbs some of the impact energy.

The following example illustrates selection of an appropriate cushion for a product:

Example: For the cushion with a characteristic cushion curve shown in Figure 13.2, calculate the load-bearing area that will provide adequate protection for a product measuring 20 cm by 30 cm, with a mass of 10 kg, and with a fragility of 50 G. The maximum expected drop height for the product is 1 m.

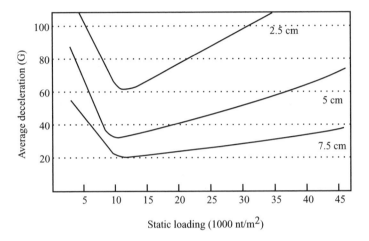

Figure 13.2 A typical cushion curve for a hypothetical packaging foam, for drops 2 to 5 from 1 m, for 2.5-cm thick, 5.0-cm thick, and 7.5-cm thick foam.

Solution:

We can eliminate the 2.5 cm cushion immediately since it does not provide decelerations below 50 G at any static loading. The 5 cm and 7.5 cm cushions are both possibilities. To determine under what conditions they would be usable, we need to calculate the static loading. The static loading must be between 7,000 and 27,500 nt/m² to keep the G under 50 for the 5 cm cushion. The static loading must be above 4,000 nt/m² and can be as high as 45,000 nt/m², the maximum extent of our information, for the 7.5 cm cushion. With a 10 kg object, the loading will be 7,000 nt/m² with a cushion area of 140 cm² (see sample calculation below). The loading will be 27,500 nt/m² with a cushion area of 35.7 cm². Since the footprint of our item is 600 cm², either of these or anything in between is feasible. Thus our cushion can provide any contact area between 35.7 and 140 cm², if we are using a 5 cm cushion, but should not provide more or less than this to get adequate performance. If we choose to use a 7.5 cm cushion, by the same process we determine that the cushion area should be at least 21.8 cm² and not more than 245 cm² to ensure adequate performance. The precise cushion geometry can be anything that is practical for the particular product.

Static loading = force/area = (mass × g)/area
So, area = (mass × g)/static loading
area = (10 kg × 980.665 cm/s² × m/100 cm × nt/(kg m/s²))/7000nt/m²) = 0.014 m²
0.014 m² × (100)² cm²/m² = 140 cm²

13.4   Thermal Insulation using Foams

Foams, in addition to being useful as cushioning, can be used to provide thermal insulation for products. A frozen product, for example, might be packaged with ice (or dry ice) to provide cooling, encased in a foam container to help reduce the conduction of heat from the surroundings into the container. Often the temperature inside and outside the container can be regarded as relatively constant, and the heat transfer process can considered essentially one dimensional. In such cases, Fourier's law of heat conduction reduces to its one-dimensional steady-state form:

$$q = k \frac{A(T_1 - T_2)}{x}$$
(13.1)

where q is the rate of heat conduction, T is temperature on the two sides of the material, A is the area available for heat transfer, x is the thickness of the material in the direction of heat transfer, and k is the thermal conductivity of the material. As can be seen, the higher is the thermal conductivity of a material, the more heat will be transferred through it in a given amount of time, at a specific temperature difference.

Plastics are, in general, good insulators. Gases are much better insulators than solids, so in a foam, the combination of the plastic and the gas pockets within it provides very low thermal conductivity. Each phase, gas and solid, contributes an amount roughly proportional to its volume fraction. Thermal conductivities of some common packaging foams are given in Table 13.1.

Table 13.1 Thermal conductivities and densities of selected packaging foams [2]

Material	density (kg/m^3)	k (w/m K)
polystyrene	16-160	0.030-0.037
polyethylene	26-43	0.036-0.053
polypropylene	10-96	0.039

These values can be used to calculate the length of time that a product in a given package can be expected to remain at a safe temperature, as shown in the following example.

Example: A 1-kg product which must be kept at a temperature of 0 °C is packaged with 2 kg of ice in a 1.5 cm thick expanded polystyrene container, with inside dimensions measuring 0.4 m × 0.4 m × 0.6 m. How long will the product be protected? The thermal conductivity of the polystyrene foam is 0.030 w/mK. The outside temperature is constant at 25 °C.

To simplify the problem, we will assume the heat conduction can be represented with reasonable accuracy by one-directional heat transfer throug h the total inside area of the container, and that the container provides the only resistance to heat transfer. We will also assume the product and the ice start out at a temperature of 0 °C. They will then remain at 0 °C until all the ice is melted. Thus we need consider only the heat of fusion of the ice, 6.01 kJ/mol.

The total amount of heat which can be conducted without harming the product, then, is

(6.01 kJ/mol)(2 kg)(mol/18 g)(1000 g/kg) = 668 kJ

The rate of heat transfer can be calculated from Eq. 13.1:

$$q = 0.030 \, w / mK \frac{(1.28 \, m^2)(298K - 273K)}{1.5 cm(m / 100 cm)} = 96 \, w$$

Since 1 w = 1 J/s, we can then calculate the time as

$$t = \frac{668,000 \, J}{96 \, J / s} = 6958 \, s = 116 \, min$$

If we need a longer lifetime, we could use more ice, a thicker foam, or make other adjustments.

One of our assumptions in this problem, that the foam container provides the only resistance to heat transfer, is, in many cases, a poor assumption. We have also implicitly assumed that conduction is the only mechanism for heat transfer - that radiation and convection do not play a role. This assumption is accurate in many instances, but inaccurate in others. We must, therefore, regard this calculation as only a rough approximation to the actual performance of the package.

13.5 Plastic Pallets

Increasingly, plastic pallets are being used in place of wooden pallets for product distribution. Although the cost of a plastic pallet can be four times the cost of a wooden pallet, its lifetime is usually much longer, typically five to nine years. Repair of broken plastic pallets is not usually feasible, but they are recyclable. Consequently, there is frequently an overall economic advantage to using plastic pallets. Because plastic pallets represent a substantial investment, their use is most suitable in situations where the distribution environment is relatively controlled, so that loss can be minimized.

HDPE is used most often in pallet manufacture, with PS, PP, and fiberglass-reinforced plastics also being used. Steel reinforcements can be added to increase the pallet's load-bearing capacity. Structural foam molding is common, with low-pressure injection molding producing pallets with a solid skin and a foamed core. Wall thickness is generally 0.5-2.5 cm (0.19-1.0 in). Ordinary injection molding is also used for pallet manufacture. Walls in this case are thinner, typically 0.08-0.9 cm (0.03-0.375 in). On occasion, rotational molding is used, producing pallets with 0.3-0.64 cm (0.125-0.25) in walls. Thermoforming, especially twin sheet thermoforming, is also a common forming method, with wall thicknesses also in the 0.3-0.64 cm (0.125-0.25 in) range. Polyurethane pallets can be produced by reaction injection molding (RIM), and have walls from 0.3-5 cm (0.125-2.0 in) thick.

Some designs for plastic shipping containers combine the functions of pallet and container. These packages are often designed to be collapsible for economy in storage and shipping when empty.

13.6   Plastic Drums and Other Shipping Containers

High molecular weight HDPE is increasingly being used for drums, pails, totes, and similar shipping containers, as an alternative to steel and fiberboard. Plastic drums are often used for chemicals, and are also used frequently in the food-processing industry. Standard drum sizes in North America are 57, 76, 114, 132, 208 and 216 L (15, 20, 30, 35, 55, and 57 gal), with 208 L (55 gal) used most often. Drums are categorized as closed-head (or tight-head) or open-top. Open-top drums are used mostly for water-based products, and are easier to clean and reuse. Plastic drums are permissible for shipment of hazardous materials provided the appropriate testing has been done and the containers are properly labeled.

Another use of plastic in drums is as a polyethylene liner in a steel drum. The liner adds chemical resistance and simplifies cleaning and reuse of the drum, since it can be discarded and replaced with a new liner.

Plastic pails, usually produced by injection molding from HDPE containing a butene comonomer, are available in sizes from 4 to 23 L (1 to 6 gal). Open top containers, which account for about 75% of the market, are referred to as pails. Closed-head containers are called either pails or jerrycans. Pails generally include a handle. Pails used for shipping hazardous materials must have passed required performance tests.

Plastic crates and boxes are usually injection molded from HDPE or PP. They are available in a variety of sizes and designs. Most of these are designed to be reusable, and often will either collapse or nest to minimize space requirements when they are not in use.

Plastic boxes can also be made from plastic corrugated, usually a PP copolymer or HDPE. Polycarbonate is used in some special applications. The corrugated board can be made directly using extruded profiles, or it can be assembled by laminating together three separate sheets, the two liners and the inner fluted layer. The boxes can then be made by die-cutting, scoring, and folding the board, similar to the process used for making corrugated fiberboard. The main difference is in joining the seams. High frequency welding generally gives the best results. Silicone-based or hot melt adhesives can be used if the board has been corona-treated to enhance adhesion. Metal staples (stitching) can be used, but are not as desirable since they create weak spots in the container around the staples. Plastic corrugated is particularly suited for manufacture of conductive containers for electronic parts that require protection against electrostatic discharge (ESD). Carbon films printed onto corrugated paperboard can be used, but there are problems with wear and fiber generation. These are avoided with conductive plastic corrugated, since the conductive material can be introduced into the polymer before extrusion.

13.7   Packaging for Electrostatic Discharge Protection

As mentioned above, sensitive electronic products require protection from electrostatic discharge (ESD). In unmodified plastics, ESD is a significant problem, since the nonconductive nature of plastics causes them to easily accumulate static charges. Consequently, when plastic packaging is used for sensitive devices, it must be modified to make it less susceptible to charge buildup, and to provide for dissipation of any charges that are created. Both of these are accomplished by providing a conductive path for electrons.

The most common way to alter the susceptibility of the plastic to static charge accumulation is to add to it some component that attracts and retains a thin layer of surface moisture, and at the same time weakly ionizes it. The ionized water provides the needed conductive path so that static charges will dissipate. The additive can be applied to a surface layer in the material, can be distributed through the bulk of the material, or can even be confined in an inner layer of a coextruded material. Antistatic additives that are commonly used include ethoxylated amines, quaternary amines, and ethoxylated amides.

Antistatic packaging materials come in a variety of forms, including film, foam, bubble wrap, and containers. ESD films in heat-shrinkable form are designed to minimize static generation by friction within the bag, as well as transmission of static charges from outside the bag. These materials are often tinted pink as a visible signal of their antistatic properties.

An alternative to the use of additives is selection of plastics that are inherently static-dissipative. While such materials are still largely experimental, one promising alternative is polyamide ethylene oxide block copolymers.

References

. Selke, Susan, *Understanding Plastics Packaging Technology*, Carl Hanser Verlag: Munich, 1997.
. Suh, K. W. and Tusim, M. H., *Foam Plastics*, 2nd ed., Wiley, New York, pp. 451-458.

Study Questions

1. Think of examples of the use of: (a) loosefill, (b) molded foam cushions, and (c) foam-in-place cushioning systems.

2. For a cushion with the characteristic cushion curves shown in Figure 13.2, calculate the load-bearing area that will provide adequate protection for a product measuring 10 in square, weighing 30 lbs, and with a fragility of 45 G, if the maximum expected drop height for the product is 36 inches.

3. What is the minimum fragility required for a product to be sufficiently protected if it is cushioned with 1-inch foam with cushion curves shown in Figure 2.2, and the static loading is 0.5? What if a larger cushion is provided, so the static loading decreases to 0.4? A smaller cushion, increasing the static loading to 0.8?

4. Explain the difference(s) between extruded and expanded PS foam.

5. There are significant differences in properties between PS foams and PE foams. How do these relate to the chemical structure/property relationships of the base polymers?

6. Why do you think closed cell foams are used more often in packaging than open cell foams? What kind of foam is used for dishwashing sponges? Why?

7. In molding with foam beads, the mold must be over-filled to avoid gaps in the finished products. Explain why this is so.

8. In making extruded foams, what factor is most influential in determining the density of the foam? Explain.

9. In making extruded foams, what factor is most influential in determining the size and number of cells in the foam? Explain.

10. Based on what you learned about polymer properties, why would you expect SAN foam to offer better performance in repeated drops than EPS foam?

14 Mass Transfer in Polymeric Packaging Systems: Sorption, Diffusion, Permeation, and Shelf Life

14.1 Introduction

For many products, an ideal packaging material, in addition to containing and enclosing a product, would provide an inert, perfect barrier between the product and the environment. With an inert perfect barrier, there would be no exchange of molecules (such as oxygen, CO_2, water, ions, product ingredients, or packaging material components) between the package and the materials confined within it. Such a material is not available in the real world. Some materials are nearly inert, such as glass, but under the right conditions, certain ions can be leached from a glass package. Metal packaging, while like glass an excellent barrier, is not inert. All types of metal packaging can be dissolved with the right solvent. Further, hydrogen can diffuse through metals and glass. Polymeric packaging materials are much less inert than metal and glass, and are certainly permeable to gases. Exchange of substances between the product and the package, along with the ability of a plastic material to transmit substances between the environment and the product, are very often significant determinants of the shelf life of a product. Collectively we refer to these as interactions.

In this chapter, we will explore the types of product/package interactions that can occur between a plastic material and a product, as well as the factors that affect the sorption, diffusion, and permeability behavior of polymers. We will then examine how polymer and product mass transfer characteristics can be used to predict product shelf life, based on moisture and oxygen transport.

14.2   Physical and Chemical Basis for Interactions

Molecules of gases, vapors, and other low molecular weight substances can dissolve in polymers, diffuse through the polymers, and then travel to a contacting substance. The extent and speed of these occurrences depends on the chemical and physical structure of the polymer, and on the nature of the traveling molecule. The driving force for net transfer of penetrant molecules from one location to another is the natural tendency to equilibrate the *chemical activity* of the species. Chemical activity will be discussed in more detail in Section 14.4.

For transfer of the penetrant through a polymer to occur, the penetrant must have the ability to move within (and through) the polymer. If the penetrant cannot enter the polymer, or cannot move within it once it has entered, the polymer will provide a perfect barrier. The ability of the penetrant to move within a polymer is greatly influenced by the *free volume* of the polymer. The free volume is the space that is not occupied by the polymer molecules - the voids between polymer molecules or between segments of molecules. The physical structure of the polymer and its temperature are important determinants of the amount of free volume within the plastic material. The effect of temperature is observed as thermal expansion of the material. As temperature increases, the spacing between adjacent atoms and molecules increases, so the free volume increases. The ability of a penetrant molecule to move within a polymer is dependent, among other factors, on its size, particularly on its size in comparison with the size of the voids in the polymer. Smaller molecules can move more readily than larger molecules. Similarly, the larger the free volume, the larger the size of the voids, and the greater the ability of the penetrant molecule to move.

An amorphous polymer that is below its glass transition temperature, T_g, such as polystyrene at room temperature, has free volume within its mass, in amounts determined by its chemical nature and its temperature. The size and location of the voids is fairly static. In other words, voids are not forming or being destroyed, except in response to changes in temperature. Above T_g, the situation is very different. Under such conditions, as discussed in Chapter 3, the polymer has enough energy for motion of polymer segments. As the segments move from one conformation to another, they successively form and destroy free volume. Therefore, the free volume in an amorphous polymer above T_g is in a very dynamic state. As will be discussed in more detail later, in a given polymer, the ability of a penetrant molecule to move within a given plastic is much higher above T_g than it is below T_g.

In semi-crystalline polymers, as discussed in Chapter 3, there are both amorphous and crystalline regions within the polymer. If the polymer is above its T_g, the chain segments in the amorphous regions have mobility, as in the case of a completely amorphous polymer. If the polymer is below T_g, there is no significant mobility. In either case, the crystalline regions have very little mobility, and have less free volume than the amorphous regions. Therefore, there is much less ability for a penetrant molecule to travel through the crystalline regions. In essence, all mass transport through semicrystalline materials can be considered to occur through the amorphous regions of the material.

In addition to crystallinity, other factors that increase interactions between polymer chains can lead to tighter packing of polymer molecules (less free volume), and can restrict segmental motion (less dynamic free volume). Increasing polarity, hydrogen bonding, and crystallinity all reduce segmental mobility and decrease free volume, therefore increasing the barrier ability of a polymer. Increasing chain stiffness and cross-links decrease segmental mobility, but can either increase or decrease free volume. Consequently, their effect on barrier ability of a plastic is more difficult to predict.

14.3   Types of Interactions

Scientists and engineers have been interested in mass transfer of low molecular weight compounds since 1829, when Thomas Graham performed the first diffusion experiments with natural membranes. In 1855, Adolf Fick derived the fundamental equations of diffusion. Mass transfer processes in modern packaging systems are of great theoretical and practical interest, since they are very important in determining the quality of packaged products. Molecular interactions in packaging systems start from the moment the package contacts the product during production, and extend throughout the product's shelf life. Packaging interactions involve changes in the product that are affected by mass transfer processes occurring within the packaging material. Changes that take place in the product which are not consequences of mass transfer, such as crystallization, are not considered packaging interactions. These packaging interactions can be categorized as permeation of compounds through the plastic, sorption of compounds into the plastic, and migration of compounds from the plastic. Figure 14.1 illustrates diffusion, sorption, and permeation occurring in a packaging material. Diffusion coefficients and solubility coefficients are the fundamental parameters controlling the mass transfer in packaging systems.

14.3.1   Permeation

Permeation is the movement of gases, vapors, or liquids (called permeants) across a homogeneous packaging material. It excludes the travel of materials through perforations, cracks, or other defects in the package. Permeation may significantly impact the shelf life of a product, since the product may gain or lose components, or undergo unwanted chemical reactions with permeating substances. The loss of water and carbon dioxide, the uptake of moisture by dry products, or the oxidation of oxygen-sensitive products all affect the product's composition, and therefore affect its shelf life. Other consequences of permeation include the transfer of airborne contaminants or volatile components into the product, which may produce off-flavors or off-odors, especially in certain flavor-sensitive foods.

While the permeation process may have a negative impact on product quality, there are instances where permeation is required to maintain product quality. For example, in modified atmosphere packaging of fresh fruits and vegetables, the package is designed so that O_2 and CO_2 will permeate through the package at a desired rate, which depends on the product characteristics.

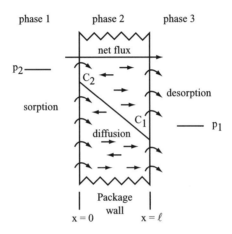

Figure 14.1 Sorption, diffusion, and desorption processes across a plastic sheet.

14.3.2 Migration

Migration is the transfer of substances originally present in the plastic material into a packaged product. The compounds that transfer are called migrants. Examples of migrants include residual monomers, solvents, residual catalysts, and polymer additives. Migration may affect the product's organoleptic (sensory) quality as well as its toxicological characteristics, since it may involve the incorporation of undesirable components from the package. The presence of potential migrant substances in a packaging material for food, cosmetics, or drugs is subject to legal control by the Food and Drug Administration, FDA (for detailed information see Sec. 15.4). Vinyl chloride in PVC, styrene in PS, and acrylonitrile in PAN are well known examples of migration of residual monomer that were the object of intense investigations in the 1970s. As a result, these resins, as well as all plastics for food applications, are commercially produced with residual monomer contents below the maximum allowed by the FDA.

However, even if virgin resins contain very low level of monomers and additives, migrating substances that may potentially impart off-flavors to products can develop during converting operations. Excessive time-temperature processing conditions may oxidize and degrade the resin. This may occur during extrusion, coating, blown film production, injection molding, blow molding, or even heat-seal operations. Adhesives, inks, pigments, printing solvents, and printing pretreatment ingredients are potential sources of packaging contamination. Residual solvents like toluene, hexane, or pentanol from packages and coupons in direct contact with the plastic can produce unwanted odor and taste in food products. The presence of recycled and regrind resins are potential sources of migrants in the finished package if not sufficiently clean. Other sources of migrants may arise when contaminated equipment and lines are used to process the resin. Similarly, impure air used to blow films can result in a contaminated package.

The sensory threshold for detection of an off-odor or flavor varies widely, depending on the substance and the product, as well as on the individual doing the sensing. It is not unusual for amounts as small as a few parts per billion to be sufficient reason to reject a product.

While migration is usually to be avoided, there are cases where migration of a substance, such as an antioxidant, from the package to the product is desired; such is sometimes the case with the antioxidants BHT and vitamin E in cereal packages.

The ultimate extent of migration of a particular substance from a package to a product depends on the initial amount of the migrating substance in the plastic, and on the partition coefficient between the plastic and the food. The partition coefficient determines the equilibrium distribution of the compound in closed systems. The actual amount of migration occurring in a given time depends on those factors, as well as on the rate of transfer.

14.3.3   Sorption

Sorption, such as flavor scalping, is the uptake of product components, such as flavor, aroma, or colorant compounds (called sorbates) by the package. European literature sometimes calls this "negative migration," but the use of this term is unfortunate since it may lead to confusion. Many studies of sorption of food components have been reported in the literature.

Similarity in chemical structure between the sorbate and the polymeric packaging material enhances sorption. As the molecular weight distribution (MWD) of the polymer broadens, sorption also tends to increase. LLDPEs produced with metallocene catalysts, which are characterized by a narrow MWD, have a reduced tendency to sorb volatile compounds, compared to LDPE. Sorption studies on benzaldehyde, citral, and ethyl butyrate revealed that sorption by ionomers is similar to sorption by LDPE. As in the case of migration, the extent to which the sorption process occurs depends on the initial concentration of the sorbate and the partition coefficient between the polymer and the product, as well as on the rate of transfer.

14.4   Thermodynamic Equilibrium

The fundamental driving force that prompts a molecule to diffuse within a polymer or transfer between a polymer and a surrounding phase is its chemical potential. Like the electrical potential of a battery causes electrons to flow through wires, chemical potential is the driving force in physical chemical phenomena. Substances will naturally tend to move from a higher chemical potential to a lower one. The equation for chemical potential is:

$$\mu_i = \mu_i^{\,o} + RT \ln a_i \qquad (14.1)$$

where

μ_i = chemical potential of substance I
$\mu_i^{\,o}$ = chemical potential of substance I at a standard state
R   = universal gas constant
T   = temperature, K
a_i = chemical activity

The drive of a compound to move from a point of high chemical potential to a low one is equivalent, in most cases, to the tendency to equilibrate the compound's chemical potential in a phase, such as within a plastic container or a product, where mass transfer is physically able to occur, in other words, the compound tends to reach thermodynamic equilibrium.

The chemical potential, and thus the chemical activity, of a substance is defined with respect to some standard, or reference state. Typically, the reference state chosen for a liquid or solid is the pure substance at the temperature under consideration. For a gas, an ideal gas at the temperature under consideration and one atm pressure is the usual reference state. With this choice of reference states, the chemical activity is approximately equal to the concentration of the substance. For equilibrium to be reached, then, there will be a net transfer of the substance from regions of high concentration to regions of low concentration. This simple concept becomes somewhat more complicated when transfer of a substance from one phase to another is involved, as in the case of packaging systems.

In packaging applications, polymers are often in contact with gas (usually air) and liquid phases. In multilayer structures, where several layers of polymers are in direct contact, a polymer is in contact with at least one other solid phase. In packages containing solid products, there are typically only small points of actual contact at the molecular level between the product and the package, with most of the apparent physical contact between the solids involving an intervening thin layer of air. Any mobile substance that is not in thermodynamic equilibrium within and between phases will tend to transfer in directions that lead to equilibration, as indicated by Figure 14.1. In sorption and migration, mass transfer involves two adjacent phases: the polymer phase and a surrounding liquid or gas. The mobile substance must diffuse within each phase and also move across the interface between the phases. In permeation, the permeant molecules need to move across the two wall interfaces (one at each side of the packaging wall) as well as diffusing within the polymer, the product, and the external air.

Therefore, in order to understand mass transfer behavior in packaging applications, we have to be able to relate the activity of a substance in one phase to its activity in any contacting phase. As will be seen in the next section, for most packaging applications the activity can be reduced to concentration.

14.4.1   Gas Phase Chemical Activity

The chemical activity of a substance I is proportional to its concentration and can be represented as follows:

$$a_i = \gamma \, c_i \tag{14.2}$$

where γ is the activity coefficient. In a gas phase at atmospheric pressure or below, the activity coefficient is approximately 1, so activity is approximately equal to concentration. For gas phases, the concentration can be expressed as partial pressure, p_i, because of the ideal gas law

$$p_i = c_i \, R\,T = \frac{n_i}{V}\,R\,T \tag{14.3}$$

This relationship will hold exactly for ideal gases, those that do not have any interaction between molecules. Most gases behave as nearly ideal gases as long as the pressure is not high, so this is usually a good approximation.

14.4.2   Solubility

The act or process by which a compound such as oxygen is molecularly mixed with a liquid (such as water) or a solid (such as a polymer) is called dissolution, and the result of the mixing is a solution. If a solution is very dilute, as is commonly found in packaging, it behaves as an ideal solution, and again the activity coefficient is approximately 1, so concentration can be substituted for activity in thermodynamic relationships. In order to describe the solubility of a compound present in a gas phase that is in contact with a solid phase, as may be the case of oxygen in air contacting a polymer, we need a relationship between the concentration in the liquid (or solid) phase and the concentration (or partial pressure) in the gas phase. In other words, we need an expression for the solubility of the substance at equilibrium, as a function of the partial pressure of the gas or vapor in the contacting gas phase.

William Henry found in 1803 that the equilibrium vapor pressure of a solute above an ideal solution was proportional to its concentration (see Fig. 14.2):

$$p_i = k\,c_i \quad \text{or} \quad c_i = S\,p_i \tag{14.4}$$

The Henry's Law proportionality constant is called the solubility coefficient, often represented by S, and is a function of temperature. Henry's Law holds exactly for ideal solutions, and is a good approximation for most real solutions, as long as they are dilute. In particular, it works well for oxygen and carbon dioxide up to about one atmosphere, and for many organic vapors at low concentrations.

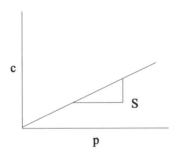

Figure 14.2 Henry's law solubility behavior.

For non-ideal solutions, other equations can be used to describe the solubility relationship. The choice of governing equation depends on the nature of the non-ideal behavior, in other words, on the types of interactions between molecules that cause the behavior to deviate from ideality.

If there is strong interaction between an organic solvent and a polymer, then a Flory-Huggins type of equation can be used, containing an interaction parameter χ representing this interaction. At moderate to high concentrations, for example, organic vapors may act as plasticizers and begin to swell the polymer, as well as clustering together, creating non-ideal interactions. An example is the interaction between d-limonene and polyethylene. The swelling of the polymer also increases the free volume in the polymer. When this type of behavior occurs, Henry's Law does not apply, and the solubility must be treated as a function of concentration. The Flory-Huggins equation correlates the vapor activity of the solute in the gas phase with its volume fraction in the polymer as

$$\ln a_1 = \ln v_1 + v_2 + \chi\, v_2^{\,2} \tag{14.5}$$

where a_1 is the sorbate activity in the fluid phase, v_1 is the volume fraction of sorbate in the fluid phase, v_2 is the volume fraction of the polymer, and χ is the interaction parameter, given by

$$\chi = \frac{\Delta H_M}{kTN} \tag{14.6}$$

where ΔH_M is the heat of mixing, k is the Boltzman constant, T is absolute temperature, and N is the number of moles of sorbate.

To describe the sorption of high pressure CO_2 in polymers such as PET, the Langmuir-Henry's law equation is useful:

$$c = k\,p + \frac{C'_H\, b\, p}{1 + b\, p} \tag{14.7}$$

where C'_H is the Langmuir capacity constant and b is the Langmuir affinity constant. This model is useful for glassy polymers and high-pressure penetrants.

14.4.3   Partition Coefficient

When a solid and a liquid phase or two solid phases are in contact, a useful way to describe the equilibrium distribution of the penetrant between the two phases is to use the partition coefficient, K, defined as:

$$K = \frac{c^*_f}{c^*_p} \tag{14.8}$$

where c^*_f and c^*_p are the sorbate or migrant equilibrium concentration in the two phases, often the product and polymer, respectively.

14.5   Diffusion

The concepts discussed in Section 14.4 describe the situation that will eventuate when a multiphase packaging system reaches thermodynamic equilibrium. However, the rate of mass transfer of permeant, sorbate, and migrant molecules in the polymer is not addressed by these equilibrium considerations. For example, if we consider a potential migration process, we know that eventually the migrant will be transferred to the food and it will finally reach equilibrium, but based on the equations presented in Section 14.4 we cannot predict how long the process will take. Similarly, these relationships will not allow us to estimate the shelf life of a product in a particular package system. For this, we need to look at diffusion.

Adolf Fick in 1855 developed equations to describe diffusion that are now called Fick's First and Second Laws of Diffusion. In an isotropic material (where the properties do not vary with direction), and when diffusion occurs along only one dimension, which is approximately the case in most packaging systems, Fick's first law can be written:

$$F = -D\frac{\partial c}{\partial x} \tag{14.9}$$

where F is flow rate, c is the permeant concentration *in the polymer*, x is the distance (in the direction of the diffusion), and D is the diffusion coefficient for the diffusant molecule itself. The diffusion coefficient is a function of temperature, and may be a function of concentration.

Fick's first law provides a method for calculation of the *steady state* rate of diffusion when D can be regarded as constant during the diffusion process, and the concentration is a function only of the geometric position inside the polymer. However, concentration is often a function of time as well as of position. We said Eq. 14.9 describes a steady state flow, but how does the system reach this steady state? The unsteady state flow, or transient state, is described by Fick's Second Law. For a one-dimensional diffusion process, this can be written as:

$$\frac{\partial c}{\partial t} = D\frac{\partial^2 c}{\partial x^2} \tag{14.10}$$

where t is time.

In systems where the diffusant concentration is relatively low, as is typical in many packaging applications, the diffusion coefficient in Equations 14.9 and 14.10 can be assumed to be independent of both diffusant concentration and polymer relaxations. It is also usual for the diffusion to occur almost exclusively in a direction perpendicular to the flat surface of the package, with negligible amounts diffusing through the edges of the package. In that case, Equations 14.9 and 14.10 hold, and solutions to these equations, with the relevant initial and boundary conditions, are one-dimensional.

14.6   Steady State Diffusion Across a Single Sheet: Permeability

Let us consider the case of a plane plastic sheet of thickness ℓ that is contacted on both sides with a penetrant at different concentration values, as indicated in Figure 14.1. At the surface $x = 0$, the penetrant concentration $c = c_2$, and at $x = \ell$, $c = c_1$. Applying these conditions to Equation 14.9, the penetrant flow rate F across any section of the sheet is given by:

$$F = -D\,\frac{dc}{dx} = D\,\frac{c_2 - c_1}{\ell} \qquad (14.11)$$

In permeability studies, however, the partial pressure in the gas phase surrounding the sheet is easier to measure than the penetrant concentration c in the polymer. If we are at low concentrations, Henry's Law applies, and we can substitute for c using Eq. 14.3. Equation 14.9 is written now as:

$$F = DS\,\frac{p_2 - p_1}{\ell} \qquad (14.12)$$

Since the flow rate F is given by the quantity q of permeant transferred through a unit of area A in a time t, i.e. $F = q/At$, rearranging Eq. 14.10 gives us

$$P = DS = \frac{q\,\ell}{A\,t\,\Delta p} \qquad (14.13)$$

where $\Delta p = p_2 - p_1$. We have introduced a new parameter, the permeability coefficient, P. P is defined as the product of the diffusion coefficient and the solubility coefficient. Since P combines the effects of the diffusion coefficient and the solubility coefficient of the permeant/plastic system, it is an indicator of the barrier characteristics of the polymer for the permeant under consideration. A material having a low value of P for a particular permeant is a good barrier, i.e., only a small quantity of permeant will be transferred through it. Conversely, a high permeability value indicates a material with poor barrier properties.

Equation 14.13 is a simple but very useful equation for the design of packages with desired steady state barrier capability, as we will see in Section 14.11.

Table 14.1 summarizes the notation we are using for permeability. It is an unfortunate common practice, in industry, for the units associated with these variables to not be consistently English, metric, or SI. Common units are given in Table 14.2.

In addition to the permeability coefficient, other parameters are also used to express the barrier characteristics of plastic materials. These include permeance (R), gas transmission rate (GTR), water vapor transmission rate (WVTR), and thickness normalized flow. The relationship between these parameters is shown in Figure 14.3.

Table 14.1 Notation Related to the Permeation Process

P	Permeability coefficient
t	Time during which permeation takes place (at constant conditions)
ℓ	Thickness of package wall
p	Partial pressure of gas or vapor
Δp	Partial pressure difference across the package wall
A	Package area over which permeation takes place
q	Quantity of permeant transferred through area A in time t (mass or volume)
c	Concentration of permeant in the polymer
S	Solubility coefficient
D	Diffusion coefficient

Table 14.2 Units Used in Permeability Calculations

Variable		Common Units	SI Units
q	quantity	g, cm^3 (STP), mol	kg
ℓ	thickness	cm, mil	m
t	time	hour, day	s
A	area	cm^2, in^2, 100 in^2	m^2
p	partial pressure	atm, psi, cm Hg, mm Hg	Pa

The permeability coefficient, P, is expressed in a combination of the above units, e.g.,

$$\frac{cm^3 \text{ (STP) mil}}{m^2 \text{ day atm}} \qquad or \qquad \frac{kg \text{ m}}{m^2 \text{ s Pa}}$$

The gas transmission rate, GTR, is the quantity of permeant flowing per unit area per unit time. The water vapor transmission rate, WVTR, is a GTR defined specifically for water vapor. The two together are permeant transmission rates, identified as F in Figure 14.3.

WVTR is related to P as follows:

$$P = \left(\frac{q}{A\,t}\right)\frac{\ell}{\Delta p} = WVTR\,\frac{\ell}{\Delta p} \qquad (14.14)$$

The relationship between the permeability coefficient and gas transmission rate is defined analogously. Thus to convert a measured WVTR or GTR to P, all that is necessary is to multiply by the film thickness that was used, and divide by the partial pressure difference used to make the measurement of WVTR or GTR.

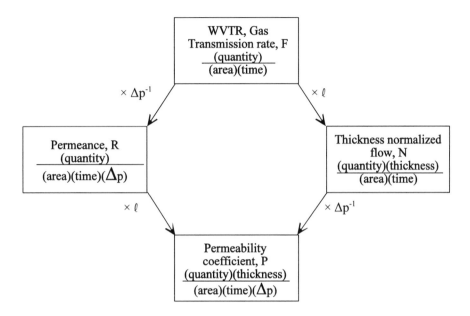

Figure 14.3 Relationship between permeability coefficient, permeance, WVTR, GTR, and thickness normalized flow.

Example 1: The WVTR of a film 25 microns thick measured by the ASTM dish method at 100 °F and 90% RH is 0.1 g/day m² (see Fig. 14.4). Calculate P.

Solution: The saturation vapor pressure of water at 100°F is 49.7 mm Hg. Table 14.3 contains saturation vapor pressures for water at various temperatures. Since $p_1 = 0$ because there is a desiccant inside the cup, $\Delta p = p_2 - p_1 = p_2 = 49.7$ mm Hg x 0.9 = 44.73 mm Hg.

P = WVTR $(\ell/\Delta p)$ = 0.1 g/day m² x 25 microns/44.73 mm Hg = 0.0559 g·microns/m²·d·mm Hg

We can convert units to S.I., if desired:
$P = 4.85 \times 10^{-18}$ kg·m/m²·s·Pa = 4.85×10^{-18} s = 4.85 as (attosecond).

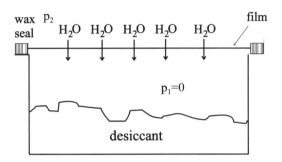

Figure 14.4 Dish method for determining water vapor transmission rate.

Example 2: Assume that the GTR of oxygen through a 1-mil PE film,
$$GTR = 3.5 \times 10^{-2} \text{ g/h m}^2,$$
when the partial pressure difference through the film Δp is 30 mm Hg. Calculate the permeance and the permeability coefficient.

Permeance $= GTR \times \dfrac{1}{\Delta p}$

Permeance $= \dfrac{3.5 \times 10^{-2} \text{ g}}{\text{h m}^2 \; 30 \text{ mm Hg}} = 1.17 \times 10^{-3} \; \dfrac{\text{g}}{\text{h m}^2 \; \text{mm Hg}}$

Permeability $= \dfrac{GTR}{\Delta p} \times$ thickness $=$

$\dfrac{1.17 \times 10^{-3} \text{ g} \times 2.54 \times 10^{-3} \text{ cm}}{\text{h m}^2 \; \text{mm Hg}} = 2.96 \times 10^{-6} \; \dfrac{\text{g cm}}{\text{h m}^2 \; \text{mm Hg}}$

It should be emphasized that permeation through a polymer involves three steps, as shown in Figure 14.1: (1) the permeant dissolves at the polymer interface, (2) the permeant diffuses within the polymer film from the side of high concentration toward the low concentration side, and (3) the permeant diffuses out from the opposite polymer interface. These steps are always present in any system regardless of whether D and S follow Fick and Henry's law, respectively, or not. For instance, the solubility of CO_2 in PET follows the Langmuir-Henry's law model, Equation 14.7, and P is given by .

$$P = k\, D_D + \frac{C'_H \, b\, D_H}{1 + b\, p} \tag{14.15}$$

where D_D and D_H are the diffusion coefficients for the Henry's and Langmuir's law populations, respectively.

As indicated, a material that is a good barrier has a low value of the combined diffusion coefficient and solubility coefficient values for a particular penetrant.

Preferably, both D and S should be low. For instance, polyethylene is an excellent barrier to water because water has very low solubility and diffusion coefficient values in polyethylene, but it has a relatively high P for oxygen because O_2 has higher solubility than water. A permeability coefficient value is valid only for a particular polymer/permeant pair, and as in the case of polyethylene, a structure may be a good barrier for a certain permeant and only a fair or poor barrier for a different one.

Table 14.3 Saturation Vapor Pressure of Water

Temp °C	mm Hg	Temp °C	mm Hg	Temp °C	mm Hg
-14	1.560	21	18.640	51	97.20
-13	1.691	22	19.827	52	102.09
-12	1.834	23	21.068	53	107.20
-11	1.987	24	22.377	54	112.51
-10	2.149	25	23.756	55	118.04
-9	2.326	26	25.209	56	123.80
-8	2.514	27	26.739	57	129.82
-7	2.715	28	28.349	58	136.08
-6	2.931	29	30.043	59	142.60
-5	3.163	30	31.824	60	140.38
-4	3.410	31	33.695	61	156.43
-3	3.673	32	36.663	62	163.77
-2	3.956	33	37.729	63	171.38
-1	4.258	34	39.898	64	179.31
0	4.579	35	42.175	65	187.54
1	4.926	36	44.563	66	196.09
2	5.294	37	47.067	67	204.96
3	5.685	38	49.692	68	214.17
4	6.101	39	52.442	69	223.73
5	6.543	40	55.324	70	233.7
6	7.013	41	48.34	71	243.9
7	7.513	42	61.50	72	254.6
8	8.045	43	64.80	73	265.7
9	8.609	44	68.26	74	277.2
10	9.209	45	71.88	75	289.1
16	13.634	46	75.65	76	301.4
17	14.530	47	79.60	77	314.1
18	15.477	48	83.71	78	327.3
19	16.477	49	88.02	79	341.0
20	17.535	50	92.51	80	355.1

A large number of polymeric structures covering a wide range of barrier characteristics are available to satisfy the diverse needs within the packaging industry. There are high barrier materials available to protect a product from oxygen, water or organic vapors, as well as structures with high permeability values for oxygen or CO_2 that are needed, for example, in modified atmosphere packaging of produce.

14.7   Variables Affecting Permeability

14.7.1   Chemical Structure of the Polymer

The chemical structure of the polymer's constitutional unit is the fundamental determinant of the polymer's barrier behavior. In addition to chemical composition, polarity, stiffness of the polymer chain, bulkiness of side and backbone-chain groups, and degree of crystallinity significantly impact the sorption and diffusion of penetrants, and hence permeability. Of particular significance are influences on the free volume and molecular mobility of the polymer, and influences on the affinity between the permeant and the polymer.

Table 14.4 shows some examples of the effect on oxygen permeability of functional groups attached to a vinyl polymer backbone, with structure:

$$-(\underset{\underset{\text{H}}{|}}{\overset{\overset{\text{H}}{|}}{\text{C}}} - \underset{\underset{\text{H}}{|}}{\overset{\overset{\text{X}}{|}}{\text{C}}})_n-$$

14.7.2   Chemical Structure of the Permeant Molecule

The size of the permeant molecule, as well as the chemical affinity between the permeant and the polymer, is an important determinant of permeability. Polymers can act as molecular sieves, allowing some molecules to pass through rapidly while retarding the passage of others. This is the principle used industrially in polymer membrane separation of gas blends.

The effect of size on permeability is complex, because permeability, as discussed, is the product of diffusion and solubility. Larger permeant molecules generally have lower diffusivity than smaller ones, but higher solubility. The effect of size on solubility is related to the dependence of solubility on vapor pressure. As size increases, the energy required to vaporize the molecule increases, so its vapor pressure decreases. For gases, the lower the vapor pressure, the greater is the tendency for the gas to remain dissolved in the liquid rather than converting to the gas form, so the greater the solubility. Of course, solubility is also strongly influenced by the chemical similarity between the solvent (the plastic) and the permeant.

Table 14.4 Effect of Functional Group X on Oxygen Permeability of Vinyl Polymers

Functional Group, X	Polymer	P (cc mil/ m^2 d atm)	Comments
H	HDPE	1,550-3,100	Nonpolar, very low cohesion between chains, tiny side group, high flexibility, high crystallinity
H	LDPE	3,900-5,400	Nonpolar, branched, less crystalline
CH_3	PP	2,300-3,900	Nonpolar, larger side group, stiffer than PE, lower crystallinity
C_6H_5	PS	3,900-6,200	Bulky side group, atactic, hinders packing, noncrystalline
$COOCH_3$	PMA	265	Polarity produced by ester linkage, but bulky atactic side group hinders packing, noncrystalline
OH	PVOH	0.15	Strong polarity, hydrogen bonding between chains, crystalline
CN	PAN	0.60	Strong polarity, bulkier side group than OH, noncrystalline
Cl	PVC	75-310	Strong polarity, less than PAN
F	PVF	45	More polar than PVC, smaller side group
$CH_2CH(CH_3)_2$	Poly-4-methyl pentene-1	62,000	Nonpolar, very bulky side group, amorphous

The effect of size on diffusivity is more straightforward. The larger the molecule, the greater is the amount of energy required to move it, and the greater the amount of energy required to create the larger free volume necessary for it to have a place to move. Therefore diffusivity decreases with increasing size.

The size of a permeant molecule is related to its molecular weight, but this is an inaccurate measure. The van der Waals diameter is a good measure for isotropic molecules, but many permeants of interest, such as n-alkanes, are strongly anisotropic. How best to compare the sizes of isotropic and anisotropic molecules is an issue that has not been resolved.

Table 14.5 illustrates the effect of molecular size, as indicated by molecular weight, and of polarity, on the permeability of amorphous PET.

Table 14.5 Effect of Molecular Size and Polarity on Permeability of Amorphous PET (P in cc cm/cm^2 s cmHg)

Permeant	MW	Polarity	P x 10^{10}
He	2	nonpolar	3.28
N_2	28	nonpolar	0.013
O_2	32	nonpolar	0.040
CH_4	16	nonpolar	0.09
CO_2	48	nonpolar	0.30
H_2O	18	polar	130.0

Van Krevelen [1] has presented an empirical relationship between relative values of the permeability of a polymer to various gases, that has been found to have reasonable accuracy for a broad range of polymers. Values are shown in Table 14.6.

Table 14.6 Relative Values of Permeability Coefficients [1]

Gas	P
N_2	1
CH_4	3.4
O_2	3.8
He	15
H_2	22.5
CO_2	24
H_2O	550

14.7.3 Effect of Temperature

Both diffusion and solubility are functions of temperature, and have been found to follow an Arrhenius type of equation:

$$\Gamma = \Gamma_o \, e^{-E_\Gamma /RT} \qquad (14.16)$$

where Γ represents either D or S, Γ_o is a proportionality constant (known as the pre-exponential term), E_Γ is the activation energy, R the gas constant, and T is temperature in degrees Kelvin. This equation also generally holds for P, which, as we have discussed, is the product of D and S. Equation 14.16 is valid over a relatively small range of temperatures. When a polymer passes through a transition, such as the glass transition temperature, there is a discontinuity, so a new relationship is needed.

Equation 14.16 can be used to estimate the permeability coefficient at a desired temperature from known values determined in the laboratory. A typical plot is shown in Figure 14.5. From this plot, the slope and intercept can be determined. The equation can then be used to predict the permeability at any temperature within the range of applicability.

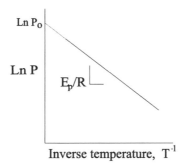

Figure 14.5 Permeability as a function of temperature.

Activation energy is given in units of energy/mole, such as calories/mole or Joule/mole. The gas constant, R, has a variety of values depending on the units chosen. Some useful values are given in Table 14.7.

We can write Equation 14.16 specifically for change in permeability with temperature as follows:

$$P = P_o \, e^{-E_p /RT} \qquad (14.17)$$

where E_p is the activation energy, R the gas constant, P_o is a pre-exponential term and T is temperature in Kelvin. Selected values of P and E_p for various polymers are presented in Table 14.8.

Table 14.7 Values for the Gas Constant, R

R	Units
1.987	cal / mol K
8.314	joule / mol K
82.06	atm cm^3 / mol K
0.0821	bar liter / mol K
1.314	atm ft^3 / mol K

Then, we can determine the relationship between P determined at one temperature and P at another temperature as follows:

$$P_1 = P_o\, e^{-E_p/RT_1} \quad \text{and} \quad P_2 = P_o\, e^{-E_p/RT_2}$$

so by dividing P_2 by P_1 and solving for P_2, we find

$$P_2 = P_1\, e^{-\frac{E_p}{R}\left(\frac{1}{T_2}-\frac{1}{T_1}\right)} = P_1\, e^{\frac{E_p}{R}\left(\frac{1}{T_1}-\frac{1}{T_2}\right)} \tag{14.18}$$

Example: The oxygen permeability of amorphous PET at 25°C has been found to be 14 cc mil/d 100 in^2 atm. What is the value of P at 50°C?

Answer:
Temperature = 25°C = 25 + 273 = 298 K
From Table 14.8, E_p = 38 kJ/mol
E_p/R = (38.0 KJ/mol)/(8.314 J/mol K) = 4,570.6 K

$$\frac{1}{T_1} - \frac{1}{T_2} = \frac{1}{25+273} - \frac{1}{50+273} = 0.0002597 \text{ K}^{-1}$$

$$e^{\frac{E_p}{R}\left(\frac{1}{T_1}-\frac{1}{T_2}\right)} = e^{4570.6(0.0002597)} = e^{1.18699} = 3.28$$

$P_2 = P_1 \times 3.28 = 14 \times 3.28 = 45.9$ cc mil/d 100 in^2 atm

---

Equation 14.18 can be written as

$$P_2 = P_1\, f \quad \text{where} \quad f = e^{\frac{E_p}{R}\left(\frac{1}{T_1}-\frac{1}{T_2}\right)} \tag{14.19}$$

The factor f gives the ratio of P_2/P_1, i.e. by what factor the permeability coefficient changes when temperature varies from T_1 to T_2.

Table 14.8 Values of Permeability Coefficients at 25°C (except where indicated) and Activation Energy (P in cc (STP) cm/m^2 d atm, E$_p$ in kJ/mol)

Polymer	Permeability				E$_p$			
	O$_2$	CO$_2$	N$_2$	H$_2$O	O$_2$	CO$_2$	N$_2$	H$_2$O
LDPE	19	83	6.4	600	43	39	49	34
HDPE	2.6	2.4	0.96	220	35	30	40	
PP	11	47	2.0	340	48	38	56	42
PS, biaxially oriented	17.5	69	5.2	6300-7350				
PAN (Barex™)	0.036	0.105		4300				
PVOH at 0% RH	0.058	.081						
EVOH at 0% RH (32% ethylene)	0.06*	0.12*	0.005*					
EVOH at 0% RH (44% ethylene)	0.24*	0.88*	0.02*					
PVDC (Saran™)	0.022	0.14	0.0039	61	67	52	70	46
PTFE (Teflon™)	28	66	8.8	53	19	14	24	
PVC, unplasticized	0.30	1.1	0.078	1800	56	57	69	23
PET amorphous	0.22	0.80	0.39	850	32	18	33	2.9
PET, 40% crystallinity	0.38	2.0			38	28		
PC (Lexan)	9.1	5.2	1.9	9100	19	16	25	
Nylon 6	0.19	0.76	0.046		44	41	47	
Cellophane (76% RH)	0.058	0.47	0.049					

* T=20°C

Example: Calculate the value of f for a polymer permeant having $E_p = 13.4$ kcal/mol when the temperature changes from 45°C to 4°C.

Answer:
$E_p/R = 13,400/1.987 = 6.743.8$ K

$$f = e^{6.743.8\left(\frac{1}{318.2} - \frac{1}{277.2}\right)} = e^{-3.1347} = 0.044$$

As a rule of thumb, a change of approximately 10°C will affect the permeability by a factor of 2 for increasing temperature and 0.5 for decreasing temperature.

14.7.4   Effect of Humidity

Hydrophilic polymers such as polyamides and EVOH, which contain polar groups and hydrogen bonding capability, strongly absorb water from humid air. Therefore, one can determine a water sorption isotherm for the polymer; that is, the equilibrium moisture content at any temperature and humidity condition. The presence of the water vapor in the polymer changes the permeation of other gases and vapors through the polymer. In most cases, the permeation rate increases with higher water sorption (see Figure 14.6) because the water acts as a plasticizer and increases the free volume of the polymer. DuPont found with some Selar™ polymers, which are nylons made from copolymers of hexamethylene diamine and isophthalic and terephthalic acids, that the permeation rate dropped and then became nearly constant as humidity increased, as shown in Figure 14.7. Hernandez [2] explained this behavior as the water not swelling the polymer, but occupying some of the sorption sites. This reduces the permeation of other gases through the competition for solubility (sorption) sites.

14.7.5   Physical Structure of the Polymer

Both crystallization and orientation affect the physical structure of the polymer, its morphology, and influence its permeability. The equations for permeation assume the polymer is a homogeneous isotropic structure. The semi-crystalline polymers often used in packaging do not match this assumption. The behavior of the crystalline regions is significantly different from the behavior of the amorphous regions, as has been discussed. Nearly all of the permeation takes place through the amorphous regions. The crystalline regions serve as blocks, around which the permeant must find a path. Thus the effective thickness of the sample, in terms of its barrier to permeant travel, is larger than its physical thickness. In addition, crystallinity may affect segmental motion in the amorphous regions of the polymer.

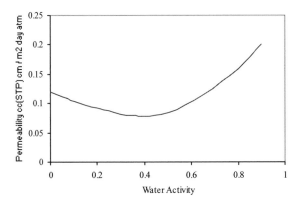

Figure 14.6 Permeability coefficient of oxygen in nylon 6 as a function of water activity at 23 °C [3].

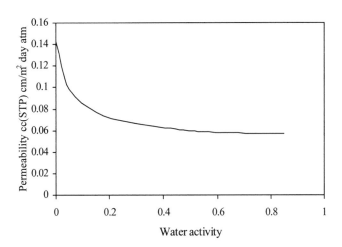

Figure 14.7 Permeability coefficients of oxygen in amorphous nylon as a function of relative humidity [2].

The solubility S of a semicrystalline polymer containing a crystalline volume fraction of α and with an amorphous phase solubility S_a is given by:

$$S = (1 - \alpha)S_a \tag{14.20}$$

The diffusion coefficient in the amorphous phase, D_a, has been shown to be related to the diffusion coefficient in the semicrystalline phase D by:

$$D_a = D \beta \tau \qquad (14.21)$$

where β is a "chain immobilization factor", and τ is a "geometric impedance factor". Both β and $\tau > 1$. Michaels et al [4] applied Equation 14.21 to semicrystalline PET. However, the equation does not work well for annealed polymers.

Table 14.9 shows the effects of crystallinity on oxygen permeability.

Table 14.9 Effect of Crystallinity on Oxygen Permeability

Polymer	Density (g/cm^3)	% Crystallinity	P (cm^3 mil/100 in^2 d atm)
PE	0.92	50	480
PE	0.96	75	110
PET		10	10
PET		50	3

Orientation, as well as crystallinity, affects the barrier properties of polymers. During orientation, in a crystalline material, the crystalline lamellae are realigned in the orientation direction, and the molecular chains in the amorphous regions are also realigned. This has two consequences. First, the chain separation in the amorphous regions is decreased, increasing the intermolecular attractions and resulting in decreased chain mobility. Second, the rotation of the crystallites in the orientation direction tends to cause their broader side to be perpendicular to the diffusion direction, increasing their effectiveness as barriers to diffusion of permeant molecules. The effect of orientation on oxygen permeability is shown in Table 14.10.

Table 14.10 Effect of Orientation on Oxygen Permeability

Polymer	% Elongation	P (cm^3 mil/100 in^2 d atm)
PP	0	150
	300	80
PET	0	10
	500	5

In semi-crystalline polymers, like HDPE, where processing conditions can strongly affect crystallinity and orientation, often permeance is not linear with thickness because thickness affects crystallization during cooling in the film formation process. Data reported by Krohn, et al [5] and shown in Figure 14.8 demonstrated that WVTR of blown films depends on the type of polyethylene used and the film thickness. Here one sees that thinner films have a proportionally better WVTR, and that medium and high molecular weight HDPE have better WVTR. These variations are caused by cooling rates and stresses imposed during blown film manufacture.

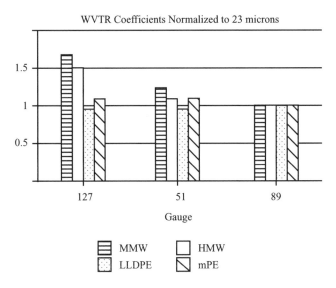

Figure 14.8 Variation in permeability coefficients of PE film

The presence of fillers can also affect permeability. Common fillers are inexpensive inorganic minerals such as talc, $CaCO_3$ or TiO_2. They can be used in concentrations as high as 40%, and are primarily intended to lower polymer cost. Sometimes coupling agents such as titanates are added to improve the interfacial bond between polymers and fillers. Table 14.11 illustrates the effect of calcium carbonate and coupling agents on the permeability of oxygen through PE.

14.7.6   Effect of Permeant Concentration

The concentration of permanent gases below one atmosphere of pressure generally does not affect the permeability coefficient. However, strong effects have been observed on the permeability of organic compounds. The permeability of organic vapors such as aromas, flavors, and solvents is usually strongly dependent on concentration. Readers interested in this topic may consult references Crank and Park [6] and Giacin and Hernandez [7].

Table 14.10 Effect of Calcium Carbonate as a Filler on the Oxygen Permeability of PE

% CaCO$_3$	Coupling agent	P (cm^3 mil/m^2 d atm)
0		5,000
15	no coupling agent	10,000
25	no coupling agent	20,000
15	with coupling agent	2,600
25	with coupling agent	1,600

14.8   Experimental Determination of Permeability

Two basic methods are used for determination of permeability, isostatic and quasi-isostatic. In both methods, experiments are run at constant temperature. Isostatic methods use continuous flow on both sides of the film to provide constant permeant concentration (Fig. 14.9). This system is employed in the Mocon Oxtran oxygen apparatus.

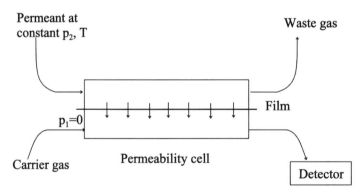

Figure 14.9 Schematic representation of a permeation cell used in the continuous flow, isostatic system.

Quasi-isostatic methods use flow to provide constant permeant concentration on the high partial pressure side, but allow accumulation on the low partial pressure side (Fig. 14.4 and 14.10). As long as accumulation is limited so that the concentration stays low, the partial pressure difference can be treated as constant.

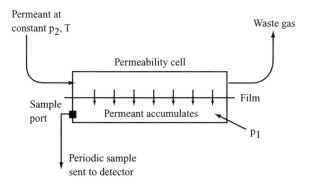

Figure 14.10 Schematic of quasi-isostatic permeation cell for gases.

The calculation of permeability and diffusion coefficients from the continuous flow method is illustrated in Figure 14.11. P and D are calculated as follows:

$$P = R_{ss} \frac{\ell}{A \times \Delta p} \tag{14.22}$$

$$D = \frac{\ell^2}{7.2 \, t_{1/2}} \tag{14.23}$$

where R_{ss} is the steady state flow rate in amount/time, and $t_{1/2}$ is the time to reach a flow rate equal to half of the steady state flow rate.

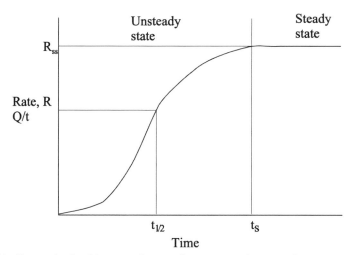

Figure 14.11 Curve obtained in a continuous flow permeation experiment.

Calculations from the quasi-isostatic method are illustrated in Figure 14.12. D and P are calculated as follows:

$$P = R_{ss} \frac{\ell}{A\,\Delta p} \tag{14.24}$$

$$D = \frac{\ell^2}{6\Theta} \tag{14.25}$$

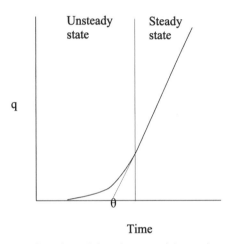

Figure 14.12 Quantity q as a function of time in a quasi-isostatic permeation experiment.

14.9   Multilayer Structures

In many packaging applications, multilayer structures are used. The calculation of permeation across such a structure can be done by examining transfer through the individual layers of the structure. Figure 14.13 and the following analysis show the situation for a three-layer material. We assume the material is at steady state. This means the amount of permeant passing through each layer is identical, and equal to the amount passing through the total structure.

The partial pressure differences across the layers are as follows:

$$\Delta p = p_2 - p_1$$

$$\Delta p_1 = p_i - p_1$$

$$\Delta p_2 = p_{ii} - p_i$$

$$\Delta p_3 = p_2 - p_{ii}$$

so, $\Delta p = \Delta p_1 + \Delta p_2 + \Delta p_3$

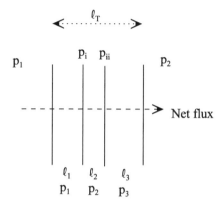

Figure 14.13 Permeation through a multilayer structure.

For the overall structure, we know

$$P_T = \frac{q \ell_T}{A\, t\, \Delta p_T} \tag{14.26}$$

and we can solve this for Δp, obtaining

$$\Delta p_T = \frac{q \ell_T}{A\, t\, P_T} \tag{14.27}$$

We can do the same for each of the individual layers, as well, obtaining

$$P_1 = \frac{q \ell_1}{A\, t\, \Delta p_1} \qquad\qquad \Delta p_1 = \frac{q \ell_1}{A\, t\, P_1}$$

$$P_2 = \frac{q \ell_2}{A\, t\, \Delta p_2} \qquad\qquad \Delta p_2 = \frac{q \ell_2}{A\, t\, P_2}$$

$$P_3 = \frac{q \ell_3}{A\, t\, \Delta p_3} \qquad\qquad \Delta p_3 = \frac{q \ell_3}{A\, t\, P_3}$$

Then, adding

$$\Delta p = \frac{q}{A\,t}\frac{\ell_T}{P_T} = \frac{q}{A\,t}\left(\frac{\ell_1}{P_1} + \frac{\ell_2}{P_2} + \frac{\ell_3}{P_3}\right) \qquad (14.28)$$

so,

$$\frac{\ell_T}{P_T} = \frac{\ell_1}{P_1} + \frac{\ell_2}{P_2} + \frac{\ell_3}{P_3} \qquad (14.29)$$

Thus knowing the thickness of each layer and the permeability coefficient for that material, we can calculate a permeability coefficient (or WVTR) for the overall structure. We can generalize to an n-layer structure as follows:

$$\frac{\ell_T}{P_T} = \sum_{i=1}^{n}\frac{\ell_i}{P_i} \qquad (14.30)$$

We can solve for P_T to obtain

$$P_T = \frac{\ell_T}{\displaystyle\sum_{i=1}^{n}\frac{\ell_i}{P_i}} \qquad (14.31)$$

Example: calculate the total oxygen permeability coefficient of the following multilayer structure (P given in cc mil / 100 in^2 d atm, thickness in mils).

	polymer	thickness	permeability
layer 1	PE	1.8	150
layer 2	nylon 6	1.0	2.6
layer 3	PVC	1.2	18.0
layer 4	PP	2.0	150

Solution:

$\ell_T = 1.8 + 1.0 + 1.2 + 2.0 = 6.0$ mil

$$\frac{\ell_T}{P_T} = \frac{\ell_1}{P_1} + \frac{\ell_2}{P_2} + \frac{\ell_3}{P_3} + \frac{\ell_4}{P_4} = \frac{1.8}{150} + \frac{1.0}{2.6} + \frac{1.2}{18.0} + \frac{2.0}{150} = 0.477\,\frac{100\ in^2\ d\ atm}{cc}$$

Then

$$P_T = \frac{6.0\ \text{mil cc}}{0.477\ 100\ in^2\ d\ atm} = 12.6\ \frac{cc\ mil}{100\ in^2\ d\ atm}$$

14.10 Applications of the Permeability Equation

Equation 14.13 is a simple but powerful equation that can be used to estimate the solution to several types of packaging problems. The equation relates many variables of a packaging system: product characteristics q, p_1 and t; package parameters P, A and ℓ; and environmental conditions p_2 and temperature.

Following are selected applications of Equation 14.13.

1. Select package materials. Together with mechanical properties, permeability values for O_2, CO_2, N_2, H_2O and organic vapors, as applicable, should be used to determine the best performance/cost ratio for specific package materials. Tables and charts of permeability values can help the packaging professional in the selection. A word of caution: most values of P are approximate values and there is no substitute for the experimentally determined value of P.

2. Calculate the amount of gas or vapor taken up or eliminated by the packaged product, for specific conditions of temperature T, pressure differential Δp, and time, t, using

$$q = \frac{P t A \Delta p}{\ell} \qquad (14.32)$$

where q is quantity (volume or mass), P is the permeability constant of the package material, t is shelf life time, A is the package area available to permeation, ℓ is the package thickness and Δp is the pressure differential across the package wall, and everything is at steady state conditions.

3. Estimate the shelf life if the quantity q of permeant transferred through the package is known or is set as a target value (P, A, Δp and ℓ should also be known). The permeant may be amount of water lost or gained by the product, or total amount of oxygen reacted (O_2) for a specific packaged product, knowing thickness ℓ, permeability; area A; and Δp.

$$t = \frac{q \, \ell}{A P \Delta p} \qquad (14.33)$$

If Δp changes through the shelf life period, equation 14.33 has to be expressed as:

$$t = \frac{\ell}{A P} \int_{q_1}^{q_2} \frac{dq}{dp} \qquad (14.34)$$

The above equations (Eq. 14.33 and 14.34) give the value of the shelf life in terms of the length of time a packaged product remains in an acceptable or saleable condition under specific conditions of storage. The validity of the above equations is subject to the following conditions:

a.   there is a rapid equilibrium between the product and the package internal conditions.

b.   the delay in reaching steady state condition of permeation through the package material is not considered.

c.   the temperature is constant through the shelf life period, t.

d.   P is not affected by concentration.

We now will consider the estimation of shelf life for moisture and oxygen sensitive products.

14.11 Shelf Life Estimation

The first step in a shelf life estimation is to determine the parameters controlling the loss of product quality. Shelf life may end for a product due to moisture uptake, oxidation, spoilage from microbial action, or a combination of these factors. Therefore, one must determine what is causing the end-point to occur. Having done that, calculations to estimate when that will occur in a package can be made.

The first estimation of shelf life due to gain or loss of a volatile component is usually made using the assumption of a constant Δp across the package wall. The accuracy of this assumption will vary, depending on the product and the package. For instance, if the product was potato chips and the failure mechanism was oxidative rancidity, the assumption is fairly good. Oxygen pressure in the atmosphere is nearly constant at 0.21 atm. The oxygen concentration in the package will be nearly zero, since any headspace oxygen will quickly react with the oil in the product. If the product were a thick liquid where diffusion is slow, the assumption would not be so good. For moisture vapor over a long shelf life, the assumption is only a first approximation because the relative humidity of the atmosphere changes over time, and the relative humidity inside the package can change significantly as moisture is gained or lost in the product. For accurate estimates of shelf life, storage testing of real packages under nearly real conditions is often needed.

To determine the behavior of a product, it must be stored at known conditions for a period of time and its properties measured. In the case of oxidation, for example, some method must be available to determine the amount of reaction with oxygen that the product has undergone. This is often done by measuring peroxide values for oil-containing products, or hexanal values for products that have hexanal as the end degradation product for oxidation. For moisture sorption, the product can be stored over a saturated salt solution until moisture uptake is at equilibrium. Then taste or texture is often the measured parameter to determine the end-point of shelf life. For pharmaceuticals, the true end-point is determined by the bioavailability of the drug.

For any type of product that gains or loses water, one can measure the moisture content as a function of relative humidity, or water activity, and determine a moisture isotherm. As shown in Figure 14.14, moisture isotherms are usually sigmoid shaped curves. However, one can sometimes use only the linear portion of the curve for shelf life predictions.

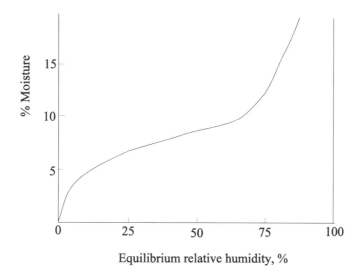

Figure 14.14 Moisture sorption isotherm.

Let us look at an example of shelf life prediction where the Δp is constant throughout the storage.

Example: Calculate the minimum thickness of PET for protection of a product that has an end of shelf life when it has reacted with 0.005% (wt/vol) of oxygen. The package design is a 500 ml container with 400 cm² area. The product is a water-based liquid. Storage conditions are 25 °C and 60% RH. The desired shelf life is six months. Also, calculate the water loss at the end of six months in this package.

Solution:
From equation 14.13 by rearrangement,

$$\ell = \frac{Pt\,A\,\Delta p}{q} \tag{14.35}$$

From the literature, PET at 25°C has an oxygen transmission rate (OTR) of 22 cm³ (STP) μm/(m² d kPa)

t = 6 months = 180 d

A = 400 cm² = 0.04 m²

Δp = 0.21 atm = 21.27 kPa (assuming p_i = 0) (p_i is oxygen partial pressure inside the package)

To determine q, we must convert the 0.005% gain over six months to a flow:

$$q = 500\ \text{ml} \times \frac{0.005}{100} \times \frac{\text{mol}}{32\ \text{g}} \times 22{,}412\ \frac{\text{cm}^3}{\text{mol}} = 17.5\ \text{cm}^3\ \text{(STP)}$$

Then

$$\ell = \frac{22\ \text{cm}^3\ \text{(STP)}\ \mu\text{m}}{\text{m}^2\ \text{d}\ \text{kPa}} \times 180\ \text{d} \times 0.04\ \text{m}^2 \times 21.27\ \text{kPa}\ \frac{1}{17.5\ \text{cm}^3\ \text{(STP)}} =$$

193 μm = 7.6 mil

We can now use the same method to calculate the amount of water loss.

$$q = \frac{Pt\,A\,\Delta p}{\ell}$$

Assume that the P_{H2O} for PET is 8.5 x 10⁴ cm³ (STP) μm/m² d kPa. The Δp is the difference in the vapor pressure in the container (100% RH) and that outside (60% RH). From a steam table, saturation vapor pressure at 25 °C = 0.4592 psi x 6.895 kPa/psi = 3.166 kPa.

Δp = 3.17 x (100 - 60)/100 kPa = 1.268 kPa

$$q = \frac{8.5 \times 10^{-4}\ \text{cm}^3\ \text{(STP)}\ \mu\text{m}}{\text{m}^2\ \text{d}\ \text{kPa}} \times 180\ \text{d} \times 0.04\ \text{m}^2 \times 1.268\ \text{kPa} \times \frac{1}{193\ \mu\text{m}} =$$

4.0×10^3 cm³ (STP) = 3.22 cm³ liquid water

Now suppose that a product is stored in a real world situation where the moisture on the inside or the outside of the package changes over time. Then one needs the external environmental conditions and a moisture isotherm for the product. The moisture on the inside of the package may change over time even if the external conditions are constant because the product is reaching equilibrium with the internal moisture content. If the external conditions vary over too wide a range of temperatures, then multiple isotherms may be needed.

Let us consider an example where the external storage conditions are constant, and a product isotherm is known.

Example: A cracker package is 1.2 mil PP with a WVTR of 2.0 g mil/100 in^2 d measured at 95°F, 90% RH. The package area is 100 in^2. Mass of the crackers is 100 g. Initial moisture content is 2%, and the critical moisture content (above which the crackers are too soggy so have reached the end of their shelf life) is 5.5%. Storage conditions are 70 °F, 70% RH. The isotherm is non-linear over the range in question, as shown in Figure 14.15. Calculate the shelf life of the crackers.

Solution:
First, the calculation will not be exact, but will be an estimation. To solve the problem we need:

    1.   Permeability constant at the storage conditions.
    2.   Sorption isotherm of the product, given in Figure 14.15.
    3.   Area of package.
    4.   Thickness of package.
    5.   Value of Δp.

We will also make several assumptions to aid in the solution:

    1.   The amount of water in the headspace is negligible.
    2.   The product reaches instant equilibrium with the headspace, and the equilibrium is reversible.
    3.   A steady state permeation rate is rapidly reached.

The first step is to convert the WVTR to a permeability coefficient. Remembering that

$$P = \frac{WVTR}{\Delta p}$$

the value of Δp can be calculated.

$$\Delta p = P_{sat} \times \frac{RH_o - RH_i}{100}$$

where RH_o is the relative humidity on the outside of the test material, 90%, and RH_i is the relative humidity on the inside of the test material in contact with the desiccant, approximately 0%. Therefore

$$\Delta p = 42.175 \text{ mm Hg} \times \frac{90}{100} = 37.96 \text{ mm Hg}$$

and

$$P = \frac{2.0 \text{ g mil}}{100 \text{ in}^2 \text{ d}} \times \frac{1}{37.96 \text{ mm Hg}} = 0.053 \frac{\text{g mil}}{100 \text{ in}^2 \text{ d mm Hg}}$$

This is the value of P at 95 °F. We need the value at 70 °F, but we do not have any information about the activation energy, E_p. Therefore, we will use this value and know that we have overestimated the permeability, and that the calculated shelf life will be shorter than the actual shelf life.

The mass flow through the package to reach the critical moisture content is

$$q = \text{mass of crackers (dry)} \times (m_c - m_i) = 100 \,(0.055 - 0.02) = 3.5 \text{ g}$$

The Δp across the packaging material is a function of the amount of moisture transmitted. We know that the outside p is

$$p_{out} = p_s \times RH_{out}/100$$

and

$$p_{in} = p_s \times RH_{in}/100$$

so

$$\Delta p = p_s/100 \,(RH_{out} - RH_{in})$$

We could assume that the RH inside the package is constant, but this would be a very poor assumption. We can see from the moisture sorption isotherm that at the beginning of shelf life, the RH inside the pouch is about 8%, and at the end of shelf life, it is about 43%. To obtain a more accurate estimate of shelf life, we can average the RH_{in} at the beginning and end of shelf life and get the following:

$$\Delta p = 18.77 \text{ mm Hg} \times \left(0.7 - \frac{0.43 + 0.08}{2} \right) = 8.35 \text{ mm Hg}$$

where the values of 43% RH and 8% RH were read from the moisture sorption isotherm as the values corresponding to 5.5% moisture content and 2% moisture content, respectively.

Then

$$t = \frac{3.5 \times 1.2 \times 100}{100 \times 8.35 \times 0.053} \text{ d} = 9.5 \text{ days}$$

A better estimate would be obtained by dividing the time up into shorter intervals, and averaging Δp across each of those shorter intervals, instead of over the whole shelf life. For example, we could use two intervals, 8% to 22.9% RH, and 22.9% to 43%, giving average RH_{in} as 15.45% RH and 33.0% RH in the two intervals, and the cracker moisture content at 22.9% RH read from the graph as 3.9%. Then the calculation for the two steps is as follows:

$$t_1 = \frac{(0.039 - 0.02) \times 100 \times 1.2 \times 100}{100 \times 18.77 \times (0.70 - 0.1545) \times 0.053} = 4.2 \text{ days}$$

$$t_2 = \frac{(0.055 - 0.039) \times 100 \times 1.2 \times 100}{100 \times 18.77 \times (0.70 - 0.33) \times 0.053} = 5.2 \text{ days}$$

and $t = t_1 + t_2 = 9.4$ days.

Shortening the interval still further will increase the accuracy of the calculation.

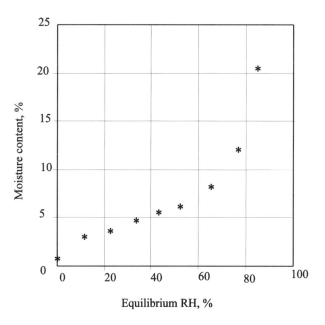

Figure 14.15 Moisture sorption isotherm for cracker example.

If the moisture sorption isotherm is approximately linear over the area of interest, an analytical solution to the shelf life equation can be obtained, using Equation 14.34.

Let the sorption isotherm be represented as

$$Y = a + bM \tag{14.36}$$

where Y is the relative humidity, and M is the moisture content of the product. Then, integrating Equation 14.34, we obtain

$$t = \frac{\ell\,W}{P\,A\,p_s\,b}\,\ln\frac{Y_o - Y_{i,t=0}}{Y_o - Y_{i,t}} \tag{14.37}$$

where Y_o is the outside relative humidity
        Y_i is the package headspace relative humidity
        W is the dry weight of product.

Example. A product with a dry weight W = 80 g and a *b* value = 9.0 kg product/kg H_2O, will be stored at 23°C and 85% relative humidity. The initial equilibrium relative humidity of the product $RH_{i,t=o}$ = 20%. The final equilibrium relative humidity of the product $RH_{i,t}$ = 70%. Permeability of the package material, $P = 4 \times 10^{-2}\dfrac{kg\,\mu m}{m^2\,d\,kPa}$. If the package thickness is 53 μm, calculate the maximum package area that will maintain the RH_i of the product at or below 70% during 100 days of shelf life.

Solution:
ℓ = 53 μm
w = 80 g = 0.08 kg
P = 4 x 10^{-2} kg μm/m^2 d kPa
p_s = 21.07 mm Hg = 2.81 kPa
b = 9.0

$$\ln\frac{Y_o - Y_{i,t=0}}{Y_o - Y_{i,t}} = \frac{85 - 20}{85 - 70} = 1.466$$

$$A = \frac{53 \times 0.08}{4 \times 10^{-2} \times 100 \times 2.81 \times 9}\,1.466 = 6.14 \times 10^{-2}\ m^2$$

References

1.   Van Krevelen, D. W., *Properties of Polymers, 3rd ed*, Elsevier Science, Amsterdam, 1997.
2.   Hernandez, R. J. *J. Food. Eng.* 22:495-507, 1994.
3.   Hernandez, R. J. and Gavara, R., *J. Polym Sci. Part B Pol. Physics*, 32:2367-2374, 1994.
4.   Michaels, A. S., Veith, W. R. and Barrier, J. A., *J. Appl. Phys.* 34:13, 1963.
5.   Krohn, J. Tate, R. and Jordy, D., *Proc. ANTEC '97*, pp. 1654-1658, 1997.
6.   Crank, J. and Park, G. S. *Diffusion in Polymers*, London:Academic Press, 1968.
7.   Giacin, J. and Hernandez, R. J., *Wiley Encyclopedia of Packaging Technology*, 2nd ed., E. Brody and K. Marsh, eds., New York:John Wiley & Sons, pp. 724-733, 1997.

Study Questions

1. You have the following permeation information about a flexible plastic structure: WVTR (obtained at 100°F, 95% RH) =1.2 g/day, area = 100 in², thickness = 3 mil. Calculate, in SI units: WVTR, permeance R, thickness-normalized flow N, and permeability coefficient P. Use SI units: kg, sec, Pa, m
   (Answer: WVTR = 2.14×10^{-7} kg/s m²; R = 3.45×10^{-11} kg/s m² Pa; N = 1.63×10^{-11} kg m/s m²; P = 2.61×10^{-15} kg m/s m² Pa)

2. You are designing a mutilayer plastic structure with 4 layers. The total oxygen permeability is required be 30 cm³ mil /m² day atm. Calculate the required thickness of the barrier layer (layer 2) in mil. (Answer: x = 2.55 mil)

layer	thickness (mil)	P (cc mil/m² d atm)
1	2.5	250
2	x	10
3	1.5	150
4	2.1	158

3. The permeability coefficient of O_2 through an unknown plastic film at 30°C is 0.1 cm³ μm/m².day atm. Estimate P at 4°C. Assume a "reasonable" value of the activation energy. (Answer: 0.0225 cm³ μm/m² d atm for E_p = 40 kJ/mol)

4. Experimental values from a quasi-isostatic permeability water vapor experiment are shown below. Plot the data and calculate D and P in g, day, atm, m², and mil.

t, hour	0	1	2	3	4	5	6	7	8	9	10
q, gram	0	.25	.5	.75	.95	2.1	3.1	3.95	5.1	5.9	7.1

   Additional information: Area of cell = 100 cm², film thickness ℓ = 1 mil, and Δp = 20 mm Hg. (Answer: P = 9.13×10^4 g mil/m² d atm; D = 9.90×10^{-15} m²/s)

5. A multilayer structure of undefined composition but containing an EVOH layer has been evaluated for its barrier characteristics to oxygen using an Oxtran continuous flow apparatus (area 100 cm²) and air. The steady state rate, R_{ss} = 2.0 cm³/day. Calculate the permeance of the structure in cm³/sec m² Pa. (Answer: R = 1.1×10^{-7} cc/s m² Pa)

6. HDPE is known to be a poor barrier for hydrocarbons such as gasoline. Using what you have learned, explain why this is the case, and why fluorination of the surface of an HDPE container (such as in an automobile fuel tank) can significantly improve its barrier ability.

7.   As discussed, migration of residual monomer from plastic packaging is sometimes a concern. However, migration of polymer molecules themselves is almost never a problem. Why?

8.   Calculate the temperature increase (and decrease) required to double (cut in half) the permeability of O_2 in amorphous PET. Reference temperature = 25°C. (Answer: T_2 = 39.1°C, T_2 = 12.1°C)

9.   To protect a moisture sensitive product you are designing a package that has the following characteristics:

Packaging material:	Area = 200 sq. cm.
Permeability:	0.44 g mil/m^2 day mm Hg
Storage conditions:	23°C, 90% RH
Thickness:	3.0 mil

Product:
Initial equilibrium relative humidity = 15%
Final equilibrium relative humidity = 80%
Product's sorption isotherm: linear with b = 16.0 g product/g H_2O
Weight on a dry basis = 150 g

Calculate the shelf life of the package, using Eq. 14.37. (Answer: t = 305 days)

10.   You are designing a bag for potato chips. This is an oxygen sensitive product that needs protection from air (21% of oxygen). Assume that the product cannot gain more than 0.02% wt/wt of oxygen. Determine the thickness of the flexible packaging material needed based on the following information:

Amount of product: 400 g
Packaging material permeability, P = 15.0 cm^3 mil/m^2 d atm
Shelf life expected: 6 months (180 days)
Area of the package: 400 in^2
(Answer: 2.6 mil)

11. A clear oxygen sensitive fruit juice (H_2O 96%) reacts with oxygen from the environment and oxidizes, turning brown. When the total uptake of oxygen is equivalent to 0.01% of its own weight, the fruit juice will no longer be saleable. The product will be packaged using a barrier film to delay the oxidation process.

Amount of juice: 200 ml, density approximately 1.0 g/cm^3
Packaging material: PET, 1 mil (semi-crystalline)
Total area of packaging material: 5" x 5"
Environmental conditions: oxygen concentration = 21%, T = 4°C, RH = 30%

1. Calculate, for the above conditions
   a. Shelf life of the product
   b. Amount of water lost by the product at the end of its shelf life.
2. Make two reasonable recommendations to reduce the loss of water by 20% while maintaining at least the shelf life calculated in Part 1.
(Answer: 4.1 months, 2.74 ml of water)

12. Your company is encountering problems with the shelf life of its fruit-flavored cereal due to an unacceptable degree of flavor scalping. One of your colleagues suggests increasing the thickness of the current pouch. Do you think this is likely to be a productive approach? Explain.

13. When a carbonated soft drink loses CO_2, it is inherently a 3-dimensional problem, as the CO_2 does not travel in just one direction, yet we can use one-dimensional mass transfer equations to estimate the shelf life for the soft drink. Explain how this can work.

14. If you bring home a party balloon filled with helium, obviously unless the balloon is a perfect barrier, over time helium will diffuse out of the balloon. Will any gases diffuse into the balloon while the helium is diffusing out? Why or why not?

15. An oxygen-sensitive product is packaged in a multilayer barrier pouch containing an inner EVOH layer, tie layers on either side of the EVOH, and outer layers of PP. The total thickness of the PP in the film is 3 mils. Does it matter whether the structure is:

/1.5 mil PP/ tie/0.5 mil EVOH/tie/1.5 mil PP/
or
/1.0 mil PP/ tie/0.5 mil EVOH/tie/2.0 mil PP/

assuming the thickness of the tie layers remains constant? Explain.

16. In question 15, does the answer change if the barrier layer is a PVDC copolymer instead of EVOH? Explain.

17. Often, when oxygen-sensitive products are packaged, nitrogen-flushing is used to remove residual oxygen from the package. Why is this important? How should you adjust a shelf life calculation if there is significant residual oxygen in the package when it is first sealed?

15 U.S. Regulations and Plastic Packaging

15.1 Introduction

There are a number of types of regulations that plastic packaging and the plastics packaging industries must meet, around the world. Most of these regulations are not directly aimed at plastics packaging, but rather are parts of more broad-based requirements relating to workplace safety, pollution prevention, product labeling, tamper evidence, and a variety of other concerns. Most of these are beyond the scope of this book. The regulations discussed in this chapter are those which limit the selection of a plastic type for a particular application, and/or limit the constituents of a plastic resin, including additives, monomer residues, and other ingredients. Typically, these regulations depend on the use of the packaging, and are most stringent for pharmaceutical or other medical products, somewhat less so for foods, still less stringent for cosmetics, and much less so for most other types of products. Our discussion in this chapter will cover only the regulatory situation in the United States. With the increasing globalization of trade, it is, of course, important to be aware of regulations in other countries around the world.

This chapter is not intended to be a complete presentation of all relevant regulatory matters; rather, it is intended to serve as a beginning point. Before making final packaging decisions for a product, experts should be consulted to make sure that all relevant regulations are satisfied.

15.2   The U.S. Federal Food, Drug and Cosmetic Act

In the USA, regulations governing food, drug, and cosmetic packaging are authorized under the Federal Food, Drug, and Cosmetic Act, first passed in 1938, and appearing, as amended, in Title 21,Chapter 9 of the United States Code. This law authorizes the Food and Drug Administration to issue regulations, which appear in the Code of Federal Regulations (CFR). As changes in regulations or new regulations are considered, they are submitted for public comment through publication in the Federal Register, which also publishes the new regulations when they are finalized. The Federal Register, which is printed daily, also contains official Notices of Federal agencies and organizations, and Executive Orders and other Presidential Documents. New regulations will appear in the next scheduled update of the CFR. Each of the 50 volumes of the CFR is updated once per year, on a rotating basis. All of these publications are printed by the Government Printing Office (GPO), which also maintains online versions that provide convenient Internet access, through the GPO website at www.access.gpo.gov. Search features are provided for each database.

For most purposes, information on the regulations is more important than information about the enabling laws. To determine if a proposed use of plastic (or other) packaging is allowed by regulations, it is important to examine the Federal Register as well as the CFR. Another point of access to these materials is through the Food and Drug Administration website at www.fda.gov, which also has a search feature. The language in the regulations can be quite complex, and it is advisable to get the help of an expert whenever possible. The FDA itself will also provide help and consultation. For food packaging, another source of information that should be consulted, if it is available, is the current Food Chemical News Guide published by FCN Publishing, which has a web site at www.fcnpublishing.com. This publication is available only on a paid subscription basis.

Since regulations for food packaging and for pharmaceutical packaging take distinctly different approaches, we will discuss them separately, beginning with pharmaceutical products.

15.3   Medical Packaging Regulations

15.3.1   Drug Packaging

The regulations for pharmaceutical products require premarket approval. This means that a manufacturer wishing to produce and sell a drug must provide the FDA with evidence that the drug is safe and effective. In particular, the package for the drug is considered part of the drug. One of the requirements is that such a product be labeled with an expiration date, and stability testing must be done to verify that the drug remains acceptable during that period. The requirements for stability testing require that the drug be tested "in the same container-closure system as that in which the drug product is marketed" (21CFR211.166). This standard nomenclature for referring to the CFR means Section 21 of the Code of Federal Regulations, Part 211, Section166. The FDA works closely with the United States Pharmacopoeia (USP) in developing regulations, as U.S. law requires that any drug sold that is listed in the USP must conform to USP standards. The USP is also a source for regulations in some countries outside the U.S.

If, after a drug is on the market, the manufacturer wishes to change the packaging in any significant way, it is necessary both to carry out stability testing in the new package, and to submit a supplementary new-drug application (NDA) to FDA for approval before the package can be used. There are provisions for accelerated testing at high temperature and relative humidity. Accelerated testing is done at severe conditions to speed up changes that may occur at a slower rate under normal storage conditions. However, even if approval to market the drug in a particular package is issued based on accelerated testing, stability tests under normal storage conditions must continue for the full labeled life of the product. Thus, changing the package for a drug product is a major undertaking.

The FDA does provide for some interchangeability of plastic resins. While in general testing "in the same container-closure system" is interpreted to mean not just the same generic plastic, but rather the same precise container, including the resin formulation, FDA has accepted procedures developed by the USP for allowing manufacturers to change packaging without requiring prior FDA approval. These test protocols provide for a determination that a new plastic resin is equivalent to the one already approved if the tests so indicate. Currently such test protocols are available for HDPE, LDPE, PET, and glycol-modified PET (PETG). If, for example, both HDPE bottles satisfy the test protocol, then they can be used interchangeably, and if one has been approved for a specific application, the other can be used as well. However, the approval is conditional on successful completion of stability testing in the new package.

15.3.2 Medical Device Packaging

While one might not expect the Food, Drug, and Cosmetic Act to regulate medical devices, the act has, in fact, covered medical devices since 1938. That regulation became more stringent with the Medical Device Amendments of 1976, which brought to some medical devices the same premarket approval needed for drugs. The amount and type of regulation of medical devices depends on their classification. All medical devices intended for human use are classified by expert panels as Class 1 - General Controls, Class II - Special Controls, or Class III - Premarket Approval. Devices in Class 1 require truthful labeling, record keeping, and other fairly standard rules, including adherence to good manufacturing practices. Premarket notification is also required, meaning that the manufacturer must notify FDA of its intent to market the product, but need not wait for FDA approval. Class III devices require the same general type of premarket approval as drugs, including approval of the packaging. The new device regulation has become more overtly a process control regulation, and specifies that the package is a component of the device.

15.4   Food Packaging Regulations

Regulation of food packaging is done on a much more generic basis than regulation of medical packaging. The primary concern with approval of plastic resins for food packaging is that the package does not adulterate the food in some way that could potentially be harmful to the consumer. Thus the main concern is migration of constituents from the plastic packaging material to the food or beverage with which it is in contact. In the 1958 Food Additives Amendment to the Federal Food, Drug, and Cosmetic Act, the U.S. Congress gave the FDA jurisdiction over both direct and indirect food additives.

15.4.1   What is a Food Additive?

The Federal Food, Drug and Cosmetic Act, in Section 321(s) (21 USC 321) defines a food additive as

> "any substance the intended use of which results, or may reasonably be expected to result, directly or indirectly, in its becoming a component...of any food...if such substance is not generally recognized...to be safe under the conditions of its intended use; except that such term does not include…..

> (4) any substance used in accordance with a sanction or approval granted prior to September 6, 1958..."

In 21 CFR 170.3(e), there is a repeat of the food additive definition. Direct food additives are those which are added to the food. Indirect food additives are those substances that are not added to the food, but can end up in the food through its contact with something containing them. Under this regulation, FDA determined that food packaging materials are considered indirect food additives and thus are subject to regulation. For plastics, the regulations address both the basic polymer resins used in food packaging and all additives, catalyst residues, residual monomer, or other components which are contained in the plastic resin. All of this applies to use of plastics for drugs, as well. Typically, approval of a new plastic for food comes first, with additional testing, including stability testing, required for approval for specific drugs.

21 CFR 177 contains positive lists of plastic resins that are permitted to be used in food packaging. These lists are written generically, not addressing products from a particular company but rather stipulating the safe uses for various plastics. Some base polymers are approved for any type of food packaging, while others are limited to certain types of applications. Some resins are regulated individually, while others, like polyolefins, are regulated as a group. The regulations frequently contain limitations on the resin, such as maximum residual monomer content and minimum molecular weight. They also often contain limits on migration of components from the plastics. In some cases, these limits are specific for a particular substance, such as residual monomer, but most often, they limit global migration, the total amount of migrating substances from either the resin or the food contact package. Tests for compliance with migration limits are generally based on short-term extraction tests at elevated temperatures with specific solvents chosen to mimic the effects of foods.

21 CFR 175 addresses indirect food additives from adhesives and components of coatings, which also generally consist of polymers.

FDA refers to substances added to a polymer in the process of formulating the resin or of manufacturing the final food package as adjuvants. Adjuvants and processing aids are regulated under 21 CFR 178. These include additives such as antioxidants and plasticizers, as well as other components. "Color additives" are excluded from the definition of food additives, but "Substances capable of imparting a color to a container for foods, drugs, or cosmetics are not color additives unless the customary or reasonably foreseeable handling or use of the container may reasonably be expected to result in the transmittal of the color to the contents of the package or any part thereof" (21 CFR 70.3). Thus coloring agents for plastic packaging are generally considered food additives and regulated under the provisions for adjuvants. Adjuvants which are permitted for use in food contact packages are, like polymers, listed separately in some cases, and grouped together in other cases. Grouping of adjuvants is based on the end use of the compound. The regulation generally specifies the amount of the adjuvant that can be used, the types of polymers in which it can be used, and often also the conditions of use of the final food package. Limitations on end uses most often address the type of food packaged, and the temperature.

It should always be remembered, of course, that the CFR contains only the regulations promulgated at the time of the last updating, so it is important to check the Federal Register for any new or changed regulations. It can also be very important to companies to check the Federal Register for official notices of regulations which are under consideration, so that FDA can take the opinions of manufacturers and users into account when the regulations are finalized.

The following examples illustrate the kind of information that can be obtained from the CFR when selection of an appropriate packaging material is being made.

15.4.1.1 Example 1: Information on Nylon Available in the CFR

The relevant section on nylon resins is 21 CFR 177.1500. Paragraph (a) defines 16 different types of nylons, including Nylon 6, Nylon 66, and Nylon 610, which are acceptable for food packaging. The identification includes the materials from which the polymers are produced, as well as, in some cases, the relative amounts of comonomers.

Paragraph (b) has specifications for each of the 16 resins, including specific gravity, melting point, solubility in a boiling acid solution, and maximum total extractibles in water, 95% ethyl alcohol, ethyl acetate, and benzene. For some nylons, such as Nylon 6/12, viscosity is also specified. In some cases, the applicable specifications depend on the characteristics of the package. For example, Nylon 6 resins used in film with an average thickness less than 25.4 μm can have a larger amount of water extractables than nylon used in other applications.

Paragraph (c) specifies a nylon modifier, its characteristics, and conditions for its use. Paragraph (d) specifies analytical methods to be used for determining whether a resin meets the specifications in paragraph (b).

15.4.1.2 Example 2: Identification of Relevant Regulations for Fabrication of an Adhesive-laminated Plastic Bag

Let us examine relevant regulations for fabrication of an adhesive-laminated plastic bag for rice, made from reverse-printed PE laminated to unprinted PE. This is a much more complex case than Example 1. In addition to examining the regulations for the base polyethylene, we must look at the printing process and ink, the adhesive, the laminating process, and bagmaking. The general standards for Good Manufacturing Process (GMP) apply, as well as individual regulations. In this example, we will examine the adhesive selection only.

The relevant regulation is Part 175 - Indirect Food Additives: Adhesives and Components of Coatings, Section 175.105 Adhesives. Paragraph (a) prescribes conditions allowing the use of adhesives in food packaging. These conditions include that the adhesive is prepared from one or more of the substances named in paragraph (c). The adhesive must either be separated from the food by a functional barrier or meet additional limitations which differ for dry and for fatty and aqueous foods.

Paragraph (b) specifies that the adhesive must be labeled as "food-packaging adhesive."

Paragraph (c) lists the substances which can be included in adhesives, including GRAS substances, prior-sanctioned substances, flavoring and color permitted for use in food (with some restrictions), and then a long list of permitted substances. Some of these are generally allowed, while others are permitted only under certain limitations, or subject to certain specifications.

15.4.2 Acceptable Amounts of Migration

Since the regulations are based on the indirect food additive provisions of the Food, Drug and Cosmetic Act, the primary determining factors in whether a resin and/or an adjuvant will be approved are its potential ability to migrate from the package to the contained food, coupled with the safety of the substance if ingested. Thus, the major focus of the pre-market safety evaluation of an indirect food additive is predicting the amount and types of migrants from the plastic material under the proposed use conditions.

The first step in determining the amount of migration that can be expected is to perform extraction studies. The FDA has formulated extraction test protocols, in which the plastic materials are exposed to food simulants under specified conditions. These protocols have been modified over the years as experience is gained. Extraction is done at elevated temperatures, with agitation, typically over a period of 10 days. Food simulants are used instead of actual foods because it is important to standardize the extraction agent, and also because detection and analysis of migrating compounds is much easier and more reliable if the medium is a simple two-component solution rather than a chemically complex food.

The next step is to convert the predicted migration into an "estimated daily intake" (EDI), which is the amount of the additive that is likely to be consumed per day by an individual if the proposed use were permitted. The EDI is calculated using "consumption factors" (CF). The consumption factors are estimates of the fraction of the diet likely to contact different types of polymers and other packaging materials, based on periodically updated data on current and projected U.S. food packaging markets. Because different types of foods have differing abilities to leach migrants from the package, food type distribution factors (f_t) are also calculated, based on the fraction of total food likely to be contacted by a plastic that is in the four general categories of aqueous, acidic, alcoholic and fatty. Then the expected migration from the packaging material for each food type, based on extraction tests with an appropriate food simulant, is multiplied by the f_t and the values are summed for the four types of foods. This estimate of migration is then multiplied by the consumption factor and then by 3000 g/day, which represents total daily food intake (including liquids), to arrive at the EDI.

The final step is to compare the EDI to the acceptable daily intake (ADI). The use of the additive can be allowed only if the EDI does not exceed the ADI. ADI values are themselves obtained using animal feeding studies to establish a no-effect level, and then applying an appropriate safety factor. If no such studies are available for the extracted chemicals, the petitioner must carry out the needed research.

15.4.3   Threshold of Regulation

In 21 CFR 170.3(e)(1) as part of the food additives definition, the regulation states "If there is no migration of a packaging component from the package to the food, it does not become a component of the food and thus is not a food additive." As with many other regulations where a zero amount of a substance was specified, the impact of the regulation as written changes with time. As analytical technologies improve, we are able to detect smaller and smaller amounts of substances, and "zero" becomes progressively harder to attain. FDA has addressed this issue through 21 CFR 170.3(e)(2), which states that no listing regulation is required for substances used in food-contact articles "that migrate, or may be expected to migrate, into food at such negligible levels that they have been exempted from regulation as food additives under §170.39."

The amount of migration under which the food additive regulations do not apply is known as the "threshold of regulation." This policy is based on the U.S. legal doctrine of *de minimis,* which holds that the law does not concern itself with trifles, so statutes should not be applied to achieve meaningless results. In the case of Monsanto vs Kennedy, involving polyacrylonitrile copolymer beverage bottles, the courts held that the FDA had latitude in applying the statutory definition of a food additive in those *de minimis* situations which clearly present no public health or safety concerns.

Based on this finding, the FDA developed 21 CFR 170.39, "Threshold of regulation for substances used in food-contact articles," which specifies the conditions under which a substance used in a food-contact article that migrates or can be expected to migrate into food will be exempt from regulation as a food additive. One requirement is that "the substance has not been shown to be a carcinogen in humans or animals, and there is no reason, based on the chemical structure of the substance, to suspect that the substance is a carcinogen." Additionally, the substance must present no other health or safety concerns because

"the use in question has been shown to result in or may be expected to result in dietary concentrations at or below 0.5 parts per billion, corresponding to dietary exposure levels at or below 1.5 micrograms/person/day (based on a diet of 1,500 grams of solid food and 1.500 grams of liquid food per person per day)..."

There is also a provision for higher limits for a substance that is currently regulated as a direct food additive. In addition, the regulation requires that the substance have "no technical effect in or on the food to which it migrates" and have "no significant adverse impact on the environment."

The regulation also specifies the information that a potential user must supply to FDA in the request for exemption from regulation of a particular use of a substance. FDA has the right to decline to grant exemptions in cases in which it determines that "available information establishes that the proposed use may pose a public health risk."

The FDA responds to requests for exemptions for specific food-contact applications by letter to the requestor. No list of these exemptions appears in the Federal Register or the CFR. However, the FDA maintains a list of exempted substances, including the name of the company making the request, the chemical name of the substance, the specific use for which it has been exempted, and any limitations on its use. This list is available from the FDA Office of Premarket Approval. If an environmental assessment is required, the finding of no significant impact and the evidence supporting that finding is also available for public inspection.

Not every user of an exempted substance must request FDA approval. If other prospective users determine that their applications meet the FDA guidelines under which an exemption was granted, they can proceed with the use. Of course, if they err and there is any hazard resulting from such use, they can be held liable. If the FDA later rescinds an exemption for a substance, in addition to notifying the requesters, a notice will be published in the Federal Register, so that other manufacturers using the exempted substances under similar conditions will be aware of the change.

15.4.4   Food Processing Equipment and the Housewares Exclusion

While we have discussed packaging materials only, the regulations for indirect food additives are written for "food contact articles" and it is clearly stated that they also apply to food processing equipment. Therefore, plastic food contact surfaces in processing equipment must meet the same requirements as plastic food packages. Migration limits, however, are often different for single-use and multiple-use items.

There is sometimes confusion within the industry about why the food contact surfaces of eating utensils, receptacles, paper towels, and other kitchen appliances do not fall under the same regulations as food packaging. These were excluded from the 1958 Food Additives Amendment from the beginning, and the FDA has reaffirmed this in many letters. This does not mean that the maker of an item that is considered a houseware can ignore the public safety issues of the plastics and additives used. It would be advisable for any manufacturer to follow similar guidelines as must food packagers, but such practice is voluntary.

15.4.5   Determining the Conditions of Use

Because migration of any material through a plastic is affected by the temperature of exposure, and the material in direct contact with the polymer surface, one must define the conditions of use for the plastic package. The CFR has implied that all approvals are based on "the intended conditions of use" and in many cases the regulations establishes specific limits on these conditions for use of a given polymer or adjuvant.

The FDA has set forth some broad categories of foods and of use conditions in 21 CFR 176.170. Use conditions include high temperature heat-sterilized, boiling water sterilized, hot filled, or pasteurized in two different temperature ranges, room temperature filled and stored, refrigerated storage, frozen storage, and refrigerated or frozen storage with reheating in the container at the time of use. There are 9 categories of types of foods, several with additional subcategories.

Frequently, an in-depth knowledge of the make-up of the product under consideration, as well as knowledge about processing and storage conditions, is required to make the determination of the appropriate category for a particular food. Further, if there are ingredients, or additives, in the product that might modify the conditions of use because they are known to be aggressive against the planned food contact surface, it may be advisable to choose a different material before proceeding further.

When the categories that fit the product and the processing and storage conditions are identified, the next step is to examine the regulations that apply to the packaging material, in order to determine whether the use is acceptable or not.

15.4.6 Multilayer Food Packages

As discussed, the indirect food additive regulations apply to substances that may migrate from the package into the food product. Therefore, the nature and composition of the food-contact layer, the layer that comes in direct contact with the product, is of prime importance. For example, if the plastic material has a coating of some type next to the food, this coating is the food-contact layer, not the plastic itself, and its composition and ingredients must be known in order to determine whether the regulations are met. The same principle holds for a multilayer package. However, there is often at least some possibility that substances contained in non-food contact layers will migrate through the food contact layer and reach the food. The thickness of the food contact layer is an important variable in determining whether and to what extent this will occur. In general, the thinner this layer, the faster materials can permeate from the bulk material to the surface and hence into the food. The rate and amount of migration also depend on the diffusion coefficient of the permeant in the plastic material, which measures how readily the migrant is able to move through the material. Another important variable is the partition coefficient, which describes the equilibrium distribution of the migrant between two phases that are in contact with each other. Of course, if there is potential for migration from non-food-contact layers, the compositions of these other layers in the package must be examined in order to determine that they also satisfy the applicable regulations.

A layer in a multilayer packaging structure that acts to retard, limit, or prevent migration of components from package layers exterior to it, into the package contents is called a functional barrier. This barrier may or may not be the food contact material itself. For example, in a multilayer structure containing an inner layer of aluminum, the aluminum can be expected to be an excellent barrier to any possible migrants from all layers from the outside of the package up to the aluminum.

15.4.7 GRAS and Prior-Sanctioned Additives

The definition of food additive, as described in Sec. 15.4.1, excludes substances

"generally recognized, among experts qualified by scientific training and experience to evaluate its safety, as having been adequately shown through scientific procedures (or, in the case as a substance used in food prior to January 1, 1958, through either scientific procedures or experience based on common use in food) to be safe under the conditions of its intended use..."

Substances meeting the first criterion are referred to as GRAS, for "generally recognized as safe." The FDA does not list all such substances, saying it is "impracticable" to do so. A number of GRAS substances are listed, however, in 21 CFR 182.1. As substances on this list are evaluated, they are, if it is determined to be appropriate, "affirmed" as safe and listed, for indirect food additives, in Sec. 186. Sec. 189 contains a listing of substances which are "prohibited from use in food." A more complete list can be found by consulting the current Food Chemicals Guide.

Any material can be added to the GRAS list if there is data to support its safety. The FDA allows a food manufacturer to affirm to the Agency that a substance is GRAS, and supply the supporting data to get the substance reviewed. This can be done and the product put into production before the review process is complete. However, the manufacturer assumes all responsibility in the event the affirmation is not verified by the Agency.

Substances that are excluded from the definition of food additive because they were "used in accordance with a sanction or approval granted prior to September 6, 1958..." are referred to as prior-sanctioned materials.

15.4.8 Use of Recycled Plastics for Food Packaging

In recent years, there has been a growing interest in the use of recycled plastics for food packaging. Here the concern about the acceptability of such uses is also largely related to the potential migration of undesirable substances into the contained food. If the material in question is clean process scrap from the manufacture of food containers that is being chipped and fed right back into the same process, the only concern is that the extra heat history not produce chemical changes that would lead to unacceptable migration (or other performance problems). Thus, this type of in-house "recycling" is routine in many processes for manufacturing plastic packaging, and is normally not considered recycling at all.

If the recycled material is manufacturing scrap that originated in some other use, perhaps even being bought and sold on the open market, it can have varying degrees of purity. There can also be varying degrees of knowledge about the identity of the components of the recycled material. Therefore, some manufacturing scrap may be suitable for recycling into food packaging, and other material may not.

Commonly, when people refer to the use of recycled plastic, for any type of application, they mean the use of postconsumer recycled plastic - material that has served its intended end use and then has been collected and processed to make some new product. With this type of recycling, the degree of uncertainty about the purity of the recycled material increases dramatically. First, the material is almost certainly a mixture of plastic resins from different manufacturers, all having their own proprietary additive packages. Depending on the type of material being collected for recycling, it may or may not be possible to verify that all the plastics coming into the system were originally of food grade. Even when this can be assured, there is some potential for the occasional inadvertent inclusion of a look-alike container. More significantly, there is no practical way to guarantee that none of the collected containers have been subject to misuse by consumers. Because of the nature of plastics, there is potential for hazardous substances, if stored in the container before it is turned in for recycling, to migrate into the container. If that occurs and the contaminants are not removed by the processing of the recycled material, they may migrate out of a finished container into a food product.

Therefore, the goal in providing for safe use of postconsumer recycled plastics in food packaging is to determine a combination of conditions that provide a reasonable degree of certainty that there will be no adverse public health impacts from such use. As was discussed earlier in this chapter, the approach for approving plastic resins for food packaging is based on determining that any migration of substances from the packaging into the food will be at levels that are within the acceptable daily intake for specific substances for that polymer, type of food, and use conditions, or that the levels are low enough to fall below the threshold of regulation. FDA's position is that it is the manufacturer's responsibility to assure that this is the case. If the manufacturer feels comfortable making this declaration, then they are free to use recycled plastics. Of course, if there is an incident where this turns out not to be the case, then the manufacturer has violated FDA regulations and will be subject to penalty. While some manufacturers have used recycled plastics on this basis, most food packagers have desired some FDA sanction of the prospective use. However, since FDA approves food packaging materials only on a generic basis, there is no provision for FDA approval of a specific use of recycled plastics.

This dilemma was solved by putting into place a procedure by which manufacturers could be issued a formal "letter of nonobjection" from FDA for a given process and/or end use of recycled plastics in food packaging. If a manufacturer seeks a letter of nonobjection, they must provide to FDA evidence that the process and/or application involved will provide the required degree of certainty that contaminants that may be present in the recycled stream will not migrate to contained food products in amounts that are potentially harmful.

This procedure provided a mechanism by which FDA could review proposed food packaging applications for recycled plastics, but a major problem still remained. Migration testing for virgin plastics is based on measured amounts of global migration or on migration of specific substances known to be present in the package. In a recycled plastic stream, as discussed, there is great uncertainty about what might be present in the plastics. It is not feasible to analyze the plastics to determine what potential migrants are present, especially since this can vary not only from day to day, but even from minute to minute. Further, global migration studies are not very meaningful, again due to the variations in the chemical constituents in the recycled stream. No matter how many measurements are made of what does migrate from the containers, these do not guarantee that some day a container won't come along with a high concentration of some hazardous substance that might migrate at unacceptable levels. Further, the number of potential contaminants is exceedingly high.

Thus, a different approach was needed. What FDA chose to do was rely on "doping" the plastic with known amounts of representative contaminant simulants. With these known amounts present, the ability of the recycling processing system to remove the simulants could be evaluated. The potential for migration of these doped contaminants into food could be evaluated using standard extraction procedures. Based on such testing, FDA has approved several processes for recycling specific types of plastic packaging for general or limited food packaging use.

Approved, or more accurately "nonobjected," ways of using recycled plastics in food packaging can be divided into three categories. The first, sometimes known as tertiary or feedstock recycling, uses a chemical process to break the polymer down into low molecular weight compounds, which are purified and then repolymerized. These processes for PET yield materials that can be used interchangeably with virgin PET, and several companies have received letters of nonobjection for these processes.

The second method for using recycled plastics involves intensive physical processing to remove potential contaminants to an acceptable degree. A few such processes for PET and for HDPE have been approved. The PET processes have received unlimited approval, while the HDPE approvals are more restricted. Closely related is approval of normally processed recycled material for "noncritical" food packaging applications. For example, the use of recycled PET for egg cartons does not result in problematic migration of substances to the eggs inside their shells, no matter what components are present in the PET. Similarly, use of recycled PET or PS is acceptable for produce trays and similar applications where there is no significant potential for migration.

The third approach to using recycled plastics in food packaging relies on the "functional barrier" concept of interposing a material between the recycled plastic and the food that will reduce or eliminate migration. One of the first such applications was the use of recycled polystyrene foam in an inner layer in fast food cartons. The food contact surface was virgin PS foam. With the very short residence time of the food in the package (shelf life of less than an hour), even a thin layer of virgin PS can provide substantial protection against migration. FDA has also issued letters of nonobjection for PET and HDPE containers for some applications using the functional barrier approach. In these cases, the FDA decision is based on both testing for migration of contaminant simulants and modeling of migration through functional barriers.

15.4.9 Role of Manufacturers and Users in Determining FDA Compliance

Both the end user and the packaging material manufacturer typically have roles to play in the securing of approval for the use of a material in food packaging. The end user will usually need to ask the packaging material supplier to provide a letter saying that the material complies with the appropriate parts of the CFR for the stated end use conditions. The packager has the responsibility of supplying the end use conditions. Since often certain details of formulation of the material are considered proprietary, the letter from the supplier will typically state that the material complies with specified paragraphs in the CFR. The packaging material supplier may, in turn (and for similar reasons), require letters from the base resin manufacturer that the polymer is in compliance. If new indirect food additives are involved, the packager, polymer or material manufacturer, and/or additive supplier are faced with the burden of submitting a food additive petition to the FDA before the material can be used. This is also the case for drug packaging.

15.5   Cosmetic Packaging Regulations

In contrast to the situation for food and drugs, FDA has no specific regulations for plastic packaging materials or their additives for use in cosmetic packaging. Such packaging regulations as there are for cosmetics deal with issues such as labeling and tamper evidence, which are outside the scope of this book.

15.6   State Laws and Regulations

There are four areas in which regulators in the U.S. have taken action that is significant to plastics packaging materials selection and manufacture primarily or exclusively at the state level, rather than at the federal level. In all of these areas, environmental concerns were the primary driving force behind the regulation. In this section, we will discuss requirements for degradable beverage carriers, the "model toxics" legislation, resin coding on plastic bottles, and recycling rate/recycled content requirements.

15.6.1 Degradable Beverage Carriers

In the 1980s, considerable attention was given to the problem of animals, especially waterfowl, becoming entangled in plastic ring connectors for beverage cans. If littered and reaching a body of water, the connectors would float, and birds, once entangled, often could not free themselves. In response, a number of states passed legislation requiring that these ring connectors, and sometimes other beverage can bundling devices as well, be formulated of a degradable plastic, defined as one that would lose a given percent of its tensile strength after a defined period of exposure to sunlight. The details of the laws differed significantly from state to state. In response, manufacturers of these ring connectors began replacing the low density polyethylene previously used for these packages with a copolymer of ethylene and carbon monoxide, which is a photodegradable plastic. At first, this was done only in states which mandated these materials, but as the number of states with this requirement grew, the photodegradable ring connectors began appearing in other states, as well.

In 1988, the U.S. Congress passed the Plastic Pollution Control Act that, among other provisions, required the Environmental Protection Agency (EPA) to issue regulations requiring the ring connectors to be degradable unless such technology was found to present environmental problems from degradation by-products, or was found not to be feasible. Since the photodegradable connectors were already in widespread use, technological feasibility was not really an issue. Therefore, these ring connectors are now required to be degradable if they are sold anywhere in the U.S.

Because state action largely preceded federal action, and the federal law did not override state laws, the individual state requirements remain law, as well. Consequently, the details of the degradation requirements differ somewhat from state to state, as some state laws are more stringent than the EPA requirements.

15.6.2 "Model Toxics" Legislation

In the 1980s, in response to declining landfill space, a number of states, particularly in the Northeast, began to rely increasingly on waste-to-energy incineration for disposal of municipal solid waste (MSW). Of course, incineration does not destroy all the waste, and the incineration facilities produced ash from the incombustible part of the waste stream, as well as that originating in the pollution control equipment. This ash had to be disposed of by landfill. However, if the ash was tested for leaching of toxic heavy metals, a significant portion of the tests showed heavy metal contents too high to allow for its disposal in ordinary MSW landfills. Instead, it was required to be treated as hazardous waste. This, of course, greatly increased the cost of ash disposal, and therefore the cost of operating the incineration facilities. One reaction was to attempt to use legal arguments that Congress did not intend MSW ash to be treated differently from MSW - after all, the heavy metals were there in the same overall amounts, just in more concentrated form. This argument failed, although the wrangling through the Courts as well as within the government took several years before it was finally resolved.

A more productive approach resulted from the action of the Coalition of Northeastern Governors (CONEG) Source Reduction Council. This was a body composed of representatives of government, industry, and environmental organizations, put together to help come up with ways of reducing the impact of packaging on problems associated with MSW disposal. In the area of heavy metals, the approach they advocated was based on the simple principle that what does not go into the incinerator cannot come out of it in the incinerator ash. Translated to heavy metals, this means keep the heavy metals out of the incinerator to the extent feasible, and the ash will be much more likely to pass the toxic leaching tests, allowing it to be disposed of as ordinary MSW. Many states, for this and other reasons, took action against lead-acid batteries, by far the largest source of lead in MSW. Disposal of such batteries in landfills or incinerators was prohibited in most states, and deposits were initiated to facilitate return of the batteries for recycling. Some states targeted other types of batteries, as well. While packaging was not the largest contributor to hazardous heavy metals in MSW, it was a significant one.

The Source Reduction Council determined that removal of heavy metals from most forms of packaging would not present insurmountable problems to industry, and would be effective in removing these sources of heavy metals not only from incinerator ash, but also from landfill leachate, and thereby would help avoid their general spread in the environment. Rather than follow the path of the ring carrier degradability legislation, where each state wrote its own law, and consequently details of the requirements differed from state to state, the Source Reduction Council produced in 1989 what became known as the "Model Toxics Law" to try to ensure uniformity in the requirements from state to state. It was essentially a "fill-in-the-blanks" law. The introducer of the legislation simply needed to fill in the appropriate state regulatory bodies and time frames, and the bill would be ready to submit to the state legislature.

This approach proved to be successful, and soon spread beyond the Northeast states. Table 15.1 contains a listing of the 18 states that have enacted the Model Toxics Law, as of August, 1999. Because the law as written applies to all packaging sold in the state, and because of the large fraction of the U.S. market residing in the listed states, this has now become a de facto U.S. requirement. Few, if any, additional states have enacted the legislation since 1999, as there are no longer many packages in the market that contain heavy metals. These laws have even affected packaging around the world, both because imported goods are covered, and because it has served as a model for regulations in other countries. However, it remains state law only, not U.S. law.

The legislation is targeted at four heavy metals: lead, mercury, cadmium, and hexavalent chromium. It has two parts. First, it prohibits the intentional introduction of these four heavy metals into packaging during manufacturing or distribution, with extremely limited exemptions. Secondly, it institutes a stepwise reduction of total incidental content of the 4 heavy metals, beginning with a maximum of 600 parts per million (ppm) two years after the legislation is signed into law, decreasing to a maximum of 100 ppm at the four year mark. Exemptions are provided for use of recycled material.

Table 15.1 States That Have Adopted the CONEG Model Toxics Law (as of August, 1999)

Connecticut	Florida	Georgia
Illinois	Iowa	Maine
Maryland	Minnesota	Missouri
New Hampshire	New Jersey	New York
Pennsylvania	Rhode Island	Vermont
Virginia	Washington	Wisconsin

While this legislation was not aimed specifically at plastics packaging, it has had some significant impacts, primarily in the area of pigments for plastics, inks for printing on plastic packages, and to a lesser extent on some stabilizers for non-food plastic packaging resins. There was not much effect on stabilizers for food grade plastics, since FDA regulations generally already did not permit use of these substances in such resins.

15.6.3 Resin Coding on Plastic Bottles

Another consequence of the solid waste concerns of the 1980s and early 1990s was a growth in plastics recycling, as will be discussed further in Chapter 16. Recycling of plastics into high value end-uses frequently depends on sorting of the plastics into resin types. However, there was no mechanism in place to enable consumers or processors to easily identify the type of plastic in an individual container. Therefore, a number of states began to express interest in legislation to require some type of identification coding on packages to identify the plastic resin in containers. The ideas discussed differed significantly among states, and some states were actually considering banning symbols that other states were considering requiring. In light of this potential for complex and conflicting regulations, the Society for the Plastics Industry (SPI) devised a voluntary plastics coding system (see Fig. 15.1), to be implemented throughout the U.S., and was successful in convincing states that desired to mandate identification to adopt this system, so that there would be consistency across the country.

As of August, 1999, 39 states had adopted laws requiring the SPI coding system. The details of the law differ somewhat from state to state. In most cases, the code is required on plastic bottles 16 oz and larger, and on rigid plastic containers 8 oz and larger. In many states, containers larger than 5 gallons are exempted, but in others that is not the case. As is the case for the Model Toxics law, few if any states have adopted such laws since 1999, as the nation-wide and often world-wide nature of the marketplace means most plastic containers are so labeled regardless of where they are sold. This decreases the perceived need for additional states to adopt similar legislation.

The SPI system has been controversial from its inception. The proposal initially called for the code for polyethylene terephthalate to be PET. However, it had to be changed to PETE to avoid trademark infringement.

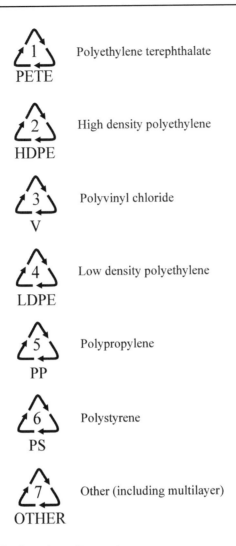

1 PETE	Polyethylene terephthalate
2 HDPE	High density polyethylene
3 V	Polyvinyl chloride
4 LDPE	Low density polyethylene
5 PP	Polypropylene
6 PS	Polystyrene
7 OTHER	Other (including multilayer)

Figure 15.1 SPI plastic resin coding system.

A major criticism levied by environmental interests is that consumers misinterpret the resin identification symbol as a claim of recyclability, and are therefore being mislead. The U.S. Department of Commerce has held that the symbol, when placed on the bottom of a container as recommended by SPI, does not constitute a recyclability claim. However, a number of manufacturers have elected to use the code on bags, lids, and other materials, although no state requires that these types of packages or package components be so labeled. If they are prominently featuring the SPI code, they may be running the risk that doing so could be found to constitute a claim of recyclability and therefore be held to be misleading if, as is likely the case, opportunities for recycling of those package components are not widely available.

Another major criticism is that the symbol does not provide sufficient information, failing to distinguish, for example, between high melt flow injection molding grades of HDPE and low melt flow blow molding grades. Mixing HDPE resins of different grades can result in production of a resin that does not have properties that make it desirable for either blow or injection molding.

In the mid-1990s, a series of talks aimed at modifying the code finally ended with no solution. There have also been, and continue to be, problems with harmonizing the SPI code with coding systems used elsewhere in the world. However, despite these objections the symbols are used not only in the USA but also in many other countries around the world.

15.6.4   Recycling Rate/Recycled Content Requirements.

In three states, Oregon, California, and Wisconsin, laws have been enacted to require plastic containers to meet certain criteria involving recycling rate, recycled content, reuse, or source reduction. A brief review of the legislation is presented next.

15.6.4.1 Oregon

The Oregon Rigid Plastic Container Law of 1994 requires plastic containers sold in the state to meet one of three 25% recycling rate options, contain 25% recycled content, or be reusable. Containers can also be temporarily exempted based on 10% source reduction. Containers for medical packaging are permanently exempted, and food packaging was exempted under a 1995 amendment. Since the law took effect in 1995, the state recycling rate for rigid plastic containers has been above 25%, so all plastic containers sold in the state have automatically been in compliance. After the law was passed, the plastics industry invested a considerable amount of money and effort in the state to facilitate and develop both collection and processing for plastic containers. This assistance was crucial in meeting the 25% goal. However, the recycling rate has been falling in the last several years, and the state has issued an official warning that the rate may soon fall below the crucial 25%, unless steps are taken to increase plastics recycling.

15.6.4.2 California

California's requirements for rigid plastic containers are very similar to those in Oregon. Here, too, medical packaging is exempted, and exemptions for food, beverage, and cosmetics were added by amendment. Both laws took effect at about the same time. Initially the overall rigid plastic container recycling rate was found to be above the 25% rate which brought all such containers automatically into compliance. However, in 1998, the state announced that the 1996 recycling rate was under 25%, and began asking companies to demonstrate compliance with the requirements of the law. In 1999, the state began to take enforcement actions against those who did not respond or were not in compliance. Expansion of the state's bottle deposit legislation to include a variety of non-carbonated beverages, which went into effect in 2000, helped increase the recycling rate to above 25%, removing the requirement for companies to demonstrate compliance in other ways. However, there have been several efforts in the state legislature to increase the requirements, in order to expand plastics recycling in the state.

15.6.4.3 Wisconsin

Wisconsin has a law that requires that all plastic containers consist of at least 10% recycled or remanufactured material. Food, beverages, and drugs are exempt if FDA has not approved the use of the specific recycled or remanufactured content in that plastic container. An exemption for cosmetics containers was added by amendment. While this law remains in effect, it appears that it has not had any significant effect on packaging decisions.

15.7   Potential Future Issues

The EPA is in the process of testing and reviewing a large number of chemical substances for potential toxicity. Some of these chemicals are found in plastics packaging, and regulations will be issued if EPA determines they are needed. In addition to effects directly related to regulation, adverse publicity may result. The past has shown that stigma associated with the presence of toxic compounds in plastics can last long after the amounts of the chemicals have been reduced to safe levels, and can even extend to plastics that never contained the suspect compounds.

A particularly volatile issue is that of hormone-mimicking or antagonizing chemicals. A recent book, articles in the news, and television programs including one on the Public Broadcasting Service (PBS) has raised the public awareness of "endocrine disruptors". Endocrine disruptors are those chemicals that affect hormonal secretions by the endocrine system of the human body. There is considerable scientific uncertainty about the effects of these substances, but public concern sometimes drives policy even without a scientific basis.

One example is phthalate plasticizers used in PVC, which are suspected endocrine disruptors. In late 1998 the U.S. Consumer Product Safety Commission (CPSC) announced that it did not find any evidence that phthalate plasticizers leaching from vinyl teething toys pose any health risk to children, but none-the-less asked manufacturers to voluntarily remove them. Reportedly at least 90% of manufacturers did so, perhaps motivated as much by consumer refusal to purchase the products as by the CPSC request.

In 1999, the concern focused on PVC bags for intravenous fluids, which also use phthalate plasticizers. The Health Care Without Harm Coalition launched a consumer education campaign claiming that the IV bags have similar problems to the PVC toys. This campaign continues at the time of this writing in 2004. One of the largest manufacturers, Baxter International Inc., said in 1999 that there is overwhelming evidence that the di(2-ethylhexyl) phthalate used in its bags is safe, but, in response to pressures from shareholders, has promised to develop non-PVC alternatives for IV bags, and foresaw a slow but steady switch away from PVC throughout its product line. Baxter currently reports that they have switched to non-PVC materials for more than a dozen products, where alternatives have proven superior to PVC and regulatory clearance has been obtained [1].

Similar concerns have been expressed about migration of bisphenol-A, another suspected endocrine disruptor, from polycarbonate. Concerns about migration of styrene monomer from polystyrene have been heard on and off for many years. Now styrene, and especially its dimers and trimers, are also suspected of being endocrine disruptors. Since styrene and its dimers and trimers tend to cause off-flavor in foods, they have for the most part already been reduced to very low levels in plastic packaging for food products.

As we can see, the regulations for plastic packaging are complex and require serious consideration. Also, this is an area where science, legislation, politics, consumer issues and business are inter-related, requiring, in most cases, the inputs of experts from different fields.

References

1. Baxter Worldwide, "Use of Polyvinyl Chloride in Medical Products: PVC and the Environment," undated, www.baxter.com/about_baxter/sustainability/our_ environment/environmental_impacts/sub/pvc.html

Study Questions

1. Locate the FDA regulations for PET as a packaging material. Summarize the regulations in your own words.

2. Why might an adhesive manufacturer not be willing to provide customers with a detailed listing of components? How might this affect the customer's ability to use the adhesive in food packaging?

3. Do you think it is reasonable to regard the package as a part of the drug when FDA approves new drugs? Explain your reasoning.

4. In your experience, how do people interpret the SPI resin coding symbols? Do you think the identification system should be changed? If so, how?

5. Explain the difference between the United States Code, the Code of Federal Regulations, and the Federal Register. What types of information will you find in each?

6. Explain the meaning of the term "threshold of regulation."

7. Why are certain materials permitted in food packaging only if they are not in direct contact with the product?

8. Why are certain materials permitted in packaging some types of food, but not permitted in packaging others?

16 Environmental Issues

16.1 Introduction

For reasons that can be difficult to understand, plastics packaging has become to some a symbol of wastefulness and pollution. A well-funded publicity campaign by the American Plastics Council, focusing on medical advances and other advances made possible by plastics has helped in improving the average consumer's view of plastics. However, criticisms such as plastics being synthetic and therefore "unnatural" and bad are still heard. It seems that since glass and steel, for example, have been around for a lot longer than plastics, they have somehow become more "natural" and thus are perceived as more environmentally friendly. More limited misperceptions, such as the belief that polyethylene creates dioxin when it is burned, also abound. This is not to say that there are not real environmental issues associated with use of plastics packaging. There are, just as there are real environmental issues associated with any type of use of any material. In this chapter, we attempt to summarize those issues, and to distinguish belief from fact. An important point to remember, however, is that in dealing with the public, what people believe to be true can be more important than the scientific facts. Beliefs, as mentioned in Chapter 15, can even drive regulation.

Another point to keep in mind is that public consciousness and concern about environmental issues tends to wax and wane. Superimposed on this cyclic pattern, however, is a general trend towards increasing expectations for cleanliness of air and water, and maintenance of natural resources. The reviving interest in sustainable development, fueled by increasing petrochemical prices and concerns about global climate change, is very much part of this overall trend.

We will first discuss solid waste-related environmental concerns, including recycling, and then turn briefly to other environmental concerns with impacts on plastics packaging.

16.2   Solid Waste Concerns

In the mid 1980s, concern about problems associated with municipal solid waste disposal became a serious issue for the U.S. packaging industry, largely because of legislation, and the threat of legislation, based on these concerns. Plastic packaging was frequently a major target of proposed legislation. To complicate the problems the U.S. packaging and plastics industries had in responding to the threats they perceived in the large amount of legislation being proposed, it was virtually all at the state and local level, rather than the federal level. Most major companies found it necessary to devote significant resources to tracking proposed legislation, testifying to legislative bodies, responding to environmentally-related inquiries from concerned citizens, and similar activities.

The basic problem was that in some areas of the U.S., there was a significant decrease in the availability of landfill space, at a time when approximately 90% of municipal solid waste (MSW) was disposed of by landfilling. The problem was generally worse in major cities than in smaller towns or rural areas, and was particularly acute in the northeastern U.S. One focus of attempts to deal with the problem was to look for ways to minimize MSW generation. Another was to seek alternatives to landfill disposal, including incineration and recycling. A third was to seek new landfill sites.

The EPA, through a contract with the consulting firm Franklin Associates, published a series of reports on the composition of MSW (Fig. 16.1). These showed that packaging, as a category, accounted for approximately one-third of MSW by weight, a fraction that has remained relatively stable over time (Fig. 16.2). The plastics packaging fraction was, of course, much less than this, but both the overall plastics fraction and the plastics packaging fraction have shown steady growth (Fig. 16.3). Further, because of the relatively low density of plastics compared to many other materials in the waste stream, the weight fraction of MSW occupied by plastics and by plastics packaging is considerably smaller than the volume fraction (Fig. 16.4). (EPA has not reported the volume of materials discarded in MSW since 1996, due to the large uncertainty in such measurements and consequently their limited value.) This to some extent legitimized a focus on plastics packaging as a particular target in the efforts to do something about MSW disposal problems. Some of the legislation resulting from these concerns was discussed in Chapter 15.

By the mid-1990s, much of the concern about MSW disposal had dissipated in the U.S. There were fewer landfills available, but those that were available were, on average, much larger, and consequently disposal capacity had returned to a relatively comfortable level, especially considering the significantly decreased overall reliance on landfill disposal (Fig. 16.5). The effort to significantly increase reliance on waste-to-energy incineration had failed for the most part, largely due to public concern about incinerator emissions and objections to siting these facilities, with the high cost of incineration also playing an important role. Recycling, on the other hand, as will be discussed in Sec. 16.3, increased dramatically during this period, and proved to be quite popular with the American public.

In some other countries, problems with inadequate facilities for solid waste disposal were, and remain, much more serious than in the United States, with its generally abundant land resources. In much of Europe, for example, landfill space is genuinely scarce, not just politically scarce. Many countries relied heavily on incineration for MSW disposal. In the late 1980s and early 1990s, concern about emissions from incineration grew considerably. Here, too, recycling was seen as the most viable way of decreasing reliance on disposal facilities, and here, too, packaging was the primary initial target in efforts to reduce disposal of MSW.

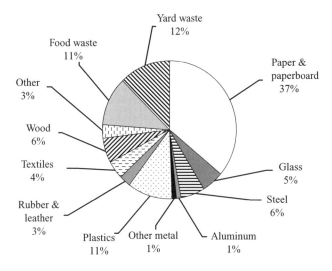

Figure 16.1 U.S. municipal solid waste, by material type, 2001[1].

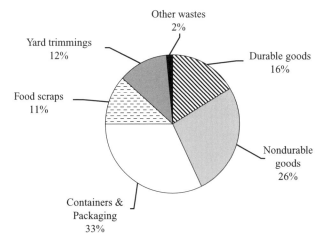

Figure 16.2 U.S. municipal solid waste, by product type, 1996 [1].

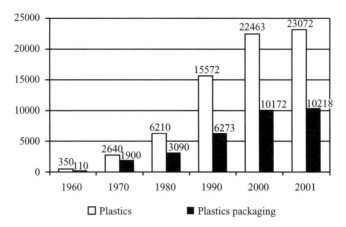

Figure 16.3 Plastics and plastics packaging in U.S. MSW, thousand tonnes [1].

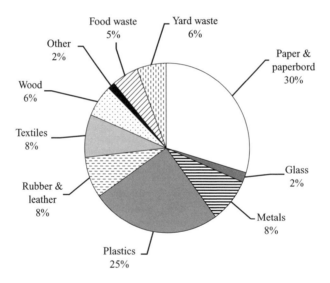

Figure 16.4 Volume of materials in discarded U.S. MSW, 1996 [2].

Germany was the first country to initiate, at least on a wide scale, what has become the preponderant philosophy governing waste disposal in Europe, and is spreading widely across the world: producer responsibility. The idea is simple: the entity that generates the material that will eventually become waste is responsible for it when it reaches the end of its life cycle. In particular, that entity has a responsibility to ensure that at least some minimum fraction is recycled, and to pay the costs of doing so. After all, the product manufacturer has the ability to make decisions that may facilitate recycling or make it more difficult. Thus was born the Duales System Deutschland (DSD), more commonly known as the Green Dot system. The government's role is to issue the mandates, requiring that producers take responsibility for the used packaging associated with their products, and to ensure that it is collected and recycled. It leaves up to industry how it will comply with those mandates. Industry's response has been, by and large, to band together, since collective action is more economical than for each company to individually manage the packaging waste from its own products.

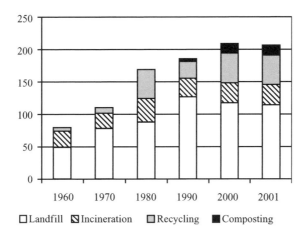

Figure 16.5 U.S. municipal solid waste generation and management, thousand tonnes [1].

While the program met with a great deal of skepticism when it was initiated, and has had some growing pains along the way, it has been successful enough that the whole European Union has mandated adoption of this approach, and is phasing in similar policies for automobiles and some consumer electronics. Charges under such programs can be substantial. In the German program, for example, participants paid about $1.76 per kg for plastic packaging in the late 1990s, which was often more than the cost of the polymer [3].

A similar philosophy has been adopted in a number of Asian countries. Ontario, Canada, has instituted a requirement that packagers or those placing goods into commerce within Ontario share in the costs of recycling through the province's "blue box" municipal recycling programs, requiring registration with the government and paying the appropriate fees. Quebec is in the process of instituting a similar policy.

The EU is also instituting a broader approach of requiring that packaging adhere to what are referred to as "essential requirements" related to source reduction, minimum presence of hazardous substances, and design and production to facilitate recovery, reuse, and recycling. At this time, not all the standards for complying with these essential requirements have been adopted at the EU level.

The general idea of producer responsibility, or manufacturers responsibility, is closely allied to the "polluter pays" principle that is at the heart of much of the hazardous waste regulation in the U.S., including the Superfund program. A valid argument can be made that this philosophy also operates in beverage bottle deposit legislation.

In much of the developing world, open dumps are still a common method of waste disposal, even in major urban areas. The appreciable amount of recycling that occurs is done almost exclusively by scavengers, who make their living by picking through the material in the dumps. In some countries, such as Brazil, establishment of organized recycling programs is proceeding hand in hand with replacement of the dumps by sanitary landfills. A number of pilot programs in Brazil are taking the scavengers and converting them into trained recycling workers, providing appropriate safety equipment, health care, living quarters, and regular pay. Many of these countries appear to be moving towards adopting policies designed to require product manufacturers to bear at least some of the burden for disposal of their packaging and products.

16.3   Source Reduction and Reuse

Plastic packaging plays an important role in two other approaches to MSW reduction. EPA considers source reduction, using less material in the first place, as the optimum approach to waste reduction. In a large number of cases, one of the main reasons for changing from some other packaging material to plastics packaging was the source reduction, and consequent cost reduction, which it provided. Plastic packages are, not always but certainly often, thinner, lighter, and smaller than their alternatives. This can reduce not only the amount of primary packaging (that which contains the product), but also the size and amount of distribution packaging. Further, the plastics packaging industry has long engaged in on-going efforts to do the same job with less material. Plastic soft drink bottles, milk bottles, pallet stretch wrap, and merchandise sacks are all, on average, thinner and lighter than they used to be.

Closely related is the use of reusable packaging rather than single-use throw-aways. While this approach has had very limited success for retail packages, use of reusable distribution packages has grown markedly. Two of the pioneering industries were office furniture and automobiles. Plastics, with their durability, light weight, and ease of cleaning, have played significant roles in adoption of reusable packaging systems for products ranging from fruit to auto parts.

16.4   Recycling of Plastic Packaging

In the U.S., recycling of plastic packaging began, on a limited scale, in the 1970s with recycling of PET soft drink bottles and HDPE milk bottles. Milk bottles were collected almost exclusively through drop-off programs, where people would bring clean empty bottles to some central location where they would be collected. Because only crude processing methods were available, labels on the milk bottles were a significant contaminant. Many programs asked residents to remove the labels, but in general met with extremely limited success. In Grand Rapids, Michigan, in one of the early programs, workers at the processing facility cut the labeled section of the bottle out with a utility knife and discarded it before sending the rest of the bottle through the grinder, in order to meet the purity requirements of the user. The recovery rate in such programs (amount of material collected for recycling compared to the amount available) was extremely low.

Recycling of PET soft drink bottles was considerably more successful. The existence of bottle deposit programs in nine states (ten if California's refund value system is included) provided a pool of collected material of consistent quality. Therefore, the early recycling efforts focused primarily on the processing and end use parts of the cycle. PET recycling also had the advantage of PET having superior properties and higher value than HDPE. By the mid 1980s, the PET recycling rate hovered around 20% in the U.S., almost exclusively due to recycling of deposit containers. There was some additional recycling of PET in non-deposit states, much of it through the Beverage Industry Recycling Program (BIRP), which was designed primarily to combat the passage of deposit legislation in additional states. Collection rates in these programs, even though some paid consumers for the containers they delivered, was comparable to those in the drop-off HDPE milk bottle programs. Deposit programs, on the other hand, generally had redemption and recycling rates approaching 90%, or sometimes even higher.

The definition of recycling rates requires some discussion. Until the late 1990s, the American Plastics Council calculated recycling rates as the amount of recovered material delivered to new markets divided by the amount of material available for collection, converted to a percentage. In the late 1990s, the APC changed the method of calculation to use the amount of material delivered to the processing center, rather than the amount coming out of the door for reuse, as the numerator. Since there are losses associated with recycling, including unwanted materials which are mistakenly included, the effect of this decision was to inflate recycling rates. APC argued that this action simply brought the plastics industry's method of reporting on recycling rates in accord with the methodology used in other industries. It certainly is true that this is the method of reporting recycling rates that is used in the paper industry. The glass industry, in addition to using quantities based on the amount of material delivered for processing, uses a formula that counts refillable containers as if they were recycled several times. In the metal industries, however, the calculation of recycling rates is based on the usable amount, rather than the total delivered amount. The change in methodologies sparked heated criticism from some environmental groups. It is likely that the fact that the change in methodology occurred at the same time as the first decrease in plastic recycling rates after a period of rapid growth, whether coincidental or not, increased the criticism. Therefore it needs to be recognized that recycling rates reported by different entities may differ in both the data that is used to determine the rate, and in the formula used for determining the rate from the data. Recycling rates coming from the U.S. EPA Office of Solid Waste are usually based on industry reports about the amount of material recycled, so the differences in methodology affect that data, as well.

16.4.1   Collection of Packaging Materials for Recycling

With the new interest in recycling spawned by the "solid waste crisis," a key issue was how to collect the material to be recycled. One approach is to take the whole garbage stream and separate out the recyclables. Since no special consumer participation is required, recovery rates are limited primarily by the efficiency of the separation processes. The U.S. Bureau of Mines supported considerable research in this area in the 1970s, most of it using technology derived from mineral processing. The end result for plastics was that processing costs were high, and the quality of the recovered material was low, so this approach was not viable. One key issue was the inability to accurately and efficiently separate various types of plastic resins from each other. This left, then, the source separation approach, in which consumers are asked to keep their recyclables separate from the rest of the waste stream. Here, obviously, the willingness of consumers to engage in the desired behavior is crucial. A number of studies examined this behavior, looking at the effects of rewards and sanctions as well as the type of preparation of the materials that the consumers were asked to do, and how and where the materials were to be delivered for recycling. It comes as no great surprise that there is a strong relationship between participation in recycling and the two key values of convenience and motivation. Simply put, individuals with very high motivation will participate in recycling even when it is inconvenient. For example, they will clean the recyclables, separate them all by material type, store them for a month or more, and then deliver them to a drop-off site located across town that is open only for a few hours on a single afternoon each month. At the other extreme, some people will be unwilling to participate in recycling no matter what is done to try to motivate them and to make recycling convenient. Most people, of course, fall somewhere in between, and will be willing to participate in recycling if the combination of motivation and convenience reaches their individual threshold.

   Different communities across the U.S. have structured recycling programs in a number of different ways. Through these experiences, some key factors have emerged. First, for single family neighborhoods, collection of recyclables at curbside, the same way garbage is collected, and on the same day as garbage collection, achieves the highest participation rates. For multi-family residences, collection sites adjacent to garbage collection locations or in some centrally used facility such as a laundry room can be effective.

Ongoing consumer education is a requirement for success. People must know how to participate, and must be reminded that their participation is important and valued. Even with a long-established program, there will be new residents who need information and people who have slacked off who need a reminder. Education of schoolchildren can be very effective not only in building the children's recycling habits, but in changing their families' behavior.

Commingled collection, in which all or most recyclable materials can be bundled together, with minimal processing expectations, achieves the highest participation. Virtually all programs require that the materials be clean, since not only is this a key to quality, but food residues can pose serious health concerns. If people are also asked to remove labels, flatten cans, etc., many will choose to bypass the effort by placing the containers in the garbage instead of the recycling bin. Providing each household with a recycling container (a blue bin is the most popular choice) can be very effective in increasing participation.

In general, mandatory recycling programs, where consumers are required by ordinance to separate target recyclables, have somewhat higher participation rates than purely voluntary programs. However, a word of caution is needed. Well-run voluntary programs can and do achieve much higher participation rates than poorly-run mandatory programs. A small number of people will react negatively to mandatory programs and totally refuse to participate; others will not totally refuse, but may try to get away with doing as little as possible. This is likely one of the reasons why a mandatory program with no enforcement activity may have a lower participation rate than an otherwise equal voluntary program.

In determining the final structure of a recycling program it is necessary, of course, to balance costs with benefits. The program attributes that maximize convenience usually increase costs, and education and motivation activities also have associated costs. A key factor is whether the amount of additional materials collected justifies the investment required.

In places where the mode of garbage collection is different from traditional curbside pickup, the most effective recycling program design will also differ. Cultural factors such as respect for authority and willingness to abide by community expectations are also important.

16.4.2   Recycling Rates for Plastics Packaging

Most curbside recycling programs for plastics packaging focus on bottles. Dropoff programs for plastic merchandise sacks and bags were relatively common in the early-to-mid 1990s, but have decreased significantly in availability. Recycling rates for merchandise sacks and bags have never been very high. Plastic stretch wrap from pallets is recycled relatively frequently. Programs for recycling polystyrene foam cushioning materials have had a moderate amount of success.

The American Plastics Council (APC) reports that 729 thousand tonnes (1.63 billion pounds) of plastic bottles were recycled in the U.S. during 2002, a 0.8% increase in tonnage over the previous year. However the recycling rate for plastic bottles fell once again, declining to 21.0% of bottles produced from the 2001 rate of 21.6%, continuing the pattern of decline that has been going on for a number of years [4].

The rigid plastic container recycling tonnage in 2001, according to EPA, was 727 thousand tonnes, for a rate of 21.8%. The flexible packaging recycling rate in 2001 was only 4.3% [1]. More recent statistics for rigid plastic containers and flexible packaging are not currently available, but the recycling rates have likely declined. The overall recycling rate for plastics in municipal solid waste was 5.5% in 2001 [1]. Figures 16.6 and 16.7 show the plastics in MSW and the plastics recycled, by resin type, for the U.S. in 2001.

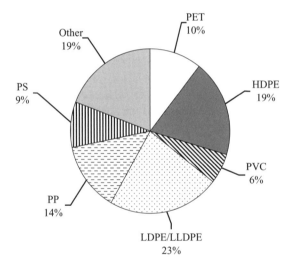

Figure 16.6 Plastics in U.S. MSW, 2001 [1].

The Association of Plastics Manufacturers in Europe (APME) reported a 14.8% recycling rate in Western Europe for mechanical recycling of plastics in municipal solid waste in 2003, up from 13.6% in 2002 The tonnage recycled increased by 24 percent between 2001 and 2003, reaching 3.13 million tonnes. An additional 350 thousand tonnes of plastic were recovered through feedstock recycling. The recycling rate for postconsumer packaging waste, through the combination of mechanical and feedstock recycling, was 23.8 percent in 2002 [5]. Rates specifically for packaging for 2003.were not reported. APME does not report recycling details by type of polymer. It should be noted that plastics recycling rates differ significantly from country to country within Europe.

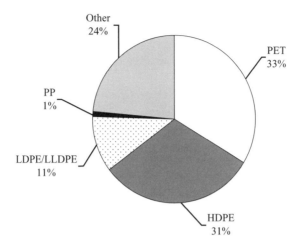

Figure 16.7 Plastics recycled in the U.S., 2001 [1].

16.4.3   Processing of Collected Plastics

Once the plastic packaging is collected, it must be processed into an appropriate form for an end-use application. While the details of the processing can vary considerably with the feedstock and the application, the major process steps typically include size reduction, cleaning and sorting, often followed by extrusion and pelletizing. Baling, which is usually done to increase the bulk density of the material if it must be transported, can also be considered a processing activity.

16.4.3.1 Size Reduction

Size reduction, as the name implies, involves reducing the size of the plastic objects that have been collected. This permits the materials, which are most often containers but may be film, to be fed into processing equipment, and also facilitates cleaning. Shredders or granulators are used, sometimes in a single step, and sometimes in two stages, to reduce the materials to the desired size.

16.4.3.2 Cleaning

Cleaning, as the term is used here, refers to removal of residual product, dirt, paper labels, adhesive residues, and other undesired contaminants from the collected material. The process actually starts with the consumer, who is typically asked to remove food residues and closures from containers and shopping receipts from plastic bags, for example. For containers, an air separation process typically follows the size reduction step, to remove label fragments and other light materials. This is usually followed by a hot detergent wash, and then rinsing and screening to remove remaining paper fibers, softened adhesives, dirt particles, and other contaminants. For HDPE and other polyolefins, a water-based sink-float separation is often employed to remove any materials with a density greater than 1 g/cm^3.

16.4.3.3 Sorting

Sorting is the term used to refer to separation of plastic packaging or packaging components by resin type. For example, when PET soft drink bottles were first recycled, all of them had HDPE base cups. Part of the recycling process involved separating the PET from the HDPE, to permit recycling of both. There is not always a firm distinction between cleaning and sorting. Is removal of aluminum caps from PET bottle streams cleaning or sorting? Fortunately, there is no need to make such fine distinctions in order to discuss some common features of sorting systems.

Sorting can occur at various points in the recycling system. In a system utilizing commingled collection of recyclables, the first step is typically sorting into the various material types. Thus in this step the plastic would be separated from the non-plastic materials. The plastics stream might then be further processed in the same facility, or might be baled and sent to a dedicated plastics processing facility.

In a system which collects a variety of types of plastics, the next sorting goal is to separate the various types of plastic resins from each other, so that each can be used in higher value applications than would otherwise be possible. The required degree of separation depends on the intended end-use. If the end market is low-end plastic lumber, for example, which can utilize a mixed, or commingled, feedstock, there may be no need for sorting. Alternatively, select high-value components may be separated out, with the remainder used in commingled form. If the system is designed to accept only one type of plastic, the goal is to remove undesired materials that have been incorrectly included. These removed items are then disposed of in some fashion.

Many plastics recycling experts have found it useful to classify sorting into three categories: macrosorting, microsorting, and molecular sorting. Macrosorting involves sorting of whole objects, and therefore takes place prior to any size reduction steps, and usually also prior to cleaning. A number of systems have been developed for quickly sorting whole plastic containers, including those that have previously been baled. Some of these systems are capable of sorting by color as well as by resin type. Typically, a combination of sensors is used, relying on reflection or transmission of IR, visible light, x-ray, or other electromagnetic radiation. The complexity of the system, of course, depends on the complexity of the sorting to be accomplished. Despite the availability of such systems, but strongly linked to their relatively high costs, much macrosorting is still done by hand. In a typical system, the materials are deposited on a conveyor belt and pass by a team of sorters, who hand-pick the materials; the sorters either select a certain type of material, or remove materials that are not the desired type. Labor costs in such systems tend to be high, despite the low wages that are usually paid. In some cases, facilities rely on special categories of labor, such as prisoners or the developmentally disabled, to reduce these costs.

Microsorting refers to sorting by resin type after the size reduction step (and generally after cleaning, as well). In these systems, plastic chips are separated by resin type, and sometimes also by color. While there are microsorting systems that are in widespread use, this technology is, on the whole, less developed than that for macrosorting. Microsorting was (and remains) an integral part of most PET bottle recycling systems. As mentioned earlier, it was originally used, in the form of a density-based separation, to separate HDPE and PET. While HDPE base cups have been phased out and PP caps have replaced aluminum, it is still used to separate PET and PP. Originally, large sink/float tanks were employed. In most facilities, these have now been replaced by much more compact hydrocyclones. In both cases, water is the separation medium, and resins with a density lower than water are separated from resins with a density higher than water.

It is more difficult to separate resins from each other if both are either heavier or lighter than water. In some cases it is possible, at least in theory, to do such separations using a liquid or a solution of the appropriate density, but the costs and sometimes the environmental hazards make this unattractive. Use of supercritical carbon dioxide is one of the most promising approaches. Its density can be manipulated readily by adjusting the pressure, it is nontoxic, and it is easily removed from the separated plastic resins.

There has been quite a lot of research on microsorting systems that rely on differences in fracture behavior, melting point, and other attributes. Cryogenic grinding and sieving have been used for PET/PVC separation. Froth flotation has also been used on a near-commercial scale. Ongoing research is also directed at applying the successful radiation-based processes designed for macrosorting to microsorting. Accuracy, cost, and speed remain the major problems in expansion of microsorting technology.

Neither macrosorting nor microsorting will be effective for resin separation if the package itself is a multi-resin material such as a coextrusion or a laminate, unless the material can be caused to delaminate. (PET/EVOH ketchup bottles are an example of a package where delamination during normal recycling processing was an important design consideration.) If the individual plastic chip contains two or more types of plastic, obviously neither macro- nor microsorting can separate the materials. For these situations, molecular sorting offers the only option if sorting is required. In molecular sorting, the plastic object is broken down into individual molecules, typically by dissolving it in an organic solvent. Selective dissolution and reprecipitation, using a variety of solvents and/or temperatures, can then be employed to yield relatively pure resin streams. In addition to high costs and environmental concerns, these systems can also result in problems with solvent retention in the processed material. No molecular sorting systems have yet been successfully commercialized.

16.4.3.4 Extrusion and Pelletizing

Following cleaning and sorting, the plastic resin may or may not be processed through an extruder and pelletized. The major advantages of pelletizing for recycled resins are that use of a filter pack in the pelletizing extruder can remove additional residual impurities, and the pelletized material is easier for end-users to handle and to mix with virgin resin, if desired. It is also possible to add additional stabilizers or other additives at this stage. The primary disadvantages of pelletizing are the added cost and the additional heat history of the recycled material, which will produce some additional degradation. This factor may or may not be significant, depending on the characteristics of the polymer and its intended use.

16.4.4   Feedstock Recycling

A relatively small amount of plastic packaging is recycled by a very different route than the physical recycling process described above. For some plastics, it is possible to use chemical treatment to break apart the base polymer structure. The treatment essentially reverses the initial polymerization process, producing a mix containing the building blocks of the polymer plus some impurities. After purification, the monomers (or oligomers in some cases) can be repolymerized, producing a polymer that is essentially identical to virgin resin. These processes are known as chemical, feedstock, or tertiary recycling. In general, they are more suited to condensation polymers than to addition polymers.

   PET is the packaging plastic that is most often recycled by feedstock recycling. Glycolysis and methanolysis processes have been developed to produce food-grade resin from recycled PET. However, at the present time resin produced in this manner is more expensive than virgin PET. This high cost significantly limits its use. Soft drink bottles with 25% repolymerized PET were available in parts of the U.S. for a short time during the mid-1990s, but were discontinued. Unless cost structures change considerably, these processes appear to be economically viable only if markets for recycled materials are mandated in some way.

   More recently, chemical recycling processes have been developed for polyurethane and for nylon 6. These appear to be somewhat more economically viable than the PET processes. However, neither have any current applications in packaging materials. The polyurethane process is targeted at automobile components; relatively little polyurethane is used in packaging. The target for the nylon 6 process is carpet. Several companies have engaged in relatively large-scale development of these processes, with mixed results; there is relatively little use of nylon in packaging.

   The only current type of feedstock recycling that is applicable to addition polymers is processes based on heat-induced degradation, generally pyrolysis. Pyrolysis involves heating in the absence of oxygen to temperatures where the molecular structure of the material breaks down. It can be applied to pure or to mixed feedstocks, including mixtures of plastics and other organic compounds. In all cases with a mixed feedstock, and in many cases even with a pure polymer, the resultant product contains a wide variety of compounds, some liquid and some gas. Separating out pure streams for use as chemical raw materials is often not economically viable. The most common use for these materials is as fuel, after separation only into gas and liquid streams.

16.5   PET Recycling

In the U.S., PET soft drink bottles were the first postconsumer plastic containers to be recycled on a large scale. As mentioned earlier, they benefitted from the existence of bottle deposit legislation in several states. In recent years, use of PET in soft drink bottles and also in a number of other packaging applications has grown at a faster rate than recycling of these containers. As a result, the U.S. PET bottle recycling rate peaked in about 1994, and by 1998 PET bottles were overtaken by HDPE bottles as the most-recycled plastic in the U.S. Figure 16.8 shows amounts and tonnage of PET bottles collected for recycling in the U.S. between 1995 and 2002.

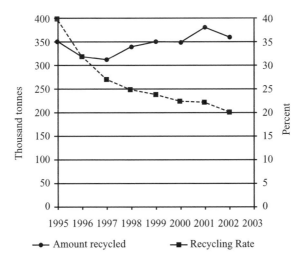

Figure 16.8 PET bottle recycling in the United States [6].

It should be noted that much, but not all, of the decline in recycling rates can be attributed to the very rapid expansion of PET bottles into new markets, most of which are not covered by deposit legislation, and the failure of collection of containers for recycling to keep pace with this growth. Even in non-deposit states, it is relatively simple to participate in PET soft drink bottle recycling. All the container requires is a quick rinse, and it can then be put in the recycling bin. In contrast, food containers such as peanut butter or salad dressing require much more intensive cleaning. Some consumers who will be willing to make the effort to recycle soft drink bottles will not be willing to clean out their peanut butter jars, so these containers will go into the garbage instead of the recycling bin. Therefore, as PET penetrates more into custom bottle markets, if all else remains constant, recycling rates can be expected to decline. The growth rate for custom PET bottles has been very high in the last several years. In fact, in 2002 custom bottles accounted for nearly 54% of all PET bottles sold in the United States [4].

Within the beverage container market, most of the growth has been in bottles for beverages such as water, juices, and sports drinks. In most states, these bottles are not covered by deposits. Even within the soft drink market, there have been changes. PET bottles have penetrated deeply into the single-serving container market. These containers, by their nature, are much more likely than the standard 2-L containers to be consumed away from home. Beverage containers of all types are more likely to be recycled if they are consumed at home than they are in the workplace, recreation site, or other locations. Most experts believe the increased use of single-serve PET soft drink containers has contributed significantly to the decline in the PET recycling rate. NAPCOR, the National Association for PET Container Resources is sponsoring PET collection containers in locations such as stadiums, amusement parks, and convenience stores, in an effort to capture more of these single-serving bottles. An additional factor that has been put forward as a factor is the declining value, in real terms, of the deposit itself, because of inflation. Five cents (the typical amount of a deposit) is a smaller incentive now than it was twenty or more years ago. The American Plastics Council also feels that the export market for recycled PET (RPET) may be significantly underestimated, artificially decreasing reported plastics recycling rates [4].

Historically, the largest end use market for recycled PET has been fiber applications of various types, ranging from fiberfill for ski jackets and sleeping bags to face yarn for polyester carpet. Other significant markets include sheet and film, strapping, non-food containers, food and beverage containers, and engineered resins and molding compounds, as shown in Figure 16.9. Both chemically recycled, or regenerated, PET and some physically reprocessed PET have been cleared by FDA for food contact applications. Recycled PET is also used in a middle layer in some multilayer PET food packages.

16.6   HDPE Recycling

HDPE bottle recycling in the U.S. exceeds PET bottle recycling both in terms of total amount. In 2002, 364 thousand tonnes (800 million pounds) of HDPE bottles were recycled, up from 341 thousand tonnes (750 million lbs) in 2001. The recycling rate increased to 24.2% in 2002 from 23.2 percent in 2001. Growth in tonnage and in recycling rate has been slow in the past several years, but HDPE recycling has not experienced the declines that have hit PET. The recycling rate is generally higher for unpigmented containers such as milk and water bottles than for pigmented containers. However, that gap has narrowed. In 2002, natural HDPE bottles had a recycling rate of 26.1%, down from 27.6% in 2001, while the rate for pigmented HDPE bottles increased from 19.3% in 2001 to 22.4% in 2002 [4]. Unpigmented containers sell for a price premium because they are suited for a wider variety of end uses. The value of pigmented bottles is increased by color separation.

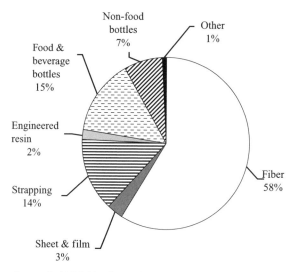

Figure 16.9 Uses of recycled PET in the U.S., 2002 [6].

In addition to bottles, there is some recycling of other HDPE containers and of merchandise sacks. The recycling rate in the HDPE bags, sacks & wraps category was 3.9% for a total of 27,000 tonnes in 2001. The overall recycling rate for HDPE packaging was 8.7% in 2001 [1].

The largest use for recycled HDPE bottles is in the manufacture of new non-food HDPE containers. For a number of years, laundry detergent bottles and containers for similar products have had a three-layer structure with a core of recycled HDPE plus bottle regrind. Many motor oil bottles are made with a blend of recycled and virgin HDPE. Pallets & lumber, drainage pipe, and film are also significant uses, as can be seen in Figure 16.10. Drainage pipe was the first large-scale use for recycled milk bottles. Pipe manufacturers now use considerable amounts of pigmented HDPE bottles. Since most of the pipes are black, mixed colors do not create any problems. However, this stream contains relatively high levels of polypropylene contamination, compared with the unpigmented HDPE stream. The PP is mostly from caps that were left on the bottles, and it limits the proportions of this feedstock that can be used by at least some manufacturers. The film market consists primarily of garbage bags and merchandise sacks.

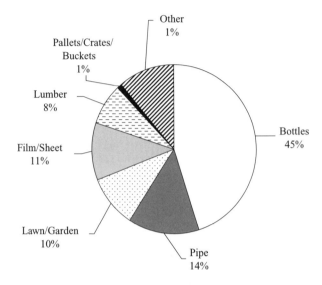

Figure 16.10 Uses for recycled HDPE bottles in the U.S., 2002 [4].

Caps also present some color contamination problems in the unpigmented HDPE stream. While consumers are requested to remove the caps before turning the bottles in for recycling, they do not always comply. The typical processing procedure for recycled HDPE does not permit separation of the PP and HDPE, since both are lighter than water. The presence of a relatively small amount of cap material can darken the reprocessed HDPE enough to lower its value by 2 to 7 cents per kilogram.

16.7 LDPE Recycling

Recycling of LDPE packaging in the U.S. is primarily directed at stretch wrap, collected from businesses, and at merchandise sacks, collected from consumers through drop-off sites located at stores. There is very little collection of plastic bags at curbside in the U.S., though there are fairly widespread curbside bag collection programs in Ontario, Canada, and to a lesser extent in Quebec. These programs, sponsored by the Plastic Film Manufacturers Association of Canada and the Environment and Plastics Institute of Canada (EPIC) have been operating for several years. EPIC published "The Best

Practices Guide for the Collection and Handling of Polyethylene Plastic Bags and Film in Municipal Curbside Recycling Programs" in 1998. This document provides valuable step-by-step guides for implementing what they have found to be the "best practices" for such systems. These programs typically collect HDPE bags as well as the more common LDPE bags. Often there is no practical way for the consumer to tell the two materials apart. Although HDPE and LDPE are not totally compatible resins, despite being produced from the same monomer, the resulting blends maintain useful properties.

There is very little recycling of LDPE containers, largely because there is little use of LDPE containers in packaging, insufficient to justify dedicated recycling programs. Some LDPE containers are collected in the relatively few curbside programs that accept all types of plastic containers, but the quantity is negligible. EPA reports that the overall U.S. recycling rate for LDPE packaging was 2.6% in 2001, for a total of 136,000 tonnes [1]. The American Plastics Council reports a 0.6% recycling rate for LDPE/LLDPE bottles in 2002, up from 0.5% in 2001 [4].

The major market for recycled LDPE is the manufacture of trash bags. Some recycled LDPE is also used in manufacture of plastic lumber, merchandise bags, bubble wrap, plastic lumber, housewares, and other products.

16.8 Recycling of PS, PP, and PVC

The remaining major packaging plastics are polystyrene, polypropylene, and polyvinyl chloride. EPA reports that the amount of PVC packaging recycled in the U.S. is negligible [1]. The American Plastics Council reports a recycling rate for PVC bottles of 0.4% in 2001, up from 0.3% in 2001 [4]. PVC recycling is much stronger in Europe, where use of PVC in packaging, especially in bottles, is also considerably higher, resulting in a larger stream of potential recyclables. In France, for example, the PVC mineral water bottle recycling rate is reported to be about 50%.

EPA reports the U.S. recycling rate for PP packaging was 0.3% in 2001. Nearly all of this material originated in the "other plastics packaging" category, which includes all packaging other than containers and films [1]. Much of this material is likely caps and fitments collected along with HDPE bottles. The American Plastics Council reported a recycling rate of 3.8% for PP bottles in 2001, and a 3.9% rate in 2002 [4].

EPA also reported a negligible recycling rate for PS packaging in 2001. The American Plastics Council reported a rate of 1.1% for PS bottles in 2001, but a negligible rate in 2002 [1,4]. Food service related PS packaging was the target of a major recycling effort in 1989 and the early 1990s, when eight PS resin suppliers joined together to form the National Polystyrene Recycling Company (NPRC), and pledged to achieve a 25% recycling rate by 1995. However, the effort was never very successful and by 1997 the NPRC had shut down all but two of its facilities and abandoned the goal. The processing costs for the heavily contaminated food service materials generally exceeded the value of the material. Recycling of PS cushioning materials has been much more successful, and accounts for nearly all PS packaging currently being recycled. The Alliance of Foam Packaging Recyclers (AFPR) reports a recycling rate for EPS transport packaging of 10%. The majority of the recycling is done by molders of EPS packaging. AFPRs website, www.epspackaging.org, contains a map showing recycling locations. There is also both reuse and recycling of loose fill EPS. The Plastic Loose Fill Council (www.loosefillpackaging.com) maintains a toll-free telephone number and a network of collection points to enable consumers to drop off their unwanted polystyrene loosefill. In addition to packaging, recycled PS packaging is used in housewares, cameras, video cassette casings, insulation, and other products.

Japan reported a 34.9% recycling rate for expanded polystyrene in 2000, but this includes use as fuel [7]. The UK reported a recycling rate for expanded polystyrene of nearly 15% in 2001, up from 10% in 2000 [8].

16.9   Recycling of Commingled Plastics

As mentioned earlier, for some applications it is possible to use recycled plastics that contain a mixture of different types of resins. Since different plastic resins are generally not compatible with each other, such mixtures tend to contain small regions of individual types of polymers within a matrix of another polymer. Adhesion between these regions can be poor, resulting in a substantial decrease in performance. Two general approaches have been taken to producing usable materials from such mixtures.

In one approach, items with relatively large cross-sectional dimensions are chosen, and the properties depend primarily on the preponderance of polyolefins typical of commingled recycled plastics. The other plastics, which generally melt at higher temperatures than polyolefins, are carried along as solid inclusions, often along with other contaminants such as paper labels, and sometimes even product residues. The products manufactured are generally in the wood-substitute or concrete-substitute categories and include such items as plastic lumber, picnic tables and benches, parking stops, and machine bases. The performance of the plastic mixture is often relatively poor, but the demands are low. In outdoor applications, these products tend to have much longer useful lives than treated lumber, since they are not susceptible to rot and other decay mechanisms. They are also easier to clean than wood. These assets come at a price, as is usually the case. The plastic items frequently sell at 3 times or so the cost of the wood items they replace. From a life cycle costing viewpoint, the money is well-spent, but that is not always the view that is taken. The attractiveness of these materials has increased over the last several years as concern about the toxic chemicals contained in treated lumber has increased, and they are no longer considered acceptable for some uses, for example in playground equipment.

The second approach to the use of incompatible mixtures of recycled plastics is to compatibilize them in some way, and thereby improve their performance. One method is to use additives, similar to adhesives, which can help join the different resins into a more uniform material. Investigators at Rutgers reportedly used this approach to make bottles from commingled plastics, in a process that has not been commercialized. Several companies are marketing or developing compatibilizing additives, often targeted to particular mixtures of plastics. Investigators at Northwestern University used a process referred to as solid-state shear extrusion pulverization to convert mixed plastics, along with scrap rubber, into a uniform fine powder. This powder was reported to be inherently compatible and usable in a variety of products [9].

16.10 Biodegradable Plastics

When the "solid waste crisis" first hit in the 1980s, one result was a surge in interest in biodegradable plastics, that is, plastics with a chemical structure that will totally break down within a reasonable period of time (weeks or months) when exposed to soil microorganisms and water, and sometimes requiring oxygen as well. After a number of plastics were marketed as "biodegradable," two things became apparent. First, some of the plastic materials being touted as biodegradable really were not, and those that were truly biodegradable were very expensive, or did not have desired performance characteristics. Secondly, in the environment of a modern sanitary landfill, with its emphasis on keeping conditions dry to minimize the potential for groundwater pollution, degradation rates are so slow that there is little, if anything, to be gained from switching from nondegradable to degradable plastics. The result was that biodegradable plastics largely fell out of favor.

In the years since then, however, there have been some significant changes. New types of biodegradable plastics have been developed, performance has improved, and costs have decreased. This continues to be an active research area, and more progress can be expected. Perhaps even more significantly, there has been considerable growth in interest in and the practice of composting. In contrast to sanitary landfilling, composting facilities are specifically designed to enhance the growth and activity of microorganisms. Use of biodegradable plastics packaging can be a significant asset if the ultimate disposal of the waste stream will be by composting.

Composting is much more prevalent in Europe than in the U.S. In several European countries, Germany and France among them, separate collection and composting of a "wet organics" fraction of the waste stream is relatively common. Biodegradable plastics packaging materials can be included in this fraction, and thereby avoid being subject to the high fees associated with other plastics in the "Green Dot" and similar systems. This financial advantage makes them more cost-competitive with nondegradable plastics. Further, composting is likely to grow significantly in the European Union with the expected adoption of policies limiting the disposal of biodegradable municipal solid waste in landfills.

Biodegradable plastics can be categorized into several major families, which will be discussed next.

16.10.1 PHAs

One of the earliest truly biodegradable non-cellulose-based plastics was polyhydroxybutyrate/valerate (PHBV), a bacterially grown polyester with properties similar to polypropylene. PHBV was first produced commercially by Imperial Chemical Industries (ICI) in Britain, sold under the name Biopol. Since then, other bacterially grown polyesters and copolyesters have also been produced. High costs have been an ongoing problem. ICI spun off the Biopol unit to Zeneca, which then sold it to Monsanto. In late 1998, Monsanto decided to discontinue Biopol production and research.

PHBV is one member of a broader family of polyhydroxyalkanoates (PHAs), produced by microbes from sugars or other biobased materials. Metabolix, of Cambridge, Mass., is engaged in attempting to commercialize a variety of PHAs for packaging and other applications. In March of this year, Metabolix was awarded a significant grant from the U.S. Navy to work on developing PHA food packaging films for the U.S. military. The company is also working on developing the ability to produce PHAs directly in plants. Depending on the particular structure, PHAs can be nearly amorphous or can be up to about 70% crystalline, and can range from soft and elastic to stiff and brittle. Metabolix PHAs are reported to be better water vapor barriers than most biodegradable plastics. WVTRs for 50 micron PHA film at 23°C and 90% RH range from 20-150 g/m^2 d [10].

16.10.2 Polylactides

Biodegradable plastics based on lactic acid have been available on a small scale for a number of years. They have been used in applications such as medical implants, but their high price was a deterrent to widespread use in lower value applications such as packaging. However, in recent years new technologies for production of lactide monomers have greatly lowered costs, making the polymers much more competitive. Generally, the lactic acid is obtained from corn or other biobased materials by a fermentation process, and then chemical synthesis is used to produce the polymer from the lactic acid or lactide monomers.

Cargill was one of the first companies to develop biodegradable polylactides (PLA) for applications such as packaging. Cargill and Dow Chemical partnered to form Cargill

Dow LLC, which now operates a full-scale production facility in Blair, Nebraska, with a capacity of 140,000 metric tons (300 million lbs) per year. Lactides are produced from cornstarch by fermentation, and then converted to PLA through a ring-opening polymerization process. Cargill Dow's NatureWorks PLA is produced in a variety of grades, including film, sheet for thermoforming, and injection molding resins. It has applications outside packaging, such as fiber for clothing, as well as packaging applications such as food containers, drink cups, blister packs, etc., and related applications such as food serviceware. The transparency of the materials is generally excellent. Mechanical and barrier properties depend on the grade [11].

Other producers of PLA resins and related copolymers have included Mitsui Toatsu, Shimatsu, CornCard International, Chronopol, Dainippon, and Neste Oy, but Cargill Dow is by far the largest producer.

16.10.3 Other Polyesters

Polycaprolactone is a synthetic polyester that has been available for a number of years. Union Carbide is the major supplier, and sells the resins under the Tone brand name. DuPont sells a family of synthetic biodegradable polyesters under the Biomax name. The degradation mechanism is a combination of microbial action and hydrolysis. Costs are reported to be only marginally higher than PET. Eastman has also developed a synthetic biodegradable and compostable polyester, with properties comparable to low density polyethylene. Other companies manufacturing biodegradable polyesters include Showa Highpolymer, Bayer, and BASF [12].

16.10.4 Starch-Based Plastics

Several companies have developed starch-based plastics. By using carefully selected starch feedstocks, and water as a plasticizer, they produce thermoplastics from nearly 100% starch, or from blends of starch with other biodegradable components. Many of these materials are water-soluble in addition to being biodegradable. The major target application has been as a replacement for polystyrene foam, including both molded cushions and loosefill.

Warner-Lambert was the first major producer of starch-based plastics, which it sold under the Novon name, beginning in about 1990. It discontinued manufacture in 1993 and sold the operation to EcoStar International, which was shortly thereafter acquired by Churchill Technology, which then declared bankruptcy. Other suppliers of biodegradable starch-based plastics include StarchTech, Inc. of Minnesota, FP International, Biotec in Germany, Mellita, and Novamont, among others.

16.10.5 Other Biodegradable Plastics

There are a number of other families of biodegradable plastics, most at an earlier stage of development than those discussed above. Examples include protein-based plastics, polysaccharides, and wood-derived plastics. Two water-soluble plastics have a much longer history.

Polyvinyl alcohol is a synthetic polymer manufactured by Air Products, ChrisCraft, Nippon Gohsei, and Italway, among others. It is biodegradable once it is dissolved in water. A major limitation to its use is that it is generally not melt processable. It is also relatively high in cost.

Polyoxyethylene is another water-soluble plastic that has been known for a number of years but has seen little use. There are conflicting reports about its biodegradability. Mitsubishi is the major supplier.

16.11 Other Environmental Concerns

16.11.1 Resource Depletion and Energy Efficiency

In the 1970s, political events and associated dramatic increases in the cost of oil brought public attention to the fact that the amount of petroleum available on this planet is finite. One result was criticism of plastics, on the basis that since they were made of oil, using these materials was diminishing these finite sources of fuel. Evidence soon emerged, however, that not only were plastics a relatively insignificant user of oil (less than 2%), but also that plastics, including packaging materials, were usually more fuel-efficient than alternative materials when the whole life cycle was examined. In particular, the light weight and easy formability of plastics gave rise to transportation and processing efficiencies that most other materials could not match. Further, it is possible to recover the energy value of the plastics themselves through waste-to-energy incineration at the end of the plastic article's useful life.

In the area of packaging, often the most competitive packages in terms of energy efficiency are refillable packages. The additional embodied energy in these heavier materials can be more than compensated by the number of use cycles that can be achieved, as long as distribution cycles are not too long. Studies of beverage bottle packaging, for example, found break-even energy costs in the neighborhood of 10 trips when returnable glass bottles were compared to throw-away plastic bottles. The exact value depends on the size of containers being compared, the distances that must be traveled during bottle return and delivery, recycling rates, recycled content, and a number of other factors.

Therefore, as time went on, the energy-efficient nature of plastics became more well-known and appreciated, and criticisms against plastics on resource use grounds diminished.

16.11.2 Pollution

Production of plastics is associated with a variety of types of air and water emissions, and in the U.S. plastics manufacturing facilities, as well as other industries, must comply with a number of strict EPA regulations. Further, there is a relatively constant tightening of the standards, requiring ongoing attention. Many countries around the world have similar regulations.

The basis of air emission regulations in the U.S. is the Clean Air Act of 1970 and its subsequent amendments. Water emission regulations are similarly based on the Clean Water Act of 1972 and its amendments. The plastics industry is also impacted by state regulations issued in order to attain the ambient air quality standards imposed by federal requirements. California is a prime example, having emission regulations that are significantly more stringent than those in most other parts of the U.S.

The plastics packaging industry has also been impacted by public pressure, such as that arising from the Community Right to Know law that requires major industries to report annual emissions to the EPA for public disclosure. Reductions in emissions have also resulted from purely voluntary industry programs designed for minimization of hazardous wastes. Many companies, including some associated with plastics packaging, have found that these programs can lead to monetary savings that increase profits, as well as to a better public image.

One issue that continues to loom, as mentioned in Sec. 15.7, is the "endocrine disrupter" concern about a variety of chlorinated organics, phthalates, styrene compounds, and some other chemicals which are suspected of being hormone mimics or antagonists, and thus of having adverse effects on humans, as well as on a variety of types of animals. This is still a relatively new area of research, and much remains unknown both about what chemical compounds have such activity, and about their effects. Therefore, it is very hard to evaluate the eventual impact of these concerns on the plastics packaging industry.

16.11.3 Climate Change

Concern that human activity is leading to a change in the planet's climate, global warming, is an issue that is very likely to become increasingly significant in the next two decades. While there are vocal dissenters, a general consensus has emerged in the community of experts on global climate that increased emissions of greenhouse gases, especially carbon dioxide, have contributed to an increase in average global temperature over the last 100 years. This increase is expected to accelerate in the next century. The consensus is also that the result will be a wide variety of changes, some of them favorable, such as lengthened growing cycles in high northern latitudes, but many of them unfavorable, such as rise in sea level and more frequent droughts and floods.

The longer action to decrease greenhouse gas emissions is delayed, the more severe the ultimate effects will be. For this reason, many countries have agreed to cooperative action to reduce emissions of these gases. A few countries, mostly in Europe, have already taken significant steps in this direction, even though the details of the international agreement are still being hammered out. There is general agreement that energy conservation efforts will be a significant part of the overall strategy, since generation of energy is the largest source of human emissions of carbon dioxide.

The effect on the plastics packaging industry will be interesting to observe. As discussed in Sec. 16.11.1, plastics packaging is generally more energy-efficient than its competitors. Therefore, in applications where plastic competes with metal or glass, these policies may benefit plastics. Competition with paper is more difficult to evaluate. Paper packaging uses trees as a major part of its energy supply. If, as is the usual practice, new trees are grown to replace those that are harvested, the net effect on greenhouse gas emissions is much less than the total amount of energy used would indicate. Therefore paper and other cellulosic materials, too, may benefit from policies designed to reduce carbon dioxide emissions.

In evaluating potential plastics packaging process modifications, one should remember that saving energy often results in cost savings. Therefore, switching to more energy-efficient systems can be wise economically, as well as benefitting the global environment. In light of the potential for future regulation, it would also be wise to document the energy savings that result, as they may become useful in demonstrating compliance.

16.12 Lifecycle Assessment

The diversity of environmentally related considerations raised in this chapter illustrates the complexity of evaluating the overall environmental impact of packaging alternatives. There is general consensus that when evaluating the environmental effects of a product or process, it is essential to consider the whole life cycle. Evaluations that are narrowly focused around a particular operation can lead to erroneous decisions if the impacts of the decision extend outside the boundary used, which is often the case. This is the motivation behind the developing technology of lifecycle assessment.

Lifecycle assessment begins with the question to be answered, for example, the choice between package A and package B. The next step is to define the system to be studied, including the choice of appropriate boundaries. There is, of course, a strong connection between the type of question to be answered and the appropriate choice of boundaries. A public policy decision, for example, will generally need to be based on a wider set of considerations than an internal company decision. Boundary choices are also influenced by availability of data, accuracy/uncertainty of information, and cost considerations. The wider the boundary, the higher will be the cost of the study. If the contribution of the information to be obtained by increasing the boundary is comparable in amount to the uncertainty of the more central data, then there is not much to be gained by including it.

Once the system is defined, including its boundaries, there are two major components of the lifecycle analysis. The first is termed the lifecycle inventory. It consists of a detailed accounting of the inputs and outputs across the system boundaries, and includes raw materials, energy, air emissions, water emissions, etc. It can even be expanded to cover, at least in a qualitative way, items such as habitat destruction, noise, etc. The next major component is the lifecycle impact analysis. This evaluates the effect of the items listed in the inventory on the environment. In other words, it asks "what does this mean?"

The impact analysis component is at an earlier stage of development than the inventory. It is, conceptually at least, relatively simple to talk about accuracy of data, listing of assumptions, peer review, etc. It is much more difficult to talk about how we convert a long listing of inventory items into some more concise description of what it all means, in a form that is useful in decision-making and not overly simplified or misleading. Different researchers have taken somewhat different approaches, and efforts to reach consensus on impact analysis approaches continue.

A third component of lifecycle analysis that is often listed is improvement analysis. The idea is that even without completion of a full lifecycle inventory and impact analysis, it may be possible to identify places where changes can be made, with confidence that the result will be a decrease in environmental impact. For example, if use of a toxic substance can be avoided by substitution of a more environmentally benign compound, and such substitution causes no other major changes in the process, that is clearly a beneficial thing to do. Researchers have paid relatively little attention to improvement analysis, likely because this is most often applied within a company, and is often in accord with a company's overall environmental goals.

Several computer programs for doing lifecycle assessments have been developed. Some European countries have even required that lifecycle assessments be done as part of certain packaging changes. The ultimate usefulness of these tools depends on the extent to which the impact analysis difficulties are finally resolved, as well as on the quality of the input data.

References

1.   U.S. EPA, *Municipal Solid Waste in The United States: 2001 Facts and Figures*, EPA 530-R-03-011, Oct. 2003.

2.   U.S. EPA, *Characterization of Municipal Solid Waste in the United States: 1997 Update*, EPA 530-R-98-007, May, 1998.

3.   Loepp, D., *Plastics News*, Oct. 26, 1998, pp. 5, 8.

4.   American Plastics Council, *2002 National Post-Consumer Plastics Recycling Report*, www.plasticsresource.com/s_plasticsresource/docs/1200/1131.pdf.

5.   APME "Plastics in Europe: An analysis of plastics consumption and recovery in Europe," available online, www.apme.org, 2004.

6.   NAPCOR, *2002 Report on Post Consumer PET Container Recycling Activity*, available online, www.napcor.org, 2004.

7.   Japan Information Network, "Recycling Rates," available online, http://web-jpn.org/stat/stats/19ENV51.html

8.   Expanded Polystyrene Packaging Group, "EPS Recycling Forges Ahead in 2000," available online, www.eps.co.uk/news/new36.html.

9.   White, K., *Waste Age's Recycling Times*, Oct. 4, 1994, p. 9.

10.   Metabolix, www.metabolix.com.

11.   Cargill Dow LLC, www.cargilldow.com.

12.   Selke, S., "Plastics Recycling and Biodegradable Plastics," in C. Harper, ed., *Modern Plastics Handbook*, McGraw-Hill, New York, ***

Study Questions

1.   What are the major advantages and disadvantages of the use of biodegradable plastics for packaging?

2.   Is recyclability an important packaging attribute, in your opinion? Explain.

3.   Much more HDPE and PET are recycled than other packaging plastics. Why do you think this is so?

4.   Do you favor "producer responsibility" laws? Why or why not?

5.   Do you favor laws requiring recycled content in plastic packages? Explain your reasoning.

6.   You are designing a new package. Will you do a lifecycle assessment of the alternatives? Why or why not? In what circumstances, if any, do you think one should/should not be done?

7.   There has been growing concern about global warming. Do you think the packaging industry should take any action to address this concern? Explain.

8.   Some medical groups have called for PVC medical packaging, such as IV bags, to be phased out. What concern(s) is this based on? What is your opinion?

9. You are serving on the board designing a new recycling program for your home community. What kind of program would you favor? Why?

10. Recycling rates for PET bottles have been falling in the last few years. What do you think this is the main reason? What are some other factors that may be playing a part?

11. Do you think the packaging industry has a responsibility to increase recycling of packaging plastics? Why or why not? If so, what should they do? If not, is there any other societal sector that does have such a responsibility? Explain.

12. Interest in the use of biodegradable plastics has increased in recent years. What do you think are they main advantages and disadvantages of such materials?

13. There has also been increased interest in the use of biobased materials, which may or may not be biodegradable. What are the main advantages and disadvantages of the use of biobased plastics?

14. Oil prices have been quite volatile, and have generally been on an upward trend. What do you think this means for the use of plastics packaging in the next decade? Explain.

15. Europe has a long history of composting, but in the U.S. there is little composting except for yard waste. However, the city of San Francisco recently began city-wide collection of wet organics for composting. If this is the beginning of a trend, do you think it will increase the demand for biodegradable plastics packaging? Discuss.

16. Plastics are a significantly larger fraction of municipal solid waste by volume than they are by weight. Is it more accurate to look at the solid waste burden by weight or by volume? Why?

17. Nearly every session of the U.S. Congress, a national deposit bill is introduced (a proposal to institute a nationwide deposit on containers for carbonated soft drinks and beer). Invariable, it seems, the bill never makes it out of committee. What are some of the pros and cons about deposit legislation? What is your position? Why?

18. Two states (Maine and California) have extended their deposit laws to a variety of non-carbonated beverages, explicitly to increase recycling. There are efforts in other deposit states to take similar action. What are the pros and cons of such expansion? What is your position? Why?

19. Plastics packaging often permits considerable source reduction, especially compared to metal and glass packaging. The greatest source reduction can often be achieved using multilayer packaging, such as coextrusions. However, these materials are more difficult to recycle. Should that be an important consideration in deciding whether to choose such a package structure? Why or why not?

20. Your company wants to use the SPI plastics identification symbol on all components of its package systems, including caps, lids, pouches, etc. What are the advantages and disadvantages of doing so. Would you favor such a policy? Discuss.

Additional Reading

Barrier and Mass Transport

Crank, J. (1975) *Mathematics of Diffusion*, London:Clarendon Press.

Koros, William J. ed. (1990) *Barrier Polymers and Structures*, Washington:American Chemical Society.

Piringer, O.-G. and Baner, A.L. (2000) *Plastic Packaging Materials for Food*, New York: Wiley.

Vieth, W. R. (1991) *Diffusion In and Through Polymers, Principles and Applications*, Munich:Carl Hanser Verlag.

Packaging

Brody, Aaron L. and Marsh, Kenneth S., eds. (1997) *The Wiley Encyclopedia of Packaging Technology*, 2nd ed., New York:John Wiley & Sons.

Hanlon, Joseph F., Kelsey, Robert J. and Forcinio, Hallie E. (1998) *Handbook of Package Engineering*, 3rd ed., Lancaster, PA:Technomic Pub. Co.

Sherman, Max, ed. (1998) *Medical Device Packaging Handbook*, 2nd ed., New York:Marcel Dekker.

Soroka, Walter (2002) *Fundamentals of Packaging Technology*, 3rd ed, Naperville, IL:Institute of Packaging Professionals.

Twede, Diana and Goddard, Ron (1998) *Packaging Materials*, 2nd ed., Surrey, U.K.:Pira International.

Packaging Plastics

Benning, Calvin J. (1983) *Plastic Films for Packaging: Technology, Applications and Process Economics*, Lancaster, PA:Technomic Pub. Co.

Brown, William E. (1992) *Plastics in Food Packaging: Properties, Design, and Fabrication*, New York:Marcel Dekker.

Jenkins, Wilmer A. and Harrington, James P. (1991) *Packaging Foods with Plastics*, Lancaster, PA:Technomic Pub. Co.

Miller, Adolph (1994) *Converting for Flexible Packaging*, Lancaster, PA:Technomic Pub. Co.

Plastics

Benelkt, George M. (1999) *Metallocene Technology in Commercial Applications*, Brookfield, CT:Society of Plastics Engineers.

Berins, Michael L., ed. (1991) *Plastics Engineering Handbook of the Society of the Plastics Industry*, 5th ed., New York:Van Nostrand Reinhold.

Birley, Arthur W., Haworth, Barry and Batchelor, Jim (1992) *Physics of Plastics: Processing, Properties and Materials Engineering*, Munich:Carl Hanser Verlag.

Brandrup, J., Immergut, E. H., and Grulke, E. A., eds. (1999) *Polymer Handbook*, 4th ed., New York:John Wiley & Sons.

Briston, John H. (1988) *Plastics Films*, 3rd ed., Harlow, Essex:Longman Scientific & Technical.

David, D. J. and Misra, Ashok (1999) *Relating Material Properties to Structure*, Lancaster, PA:Technomic Pub. Co.

Dominghaus, Hans (1993) *Plastics for Engineers: Materials, Properties, Applications*, Munich:Carl Hanser Verlag.

Elias, Hans-Geo (1997) *An Introduction to Polymer Science*, New York:VCH.

Gruenwald, G. (1993) *Plastics: How Structure Determines Properties*, Munich:Carl Hanser Verlag.

Harper, Charles A., ed. (2002) *Handbook of Plastics, Elastomers, & Composites*, 4th ed. New York:McGraw-Hill.

Harper, Charles A., ed. (2000) *Modern Plastics Handbook*, New York:McGraw-Hill.

International Union of Pure and Applied Chemistry, Macromolecular Division, Commission on Macromolecular Nomenclature (1991) *Compendium of Macromolecular Nomenclature*, Oxfod:Blackwell Scientific.

Krentsel, B. A., Kissin, Y. V., Kliener, V. J., and Stotskaya, L. L. (1997) *Polymers and Copolymers of Higher α-Olefins*, Munich:Carl Hanser Verlag.

Mittal, K. L. (1998) *Metallized Plastics: Fundamentals and Applications*, New York:Marcel Dekker.

Oswald, Tim A. and Menges, Georg (2003) *Materials Science of Polymers for Engineers*, 2nd ed, Munich:Carl Hanser Verlag.

Progelhof, Richard C. and Throne, James. L. (1993) *Polymer Engineering Principles: Properties, Processes, and Tests for Design*, Munich:Carl Hanser Verlag.

Rodriguez, F., Cohen, C., Ober, C. K., and Archer, L. A. (2003) *Principles of Polymer Systems*, 5th ed., New York:Taylor & Francis.

Rosen, Stephen L. (1993) *Fundamental Principles of Polymeric Materials*, 2nd ed., New York:John Wiley & Sons.

Schultz, Jerold M. (2001) *Polymer Crystallization*, Washington, D.C.:ACS.

Sperling, L. H. (1992) *Introduction to Physical Polymer Science*, 2nd ed., New York:John Wiley & Sons.

Staudinger, Herman (1970) *From Organic Chemistry to Macromolecules; a Scientific Autobiography Based on my Original Papers*, translated from the German with a foreword by Herman F. Mark, New York:John Wiley & Sons.

Tonelli, Alan E. with Srinivasaroa, Mohan (2001) *Polymers from the Inside Out*, New York:Wiley Interscience.

Woebcken, Wilbrand (1995) *Saechtling International Plastics Handbook for the Technologist, Engineer and User*, 3rd ed., translated and edited by John Haim and David Hyatt, Munich:Carl Hanser Verlag.

Woodward, Arthur E. (1995) *Understanding Polymer Morphology*, Munich:Carl Hanser Verlag.

Additives

Gächter, R. and Müller, H. (1996) *Plastics Additives Handbook: Stabilizers, Processing Aids, Plasticizers, Fillers, Reinforcements, Colorants for Thermoplastics*, 4th ed., Munich:Carl Hanser Verlag.

Pritchard, Geoffrey (1998) *Plastics Additives: An A-Z Reference*, London:Chapman & Hall.

Adhesives

IoPP Adhesion Committee (1995) *Adhesives in Packaging: Principles, Properties and Glossary*, Herndon, VA: Institute of Packaging Professionals.

Skeist, Irving (1990) *Handbook of Plastics Testing Technology*, 2nd ed., New York:John Wiley & Sons.

Testing

Shah, Vishu (1998) *Handbook of Plastics Testing Technology*, 2nd ed., New York:John Wiley & Sons.

Polymer Processing

Baird, Donald G. and Collias, Dmitris I. (1998) *Polymer Processing: Principles and Design*, New YOrk:John Wiley & Sons.
Osswald, Tim A. (1998) *Polymer Processing Fundamentals*, Munich:Carl Hanser Verlag.
Pearson, J. R. A. (1985) *Mechanics of Polymer Processing*, London:Elsevier Applied Science.

Blow Molding

Brandau, Ottmar (2003) *Stretch Blow Moulding: A Hands-on Guide*, Heidelberg: PETplanet.
Rosato, Donald V. and Rosato, Dominick V., eds. (1988) *Blow Molding Handbook*, Munich:Carl Hanser Verlag.

Extrusion

Butler, Thomas I. and Veazey, Earl W., eds. (1992) *Film Extrusion Manual: Process, Materials, Properties*, Atlanta: TAPPI.
Hensen, Friedhelm, ed. (1997) *Plastics Extrusion Technology*, 2nd ed., Munich:Carl Hanser Verlag.
Rauwendaal, Chris (1994) *Polymer Extrusion*, 3rd ed., Munich:Carl Hanser Verlag.
Rauwendaal, Chris (1998) *Understanding Extrusion*, Munich:Carl Hanser Verlag.

Injection Molding

Rees, Herbert (1994) *Understanding Injection Molding Technology*, Munich:Carl Hanser Verlag.

Rosato, Donald V. and Rosato, Dominick V. (1995) *Injection Molding Handbook: the Complete Molding Operation, Technology, Performance, Economics*, 2nd ed., New York:Chapman & Hall.

Molds

Stoeckhert, Klaus and Mennig, G., eds. (1998) *Mold-Making Handbook*, 2nd ed., Munich:Carl Hanser Verlag.

Thermoforming

Throne, James L. (1987) *Thermoforming*, Munich:Carl Hanser Verlag.

Throne, James L. (1996) *Technology of Thermoforming*, Munich:Carl Hanser Verlag.

Index

catalyst 2, 8, 21, 35-38, 97, 98, 101-104, 107, 128, 163, 310, 336, 378
  chromium 37
  metallocene **101-105**, 337
  single site 2, 102
  Ziegler-Natta 2, 35, 37, **97-98**, 102, 103
CB-A **146**
CBA **163**
CB-D **145-146**
CED, see cohesive energy density
cellophane 2, 3, 95, 112, 131, **133-134**
cellulose 7, 9, 86, **133**, 134, 170
  acetate 3, 13, **135**, 170, 250
  butyrate **135**
  nitrate 1, **134-135**, 170
  propionate **135**
cellulosic plastics **134-135**
CF **381**
CFCs, see chlorofluorocarbons
CFR **375-385**, 388
channel depth ratio **198**
chemical
  activity 334, **338-341**
  composition 17
  potential **338**
  resistance 31
  structure 18
child-resistant packaging 276, **278**
chill roll **202-203**
  also see film, cast
chiral 41, 42
chlorofluorocarbons 163, 321
chromium compounds **153-155**, 390
  also see heavy metals
citric acid 321
clamp 268, 269
clarity 25, 93, 96, 101, 102, 106, 108, 116, 117, 122, 124, 128, 131, 145, 151, 156, 216, 281, 298, 302, 303, 312, 314
  also see transparency
clay 150, 155, 162, 180, 234
cling 217
closures 127, 131, 240, 245, 247, 267, 272, 273, **274-279**, 280, 414, 415
  CT **276-278**
  dispensing **278-279**
  friction **275**
  linerless **276-277**

snap-fit **275-276**
  threaded **276-278**
coat hanger die **206**
coating(s) 7, 24, 96, 112, 117, 120, 130-132, 134, 160, 165, 180, 218, 225-234, 285, 314, 315, 337, 379, 384
  aluminum oxide **233-234**
  extrusion 93, 96, 97, 99, 118, 122, **225-228**, 234
  metallization 120, **231-232**
  silicon oxide 120, **233**
cobalt compounds 154, 160, 310
COC, see cyclic olefin copolymers
coding, resin, see resin coding
coextrusion(s) 97, 112, 116, 117, 119, 125, 128, 129, 131, 160, 166, 209, 210, 214, 217, **218-220**, 228, 234, 255-257, 281, 409
  also see blow molding, coextrusion
COF, see friction, coefficient of
cohesive bond strength **82**, 166, 167, **172-173**, 190
cohesive energy density **16**, 71, 135, 137, 169
cohesive forces **167**
coinjection, see blow molding, coinjection
cold cast **202**, 203
  also see film, cast
cold glue **177**
cold seal 173, **177**, 230
collapsing core **272**
collapsing frame **213-214**
collection **404-405**
  curbside 404, 414
colligative properties **49**, 52
colorants 141, **152-155**, 291, 379
  see also dye, pigment
commingled 405, 408, **416-417**
commodity plastics **19**
compatibilizer 136, 417
composting 417
compounding 37, **141-144**, 197
  also see mixing
compression molding **279-280**
compression section **196-197**
condensation polymer 12, 29, **42-46**, 117, 120, 122, 133, 173, 410

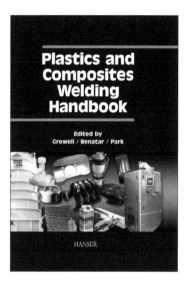